FOR REFERENCE

Do Not Take From This Room

BIRDS OF THE WORLD

Birds
of the World

Les Beletsky

Illustrated by
DAVID NURNEY
JOHN SILL
FRANK KNIGHT
BRIAN SMALL
H. DOUGLAS PRATT
JOHN GALE
DAVID BEADLE
DIANE PIERCE
DAN LANE
JOHN O'NEILL
NORMAN ARLOTT

THE JOHNS HOPKINS UNIVERSITY PRESS
Baltimore

Boca Raton Public Library, Boca Raton, FL

© 2006 Les Beletsky
All rights reserved. Published 2006
Printed in China on acid-free paper
9   8   7   6   5   4   3   2   1

THE JOHNS HOPKINS UNIVERSITY PRESS
2715 North Charles Street
Baltimore, Maryland 21218-4363
www.press.jhu.edu

Prepared by
Scott & Nix, Inc.
150 West 28th Street
Suite 1103
New York, New York 10001-6104
www.scottandnix.com

Library of Congress Cataloging-in-Publication Data

Beletsky, Les, 1956–
Birds of the world / Les Beletsky; illustrated by David Nurney ... [et al.].
p. cm.
Includes bibliographical references and index.
ISBN 0-8018-8429-2 (hardcover : alk. paper)
1. Birds. 2. Birds—Pictorial works. 1. Nurney, David. 11. Title.
QL673.B44 2006
598—dc22   2005032622

Frontispiece: Red-billed Streamertail (*Trochilus polytmus*)

Contents

Preface

BIRDS ARE BEAUTIFUL animals and many people enjoy watching them. In our contemporary societies, wild birds, in addition to the pleasures they provide with their compelling physical forms, colors, and behaviors, are perhaps increasingly significant because of the continuing connection they give us to the natural world. Among larger wild animals, birds are the ones most town- and city-dwellers still encounter frequently—so birds have assumed an essential role in allowing us, even in our usual paved and built-up surroundings, a degree of exposure to wildlife and nature. Regular exposure to wild animals, often in the guise of wild bird sightings, may also help motivate us to take part in efforts to preserve the world's threatened animal species and remaining unspoiled wilderness.

This book is for people with curiosity about the many kinds of birds, and particularly fairly new bird-watchers (birders), who are starting to master the identities of their local species and, as they do so, begin to wonder about birds around the rest of the world. When a novice birder begins learning how to identify local species, the task at first often seems enormous, the array, or diversity, of bird species in the region seemingly limitless. But after months or years of learning about, finding, and watching these birds, birders gain a mastery of the local avifauna. The birds become familiar, their diversity bounded. Yes, there are always a few species in the local region one would still like to find for the first time. But as birders begin to appreciate the different types of birds and bird families, they realize that some entire types occur only in restricted places far from home, and their birding thoughts start wandering to other parts of the world.

This book is an introduction to the birds of the world. It is to let birders and others know "what else is out there," in terms of bird families and species, and where, generally, these birds are located. It is not an encyclopedia of ornithology, nor is it a catalog of or visual guide to all the world's bird species. The main text, which tells about bird families, provides the essentials about bird groups, but only a small fraction of the rich and extensive information now available about the general ecology and behavior of a substantial proportion of the world's birds. And the illustration plates provide field-guide-like images of a good sampling of species—more than thirteen hundred total—found in each avian family. But if this book's introduction to the world's birds sufficiently raises one's interest level, then there are myriad other books that can further that interest, including bird encyclopedias, country-by-country field guides, and even volumes devoted to single bird families.

Many books of this kind offer, as an introduction, a primer of ornithology, defining the characteristics of birds and telling generally of their ecology and behavior. Instead, in keeping with the "birds of the world" emphasis, at the end of the book I discuss two basic, interrelated questions: how many types of birds are there? and where do birds occur geographically? Also, because bird diversity is increasingly jeopardized, I discuss some of the major threats faced by birds.

My hope for this book is that it stimulates some readers who have only a mild interest in birds or wildlife to further that interest and perhaps become frequent or serious wildlife watchers, and that it propels some newer birders, who have the time and resources, to begin exploring the world's birds—not just in the pages of books, but in the field.

Acknowledgments

I ACKNOWLEDGE THE ASSISTANCE OF THE PEOPLE who contributed to this book and helped put it together. First, I thank the artists who permitted their beautiful bird images to be included here. They are, in descending order of number of species illustrated, David Nurney, John Sill, Frank Knight, Brian Small, H. Douglas Pratt, John Gale, David Beadle, Diane Pierce, Dan Lane, John O'Neill, and Norman Arlott. (Artist credits for individual illustrations may be found at www.jhu.edu.) I especially thank David Nurney for painting extra bird images when they were needed at the last moment. Many of the African birds illustrated in this book (237 total bird images, by artists John Gale, Brian Small, and Norman Arlott) are reproduced from Terry Stevenson and John Fanshawe, *Field Guide to the Birds of East Africa* (London: A. and C. Black and T. and A. D. Poyser, 2002), and I thank A. and C. Black Publishers, Nigel Redman, and Paul Langridge for permission to use these illustrations. The researchers, writers, and compilers of avian data and life histories, field guides, and major reference works from which I culled much information also deserve my appreciation and acknowledgment; the main sources I used are provided in the bibliography. For various kinds of important assistance, especially for their expert ornithological knowledge, I thank David L. Pearson and Dennis Paulson. Richard Francis, Sharon Birks, and Gordon Orians read and commented on various parts of the book's text, as did Cynthia Wang, and I thank them all for their constructive criticisms and suggestions for improvements. Last, I thank my editors and the art and production staff at the Johns Hopkins University Press, Scott & Nix, Inc., Russell Galen, Vincent Burke, and, finally, Cynthia Wang, who helped me extensively with editing and artwork organization.

Introduction

Birds and Bird-watching

BIRDS ARE WONDERFUL animals to watch. Sure, we could watch mammals or reptiles or dragonflies or even worms, and each of these, except perhaps the last, has its aficionados and defenders. But the great majority of us who enjoy animal watching in the wild gravitate to birds. Many millions of people in the United States and Europe watch birds at least occasionally, and bird-watching as a participatory nature activity is on the rise in many other parts of the world, indicating the great general interest in these feathered animals. A case in point: on a recent trip to some of Thailand's national parks, I was surprised and pleased to find not only American and European birders roaming forests looking for birds, but also many Thai families, clad in khaki and tan and outfitted with binoculars and telescopes, doing the same.

What is the source of our fascination with birds? And why, when we select wild animals to watch, do we tend to choose birds? Most of us could quickly compile a list of possible reasons besides the obvious one: birds are beautiful. Such a list might include:

- Birds are mainly day-active animals; like most of us, they are out and about during daylight hours and asleep at night. Many other kinds of large animals—especially mammals—are more active at night, which makes finding and observing them much more difficult.

- Birds fly, and as a consequence of this marvelous ability, they do not always depart with all due haste after being spotted, as is common with most other types of vertebrate animals. Their ability to fly, and thus easily evade our grasp, permits many birds when confronted with people to leisurely go about their business (albeit keeping one eye on us at all times), often allowing extensive time to watch them.

• Many birds are highly or moderately conspicuous. They are especially evident during their breeding seasons, when many of them change into bright, pleasing plumages and engage in courtship displays—such as energetic dances or acrobatic flights, the spreading and waving of long, colorful tail feathers, loud vocalizations—that attract the notice not just of other birds (the presumed audience) but of people as well. Some types of birds, of course, are conspicuous all the time, including the larger, visually arresting birds such as colorful parrots, toucans, and hornbills in the tropics and large birds of prey the world over. Equally conspicuous, although plainer and smaller, are the garden, park, pond, and city birds such as many members of the duck, gull, plover, pigeon, thrush, starling/myna, crow, finch, sparrow, and blackbird families.

• Birds engage in many behaviors that are truly fascinating. A short list of these includes (1) breeding-season displays and nest-building behaviors, (2) mind-boggling migrations that may involve biannual flights of thousands of miles, from breeding sites in temperate or arctic regions to wintering sites often in tropical or subtropical areas, (3) spectacular flying prowess, such as that of hummingbirds (which, with their superbly designed and rapidly beating wings, can hover and actually fly backward), swifts (which tend to fly throughout the day), and albatrosses (which soar over oceans for extended periods essentially without beating their wings), and (4) amazingly varied breeding systems and sometimes seemingly torturous nesting efforts, such as that of hornbills, with their unique breeding during which the female of a pair is encased within a tree cavity and fed by her mate through a small hole, and bowerbirds, many males of which build large courting structures of plant materials, decorate them, and then show them to females in hopes of convincing them to mate.

- Birds are innocuous. Typically the worst that can happen from any encounter is a soiled shirt. Contrast that with too-close, potentially dangerous meetings with certain reptiles (venomous snakes, crocodiles), amphibians (frogs and salamanders with toxic skin secretions), fish (shark dangers), and mammals (bears or big cats). While it is true that an Ostrich or cassowary can injure with a swift kick, such damaging encounters are extremely rare in the wild. Safety undoubtedly contributes to birds being the most frequently watched wildlife.

Add to these attractions the benefits of an active outdoor pastime often conducted in beautiful, wild surroundings, the intellectual challenge of bird identification, and the nature-study and "collecting" elements of bird-watching, and it is not difficult to understand its popularity, or even its many near-fanatical adherents.

As people learn about birds and bird-watching, inevitably they develop an appreciation for bird diversity. This brings us to the purpose of this book, which is to introduce readers to all the world's bird types.

The primary organization of the book follows scientific animal classification, in particular, the bird species of the world separated into groups called orders and families. If you are unfamiliar with animal classification or need to refresh your memory, explanations and definitions are below. I have followed a classification that divides birds of the world into about two hundred families and about 9,750 species (generally following the classifications of the Lynx Edicions *Handbook of the Birds of the World*, and the 5th edition of J. F. Clements, *Birds of the World: A Checklist*).

At the end of the book are questions and answers on diversity, abundance, and geography of birds. Several seemingly simple questions—How many kinds of birds are there? Why are some birds abundant but others rare? Why do some occur in such small numbers that they are threatened with extinction? Where are birds located and why?—have fairly complex answers. Understanding these basics of avian diversity and biogeography is central to attaining a broad appreciation of the feathered vertebrates.

GENERAL PLAN OF THE BOOK
AND TERMS USED IN THE TEXT

THE LAYOUT OF this book is such that each bird family, or group of families, is described on a few pages of text, and several species in the family or families are illustrated in an associated color plate. For large, visually diverse families, three plates, instead of one, are associated with the text account. Some of the world's leading bird artists produced the artwork. Most of the images are of the kind used in field guides; usually the bird is portrayed in a formal perched or standing position that allows a good view of its distinguishing physical characteristics. Particular bird species were chosen for illustration for various reasons: some because they are common and widespread representatives of relevant families, some because they are rare or narrowly distributed or particularly striking. The goal is to provide the reader with a good overview of each family's diversity. The plates include common and scientific names of the illustrated birds, their body lengths, and general geographic distributions.

Most of the family accounts begin with introductory comments and then proceed to information about family classification, diversity, distribution, morphology, ecology and general behavior, breeding behavior, and conservation status.

Introductory Comments

This section identifies the group in question and relates, in a nontechnical way, what distinguishes the group from others and which aspects of its biology are especially noteworthy (if any).

Classification

Classification information is used by scientists to separate birds and other organisms into related groups. It is provided here as it often enhances our appreciation of birds to know these relationships. Below is a quick review of relevant animal classification:

KINGDOM ANIMALIA: All the species detailed in the book are members of the animal kingdom, as are mammals, reptiles, amphibians, fish, insects, crustaceans, mollusks, worms, and many others.

PHYLUM CHORDATA, SUBPHYLUM VERTEBRATA: All species in the book are vertebrates, animals with backbones, as are mammals, reptiles, amphibians, and fish.

CLASS: The book covers a single vertebrate class, Aves, the birds.

ORDER: Class Aves is divided into twenty-nine orders (in this book's particular classification scheme); the birds in each order share many characteristics. For example, one of the avian orders is Falconiformes, which encompasses all the day-active birds of prey, including hawks, falcons, and New World vultures.

FAMILY: Families of animals are subdivisions of each order that contain closely related species that are often similar in form, ecology, and behavior. The family Accipitridae, for instance, contains hunting birds that all have hooked bills, powerful legs and feet, sharp, curved claws, and that are all carnivorous—hawks, kites, buzzards, eagles, and Old World vultures. Birds of the world, in this book, are divided into about two hundred families.

GENUS: Further subdivisions; within each genus are species that are very closely related; they are all considered to have evolved from a common ancestor.

SPECIES: The lowest classification level; all members of a species are similar enough to be able to breed together and produce fertile offspring.

As an example of how birds are classified, here is the classification of a single species, the Golden Eagle, which is illustrated on p. 73

KINGDOM: Animalia, with more than 2 million species.

PHYLUM: Chordata, Subphylum Vertebrata, with more than 47,000 species

CLASS: Aves (Birds), with about 9,800 species.

ORDER: Falconiformes, with about 307 species; includes hawks, kites, eagles, falcons, New World vultures, the Osprey, and the Secretarybird.

FAMILY: Accipitridae, with 237 species; includes hawks, kites, and eagles.

GENUS: *Aquila*, with 11 species; one group of eagles.

SPECIES: *Aquila chrysaetos*, the Golden Eagle.

A main classification division among birds is that between passerines and nonpasserines. The passerine birds (order: Passeriformes) are the perching birds, with feet specialized to grasp and to perch on tree branches. These are the more recently evolved birds (the most "advanced") and include all the small land birds with which we are most familiar—flycatchers, robins, crows, wrens, warblers, blackbirds, finches, sparrows, and so on. The passerine order is the most diverse of the bird orders, including more than 50 percent of all bird species (about fifty-eight hundred species) and about half the bird families. A major subgroup within the passerines (containing about forty-five hundred species) is called the oscines, or songbirds: they all have a distinctive, advanced syrinx, the sound-producing organ in their respiratory passages. The oscines are responsible for most of the avian world's more melodic vocalizations. The remainder of the globe's birds—seabirds and shorebirds, ducks and geese, hawks and owls, parrots and woodpeckers, and a host of others—are nonpasserines, divided among the other twenty-eight orders. The nonpasserine family accounts in this book pertain to the ratite birds (Ostrich, Emu, etc.) through the woodpeckers; the passerine accounts are the ones that deal with the pittas through the New World blackbirds.

Common bird names, such as Song Sparrow or European Starling, sometimes vary from place to place. A species that occurs in Europe and Africa may have different names on the two continents, or the same species may have different names in East Africa and Southern Africa. There are ongoing attempts to standardize such English names, but not yet universal agreement. Scientific names are less variable, but they sometimes change as researchers make new classification decisions. For example, the same bird species may have different genus names depending on the source consulted. The names I have used in this book, both common and scientific, come mainly from Lynx Edicions *Handbook of the Birds of the World* or the 5th edition of J. F. Clements, *Birds of the World: A Checklist*. There is also controversy over how best to write compound bird names, for example, Wompoo Fruit Dove, Wompoo Fruit-dove, or Wompoo Fruit-Dove. I've tried to be consistent in these kinds of names, but for this book's purposes, such minute details are of minimal concern.

Diversity

The words *diverse* and *diversity* are used in different ways in various scientific and technical fields. Here, a diverse group, such as a diverse family, means one with a relatively high number of species. Thus, the loon family (Gaviidae), with a global total of 5 species, is not very diverse; the duck, geese, and swan family (Anatidae), with 157 species, has a high degree of diversity. Diversity, when used in this context, is similar to the term *biodiversity*, which refers to the different types of animals, plants, and other life forms found within a region. A group or family that is "ecologically diverse" is one that exhibits many different adaptations to the environments in which its members occur. For instance, an ecologically diverse group, such as the jay and crow family (Corvidae), might occur in several different habitat types, eat various types of foods, and employ a variety of foraging methods.

When I provide a definite number of species in a given family, such as "there are 10 species of motmots," I mean there are 10 living species; recently extinct species are not included in these totals. Often I do not give exact numbers of species; instead I give approximations, such as "there are about 118 species in the jay and crow family." For reasons I detail in the book's concluding essay, exact species numbers are sometimes problematic.

Distribution

I give the geographic distribution of each bird family, generally in terms of the continents on which its member species occur. North America, in the family descriptions and in the information provided on the color plates, includes Mexico (noted here because many bird field guides for North American birds include only species that occur north of Mexico). A distribution of land birds given as "worldwide" or "global" actually means that the group in question occurs on all continents except Antarctica, where there are essentially no terrestrial birds. A few specialized terms are used to describe family distributions, and a reading of "Zoogeographic Regions: Describing Bird Distributions" at the back of the book will be helpful in this regard. The only geographic terms used often that might be unfamiliar to many readers are *Neotropics*, which refers to South and Central America, southern Mexico, and the West Indies; *Australasia*, which refers to the region mainly encompassing Australia, New Guinea, and New Zealand (but also sometimes Polynesia and Micronesia); *Old World*, which includes Europe, Asia, and Africa (and sometimes Australia); and *New World*, which refers to South, Central, and North America.

Some key terms regarding bird distributions are

- Range: the particular geographic area occupied by a species

- Native or indigenous: both mean "occurring naturally in a particular place"

- Introduced: "occurring in a particular place owing to peoples' intentional or unintentional assistance with transportation, usually from one continent to another; the opposite of native"

- Endemic: "occurring in a particular place and nowhere else"; a species, a genus, or an entire family, can be said to be endemic. (See the section entitled "Endemics: Some Birds Occur in Very Limited Areas" on p. 493.)

On the color plates, I provide general natural ranges for each illustrated species, usually in terms of continents on which various species occur. Ranges for migratory species usually include both breeding and nonbreeding locations. A range given as "North America, South America" generally includes Central America.

Morphology

Early in each account are details of the typical family morphology—the general shape, size, and coloring of the various constituent species. Whether males and females within a species look similar or different is noted. On the color plates, in most cases, if only one individual is pictured, you may assume that male and female of that species look alike, almost alike, or have only minor differences in appearance; when there are major sex differences, usually both male and female are depicted. Bird sizes are usually given as lengths, which are traditionally measured from tip of bill to end of tail. For the passerine birds, when I use the words large, medium-size, small, or very small to describe their sizes, large usually means more than 12 inches (30 cm) long; medium-size means between 6 to 7 inches (15 to 18 cm) and 12 inches (30 cm); small means 4 to 6 inches (10 to 15 cm); and very small means less than 4 inches (10 cm). Typical bird lengths, sometimes ranges of lengths, are provided on the plates themselves. A large range may indicate that the male and female of a species are very different lengths, or the species in question has a long, variable-length tail. Abbreviations on the plates, referring to age and breeding status, are: BRD, breeding; NON-BRD, non-breeding; and IMM, immature.

Ecology and General Behavior

In these sections I describe some of what is known about the basic activities pursued by each group. Much of the information relates to the habitat types (environments) the birds occupy, when and where the birds are usually active, what they eat, and how they forage. Arboreal birds pursue life and food in trees or shrubs; terrestrial ones pursue life and food on the ground. Whether birds tend to stay in social groups is also noted.

Breeding Behavior

Here I comment on each group's breeding, including types of mating systems employed, special breeding behaviors, and nesting particulars. A monogamous mating system is one in which one male and one female establish a pair-bond and contribute fairly evenly to each breeding effort. In polygamous systems, individuals of one of the sexes have more than one mate (that is, they have harems): in polygynous systems, one male mates with several females, and in polyandrous systems, one female mates with several males. Some bird species are "promiscuous" breeders. In these species, no pair-bonds are formed between males and females. Males mate with more than one female and females probably often do the same. Males individually stake out display sites, usually at traditional communal courting areas (called leks), and repeatedly perform vocal and visual displays to attract females. Females enter leks, assess the displaying males, and choose the ones with which they wish to mate. In this type of breeding, females leave after mating and then nest and rear young alone.

Conservation Status

Detailed here, usually very briefly, is the conservation status of each group, including information on relative rarity or abundance, and factors contributing to population declines. I also provide the numbers of species within each avian family that are presently threatened. Several organizations publish lists of threatened species, and these lists do not always agree on criteria used for inclusion on the lists or on the species listed. Here I followed the listings of the comprehensive BirdLife International *Threatened Birds of the World.* The term *threatened* in my writing is simply a synonym for a bird species "in jeopardy," not a formally defined level of threat. The three formal levels of threat to bird populations used here are

1. CRITICALLY ENDANGERED: species with very small remaining populations or tiny ranges, or that are undergoing very rapid population declines. Such species face an extremely high risk of extinction in the immediate future; a critically endangered species has only an estimated 50 percent chance of avoiding extinction during the next ten years or three generations.

2. ENDANGERED: species known to be in imminent danger of extinction throughout their range, highly unlikely to survive unless strong conservation measures are taken. Such species face a very high risk of extinction in the near future.

3. VULNERABLE: species known to be undergoing rapid declines in the sizes of their populations. Unless conservation measures are enacted, and the causes of the population declines identified and halted, these species are likely to move to endangered status in the near future; such species face a high risk of extinction in the medium-term future.

Also, "at-risk" or "near-threatened" species are those that, owing to their habitat requirements or limited distributions, and based on known patterns of habitat destruction, are extremely likely to move to vulnerable status in the near future.

With the above information in mind, we can begin our exploration of the world's birds.

GREATER RHEA
Rhea americana
50 – 55 in (127–140 cm)
South America

OSTRICH
Struthio camelus
67–106.5 in (170–270 cm)
Africa

EMU
Dromaius novaehollandiae
59–75 in (150–190 cm)
Australia

SOUTHERN CASSOWARY
Casuarius casuarius
51–67 in (130–170 cm)
Australia, New Guinea

BROWN KIWI
Apteryx australis
19.5–25.5 in (50–65 cm)
New Zealand

GREAT SPOTTED KIWI
Apteryx haastii
19.5–23.5 in (50–60 cm)
New Zealand

Ratites: Ostrich, Emu, Cassowaries, Rheas, and Kiwis

OSTRICH, EMU, CASSOWARIES, and RHEAS are huge flightless birds mainly relegated to the Southern Hemisphere. KIWIS are smaller flightless birds of New Zealand. Together, these birds comprise the world's ten species of ratites, birds with flat, raftlike (ratite) sternums. Other birds have sternums with deep keels, upon which the breast flight muscles attach (think of the deep keel to which attaches the main portion of white meat in chickens and turkeys). The ratites, flightless birds that run along the ground, have no need of thick, powerful flight muscles, so have no keel. Biologists believe that Ostrich, Emu, cassowaries, and rheas lost their ability to fly because they followed an evolutionary pathway to become very large, which was beneficial for them; with their long legs, great weight, and sometimes aggressive natures, they could run rapidly and defend themselves. But as they evolved to be larger, flying became more and more difficult and was less needed to escape predators; eventually it was lost. As for New Zealand's kiwis and some other flightless birds, such as some island-dwelling pigeons and rails, they may have lost the power of flight because they evolved in what were essentially predator-free zones, where large, powerful wings and energetically expensive flight were not needed.

Ratite classification is controversial, with some authorities placing all living groups in a single order, Struthioniformes; others divide the ratites into Struthioniformes (the Ostrich of Africa), Rheiformes (the Greater and Lesser Rheas of South America), Casuariiformes (the Emu of Australia and three cassowaries: one in northeastern Australia and it and two others in

Distribution:
South America, Africa,
Australia/New Guinea,
New Zealand

No. of Living
Species: 10

No. of Species
Vulnerable,
Endangered: 5, 0

No. of Species Extinct
Since 1600: 2

the New Guinea region), and Apterygiformes (three species of kiwi, all in New Zealand).

The OSTRICH, the most widely known of this group, is the largest and heaviest of living birds, adult males standing 6.9 to 8.8 feet (2.1 to 2.7 m) tall and weighing up to 330 pounds (150 kg); females are a bit smaller (5.6 to 6.2 feet [1.7 to 1.9 m], up to 240 pounds [110 kg]). Males are mostly black and white, with either reddish pink or bluish gray necks and legs; females are generally brownish with dull white wing feathers. Ostriches occur throughout much of East and Southern Africa, and extend into Central and West Africa, wherever their preferred habitats of savanna and other dry, open, grassy habitats are found. Ostrich wings are used not for flight but in threat and breeding displays, and they are fanned to cool the body. Their legs are long and powerful, capable of propelling these birds at a steady trot of about 19 miles per hour (30 kph) and a top speed of 44 miles per hour (70 kph) during short bursts, and with this speed, they can outrun many predators. Ostriches eat leaves, flowers, and seeds. Walking slowly, their heads down and moving from side to side, they pick up suitable foods, which collect at the top of the neck. Occasionally they pause, stand alert, check for danger, and swallow the accumulated ball of food. Ostriches occur alone, in pairs, or in groups that can reach hundreds of birds. They are day-active, and at night usually roost in groups, individuals squatting on the ground near each other, typically holding their heads high.

Usually breeding in polygamous groups, one male Ostrich typically mates with two to five females. The females lay in a single communal nest, which eventually contains fifteen to twenty or more enormous eggs. The nest is a wide, shallow scrape in the ground; incubation is shared by the male (mainly at night) and the dominant female (during the day). Hatched young are quickly able to walk, run, and feed themselves. Although Ostrich numbers have been greatly reduced, mainly as a result of habitat loss to agriculture, they are still widely distributed and the species as a whole is not threatened. Commercially bred Ostriches now may be encountered almost anywhere in the world because they are widely farmed for their meat (valued as a source of low-fat protein), feathers (used as decorations), and skin (for making fine leather products).

EMUS, which roam over most of the Australian mainland except heavily settled or thickly forested areas, are enormous brownish or blackish birds, with pale or bluish bare skin on head and neck. They stand 5 to 6.2 feet (1.5

to 1.9 m) tall and weigh up to 120 pounds (55 kg); favor open woodland, savanna, and agricultural habitats. They are omnivores, taking grasses and other vegetation, fruit, seeds, flowers, and insects, and occur alone, in pairs, or in small groups of four to nine. A typical day in the life of Emus consists of early morning foraging, followed in late morning by a slow amble toward a water source, feeding as they go. Afternoons are also spent foraging, but at a slower pace; in hot weather, parts of afternoons are spent in tree shade. Emus, which can move along at a quite respectable 30 miles per hour (48 kph) when they are in a hurry, are considered nomadic wanderers, at least in some of their populations. Having no real territories, they are very mobile, following food availability. Populations in western Australia are known to move over distances of 300+ miles (500+ km) in a year. These birds are very curious and sometimes approach or even follow people. Aboriginal people would sometimes use Emu curiosity to hunt them: they would attract a bird's attention by waving objects or flashing it with a mirror, then spear it when it approached.

After mating, a female Emu lays eggs in a nest of twigs and crushed vegetation in a depression on the ground, then departs; she may later mate with other males. The male incubates the eggs and, after hatching, the male cares for the young for up to a year or more by leading them and, when they are small, brooding them at night in his feathers.

Emus did well in Australia until European colonization. They were killed with abandon by early settlers for their meat and oil (for lamps), and their eggs were collected and eaten. (They were eliminated in Tasmania by the mid-nineteenth century.) Later, farmers killed them because the birds ate their crops. However, Emus survived and eventually even benefited to a degree from development, because, with new sources of open water provided by farmers and ranchers, they could occupy many semiarid regions in the center of Australia that previously had been closed to them. Emus are now common, but they are still persecuted in agricultural areas for breaking fences and raiding crops. They are farmed in Australia and some other countries for their leather, oil, feathers, and low-cholesterol meat.

CASSOWARIES, rainforest and woodland birds, usually just a bit smaller than Emus (to 5.6 feet [1.7 m]; to 120 pounds [55 kg]), are also dark, with a bare-skin head and neck, and an unmistakable large, bony crest, or "helmet." They are chiefly fruit-eaters, taking mostly fallen tree fruit but also some pulled from plants; they also eat seeds, insects, other small invertebrates, and carrion. These large birds occur singly, in pairs, or in small groups of up to six individuals. They are fairly sedentary, not moving over long distances unless forced to by lack of food. Cassowaries can be aggressive birds, especially when breeding, and have one very long, sharp claw on each foot that can do considerable damage; they have been known to disembowel midsize mammals.

Cassowary breeding is similar to that of Emus. Two species, the Southern Cassowary and Northern Cassowary, are threatened, both considered vulnerable. The third species, New Guinea's Dwarf Cassowary, is considered near-threatened. All three species occur in New Guinea, where they are still a main source of food and other materials for some of the native peoples. Wild birds are hunted and young ones are kept in pens and fed until they are large enough to eat. A single cassowary provides a lot of meat, feathers for decorations, sharp claws for arrow construction, and robust leg bones to make daggers and other tools.

RHEAS, Ostrichlike gray-brown birds restricted to South America (central and southern portions), range up to 5 feet (1.5 m) tall and weigh up to 75 pounds (34 kg). The Greater Rhea occurs in various savanna, woodland, and scrub habitats, but often avoids very open grassland. The Lesser Rhea occurs in shrublands, grassland, and various high-elevation habitats. Both feed mainly on grass, roots, and seeds, but they will also take lizards and small rodents. They regularly graze among deer or cattle, and are usually found in small groups of five to fifteen. Long-legged, long-necked, and flightless, they can at times be very shy, and run quickly from danger.

Each rhea male defends a large territory, and tries to attract a harem of females with which to mate. The male prepares a nest (a hollow in the ground) and incubates eggs. Each of the females can lay up to fifteen eggs, and a male with a harem of six to eight females can have a nest full of ninety or more eggs. After hatching, the young stay with the father, who guards them for up to 5 months. Both of the rheas are considered to be near-threatened. They are still hunted extensively for their meat, feathers, and skins; their eggs are taken for food; and their natural habitats are increasingly converted to agricultural uses.

KIWIS, New Zealand's most famous animal residents, are medium size to large, secretive, flightless brown birds. They are now so much a part of the New Zealand national consciousness that New Zealanders traveling overseas refer to themselves as Kiwis. The three kiwis (*kiwi* is sometimes the preferred plural) are the smallest ratite birds, ranging in length from 14 to 26 inches (35 to 65 cm), and weighing from 2 to 8 pounds (0.9 to 3.9 kg). They are generally pear-shaped (relatively small head, larger lower body) and have long, slightly down-curved bills that are somewhat flexible. Their legs are short and sturdy; they lack a tail and have only rudimentary wings. They are covered in dense, shaggy brown or gray feathers that almost give the appearance of fur; bills are pinkish, brown, or off white. Females within a species are usually slightly larger and heavier than males.

Kiwis are mainly nocturnal denizens of forests, woodlands, and scrub areas. During the day, they sleep in burrows, hollow logs, or under tree roots. Emerging at night to feed, their diet includes earthworms and other forest-floor invertebrates, as well as fallen fruits, seeds, and leaves. The Great Spotted Kiwi often forages above treeline in alpine bogs and meadows. Kiwis have an excellent sense of smell, which may assist them in locating prey underground. They use their long bills to probe soft soil for earthworms. These birds are usually found in pairs and are highly territorial, both sexes defending parcels of real estate of up to 50+ acres (20+ hectares) from other kiwis. Fights between males are probably common. Adult kiwis have powerful legs and claws, so they can harm one another. When threatened, males often stand their ground while the rest of a kiwi family, if present, scurries for shelter. Kiwi calls, which are long, loud, and sometimes warbling whistles, are common and are given in aggressive and sexual contexts. Pairs sometimes duet or answer one another's calls. When alarmed or angry, kiwis growl and hiss.

Monogamous breeders, kiwis probably mate for life. Nests, lined with vegetation, are placed in burrows that the kiwis dig or in natural holes. The male only or both sexes incubate eggs; the young can feed themselves soon after hatching. All three kiwi species are threatened, mainly considered vulnerable (but the Brown Kiwi on New Zealand's North Island is endangered because of recent steep declines in population numbers, mainly due to predation and habitat loss). Kiwis have been hunted throughout historic times for their meat and feathers. Now their main enemies are introduced predators, including rats, weasels, ferrets, cats, dogs, and even pigs that people brought to New Zealand, which are especially deadly to juvenile kiwis. One species, the Little Spotted Kiwi, is now found only on a few offshore islands. Conservationists believe that breeding programs probably will ensure the survival of kiwis in captivity, but unless introduced mammal predators are controlled in the wild, natural kiwi populations may become extinct.

GREAT TINAMOU
Tinamus major
18 in (46 cm)
South America, Central America, Mexico

THICKET TINAMOU
Crypturellus cinnamonomeus
11 in (28 cm)
Central America, Mexico

SLATY-BREASTED TINAMOU
Crypturellus booucardi
11 in (28 cm)
Central America, Mexico

LITTLE TINAMOU
Crypturellus soui
9 in (23 cm)
South America, Central America, Mexico

UNDULATED TINAMOU
Crypterellus undulatus
12 in (30 cm)
South America

Tinamous

TINAMOUS are secretive, usually elusive, chickenlike birds that are occasionally seen walking along forest trails in Central and South America. They apparently represent an ancient group of birds, most closely related to the rheas of South America, large, flightless birds in the Ostrich mold. The family, Tinamidae, with about forty-five species, is confined in its distribution to the Neotropics, from Mexico to southern Chile and Argentina. Compared to most other bird families of similar size, relatively little is known about tinamous. Studying them, even censusing them is difficult. Some inhabit very remote areas, such as the Amazon forests and high Andes, and their habitats are usually densely vegetated and often marshy. Tinamous are also stealthy and superbly camouflaged.

Tinamous are perhaps best known for their songs, the loud, pure-tone, melodious whistles that are some of the most characteristic sounds of neotropical forests. They are often heard on the soundtracks of movies with rainforest settings. Sometimes resembling organ- or flute-produced notes, these vocalizations often occur all day long; the Undulated Tinamou, for example, makes one of the most common sounds of Amazonian Brazil. Some claim that one of the most delightful parts of neotropical forests is the early evening serenade of tinamou whistles.

Six to 18 inches (15 to 45 cm) long, tinamous have chunky bodies, with fairly long necks, small heads, and slender bills. They have short legs and very short tails. The back part of the body sometimes appears higher than it should be, a consequence of a dense concentration of rump feathers. Tinamous are

Distribution: Neotropics

No. of Living Species: about 45

No. of Species Vulnerable, Endangered: 5, 2

No. of Species Extinct Since 1600: possibly 2

attired in understated, protective colors such as browns, grays, and olives and are often marked with dark spots or bars. Males and females look alike, with females being slightly larger.

Tinamous inhabit a variety of environments, including grasslands and thickets, but most commonly they are forest birds. Except during breeding, they lead a solitary existence. They are among the most terrestrial of birds, foraging, breeding, and usually sleeping on the ground (some larger forest species roost in trees). Very poor flyers, they take flight only when alarmed by predators or surprised, and then only for short distances; their main form of locomotion is running. The tinamou diet consists chiefly of fruit and seeds, but they also take insects such as caterpillars, beetles, and ants, and occasionally, small vertebrates such as mice. Some South American species dig to feed on roots and termites. Tinamous avoid being eaten themselves primarily by staying still, easily blending in with surrounding vegetation, and by walking slowly and cautiously. If approached closely, tinamous will fly upward in a burst of loud wing-beating and fly, usually less than 150 feet (45 m), to a new hiding spot in the undergrowth, often colliding with trees and branches as they go.

Unusual mating systems are characteristic of tinamous, the most intriguing of which is a kind of group polyandry (one female mates with several males during a breeding season). One or more females will mate with a male and lay clutches of eggs in the same nest for the male to incubate and care for. The females then move on to repeat the process with other males of their choosing. Apparently in all tinamous the male incubates the eggs and defends the young. Nests are simple indentations in the ground, hidden in a thicket or at the base of a tree. From hatching, the young feed themselves.

Outside of protected areas, all tinamous are hunted extensively for food. Tinamou meat is considered tender and tasty, albeit a bit strange-looking; it has been described variously as greenish and transparent. Still, tinamous' cryptic coloring and secretive behavior must serve the birds well because, although hunted, many species still maintain healthy populations. Some are known to be able to move readily from old, uncut forest to secondary, recently cut forest, demonstrating an adaptability that should allow these birds to thrive even amid major habitat alterations such as deforestation. Currently five South American species are considered vulnerable. Two others, Colombia's Magdalena Tinamou and Peru's Kalinowski's Tinamou, are critically endangered and, with no recent sightings, are most likely extinct.

Penguins

PENGUINS have some paradoxical relationships with people. With their unique shape, black and white coloring, and the adoption of their image into popular culture, penguins are among the most widely recognizable types of birds, along with ducks, parrots, pigeons, and some others. However, they are relatively rarely seen in the wild; close encounters are usually limited to zoos. This is because penguins' natural home is the remote cooler waters of the southern oceans. The majority of species breed in Antarctic or sub-Antarctic regions—areas unpopulated or only lightly visited by humans. Another powerful penguin paradox is that we know them best for their land activities, their oft-photographed dense breeding colonies and their awkward waddling walk. But the majority of the penguins' lives are spent at sea; they are superbly graceful, streamlined swimming and diving ocean animals, essentially "flying" underwater, behavior that is increasingly appreciated by people at zoos and aquariums equipped with glass-walled penguin tanks. The only location people regularly encounter penguins face to face in the sea is in the Galápagos Islands, where fortunate tourists, entering the equatorial Pacific to snorkel or dive, are sometimes approached by the small, inquisitive Galápagos Penguin, the world's northernmost species.

Distribution: Antarctic, Southern Hemisphere

No. of Living Species: 17

No. of Species Vulnerable, Endangered: 7, 3

No. of Species Extinct Since 1600: 0

The seventeen penguin species (family Spheniscidae) are restricted to the Southern Hemisphere. The largest species, the Emperor, stands 3 feet (1 m) tall; the smallest, the Little, only 16 inches (40 cm). All are flightless and use their highly modified wings (flippers) for propulsion underwater. These wings are unique in that, unlike in other birds, the bones are fused together and the wings cannot be folded. Penguin feet are placed far back on the body, and their toes are webbed. The feet are used for steering and braking underwater and for clambering up steep and slippery slopes when going ashore. Penguins are well adapted to cold, having short overlapping feathers (that actually look more like large scales than proper feathers) for waterproofing, a layer of down, and beneath the skin, a thick layer of insulating fat.

Most species are highly social, great numbers of individuals foraging together at sea and breeding in colonies (a few, such as the Little Penguin, often feed alone). They eat fish, squid, and crustaceans (often shrimplike krill), which are captured in their strong, sharp bills after underwater pursuit. Their mouths are lined with rear-facing spines that help them hold onto and swallow slippery, wiggling prey. Some large species in the Antarctic can dive to depths of nearly 900 feet (275 m) and stay under for almost twenty minutes; smaller penguins stay under for briefer periods and dive only to about 200 feet (60 m). Larger marine predators such as Killer Whales and Leopard Seals include penguins in their diets. Penguins are monogamous; they nest in burrows or natural cavities such as holes under rocks; two species, King and Emperor Penguins, make no nest.

Some penguins receive more attention than others because they are sometimes encountered by or, indeed, marketed to, tourists and bird-watchers. These include the Galápagos Penguin; the Jackass Penguin, endemic to the coast of southern Africa and named for its strange braying call; New Zealand's endemic Yellow-eyed Penguin; the four Antarctic species (Adelie, Emperor, Chinstrap, and Gentoo), larger penguins whose breeding colonies are often destinations for nature tours; and the Little Penguin, endemic to Australia and New Zealand, mainly because of the famous "Penguin Parade" on southeast Australia's Phillip Island, a daily event in which tourists line up to watch the penguins emerge from the ocean and waddle to their nearby burrows.

Many penguins are secure, with several species, including Magellanic, Macaroni, Royal, Chinstrap, Rockhopper, King, and Adelie Penguins, having total populations of more than a million individuals. However, seven species are considered vulnerable owing to declining or already small populations. Three others are endangered, due to highly restricted ranges and, for two of them, tiny populations: the Erect-crested Penguin, which breeds only on two small islands off New Zealand; the Yellow-eyed Penguin, which breeds only over small areas of New Zealand and has a population of fewer than five thousand; and the Galápagos Penguin, which is down to between two thousand and five thousand individuals.

HUMBOLDT PENGUIN
Spheniscus humboldti
25.5–27.5 in (65–70 cm)
South America (Chile, Peru)

LITTLE PENGUIN
Eudyptula minor
15.5–17.5 in (40–45 cm)
Australia, New Zealand

IMM

ROCKHOPPER PENGUIN
Eudyptes chrysocome
21.5–24.5 in (55–62 cm)
Southern oceans

ADELIE PENGUIN
Pygoscelis adeliae
28 in (71 cm)
Antarctica

YELLOW-EYED PENGUIN
Megadyptes antipodes
26–30 in (66–76 cm)
New Zealand

CHINSTRAP PENGUIN
Pygoscelis antarctica
27–30.5 in (68–77 cm)
Antarctic and South Atlantic waters

EMPEROR PENGUIN
Aptenodytes forsteri
44–45.5 in (112–115 cm)
Antarctica

JACKASS PENGUIN
Spheniscus demersus
27.5 in (70 cm)
Southern Africa

BRD

BRD

NON-BRD

NON-BRD

RED-THROATED LOON
Gavia stellata
25 in (64 cm)
North America, Eurasia

PACIFIC LOON
Gavia pacifica
26 in (66 cm)
North America, Asia

BRD

BRD

NON-BRD

NON-BRD

YELLOW-BILLED LOON
Gavia adamsii
34 in (86 cm)
North America, Eurasia

COMMON LOON
Gavia immer
32 in (81 cm)
North America, Europe

Loons (Divers)

LOONS, considered by many bird enthusiasts to be the earth's most beautiful waterbirds, are divers specialized for a life spent foraging underwater. Their elegant black and white patterned plumage makes them a favorite subject of wildlife artists and photographers. But loons are best known for inhabiting waters of the far north and for their loud, long, haunting vocalizations given during breeding seasons, often at night, from otherwise quiet lakes of northern New England, northern Minnesota, Canada, and Alaska. The five loon species, called divers in Europe, comprise family Gaviidae (placed in its own order, Gaviiformes). They breed across the northern reaches of Eurasia and North America and winter as far south as the temperate waters off Mexico, Florida, Spain, and China.

Large and heavily built, loons range in length from 21 to 36 inches (53 to 91 cm) and weigh up to 13 pounds (6 kg). Although bulky, their bodies are streamlined, facilitating speedy swimming. They have thick necks, long, sharply pointed bills, relatively small wings, short tails, and short legs. Their feet are modified for underwater propulsion, set far back on the body and with webbed toes. All have dark bills (except the Yellow-billed Loon) and reddish eyes. They have bright, contrasting breeding plumage and dull, brown to black nonbreeding plumage. Like many animals that pursue prey underwater, they are countershaded, dark above and light below. With light coming from above, the back of a countershaded animal is illuminated and the underside shadowed, causing them to blend in with the water, thus rendering them harder for prey to detect. Being specialized for living on and under the water

Distribution: North temperate and Arctic regions

No. of Living Species: 5

No. of Species Vulnerable, Endangered: 0, 0

No. of Species Extinct Since 1600: 0

makes life in the terrestrial world difficult for loons. Because of the structure of their legs and feet, they essentially cannot walk; they move about on land, when necessary, by messily hopping and flopping about. (The word *loon* may come from an old Scandinavian word meaning "clumsy.")

Loons spend much of their year on saltwater, mostly in coastal areas and bays, where the surface remains unfrozen in winter; they breed on freshwater. Loons are amazingly fast and accomplished underwater swimmers, where they pursue and capture fast-moving fish. Most fish are captured 5 to 30 feet (2 to 10 m) down. Loons disappear below the surface with a thrust of their powerful feet and typically reappear far away. When in deeper water, they typically jump into the air at the beginning of a dive, to provide added momentum. In addition to fish, they sometimes eat crustaceans, mollusks, frogs, and even occasionally some plant material. Small fish are swallowed underwater; larger ones and spiny ones are brought to the surface to be disabled and positioned advantageously prior to swallowing. Outside the breeding season, loons gather to feed and roost in flocks of hundreds or even thousands. They are strong fliers. Sometimes they nest on fishless lakes, so parents must make continual flights to other lakes to find food for themselves and their young.

Loons are monogamous and usually pair for life. Mates may remain together during winter or, if separated, typically reunite for the next breeding season when they encounter one another at their breeding lake. They have elaborate courtship behaviors, pairs displaying in flight and moving synchronously on or under the water. Nests are piles of aquatic vegetation on the water or at its edge. Pairs share in nest building, egg incubation, and feeding young, which are provisioned until they are able to fly. These large birds consume a lot of fish; one estimate is that a typical pair and their two chicks consume about a ton (900 kg) of fish during the 15-week breeding season. Eventually the young follow the adults in migration from breeding lakes to the ocean.

None of the loons are currently considered at risk. However, wildlife conservationists worry about them because their preferred habitat of remote, little-disturbed northern lakes is likely to be reduced in the future, as people penetrate ever deeper into wilderness regions. Loons are also very susceptible to water pollution. Large numbers of Yellow-billed Loons, for instance, were killed during Alaska's 1989 Exxon *Valdez* oil spill.

Grebes

GREBES are among the most frustrating birds to watch: no sooner do you fix them in your sight than they dive under water, only to pop up again some distance away. These graceful, slender-necked, fully aquatic diving birds are often mistaken for ducks, but are actually members of a different family (Podicipedidae), one only very distantly related. The nineteen grebe species are distributed nearly globally, with most in the New World. They are 9 to 29 inches (23 to 74 cm) long and built to dive and swim well underwater. They have compact, streamlined bodies, short wings, very little tail, legs placed well back on their bodies, and lobed toes that provide propulsion underwater. Bills are usually sharp-pointed, not ducklike. Grebes in nonbreeding plumage are generally drab-looking, dull brown, gray, or blackish above, whitish below. Many are brighter during breeding season, and some at this time have small color patches on the head. Some species have crests or conspicuous fan-shaped spreads of feathers, often white or yellow, behind the eye. Male and female grebes look alike or nearly so.

Grebes are probably known best for their foraging behavior, called dive-hunting. They dive from surfaces of lakes, rivers, and coastal seawater, locating and then pursuing their prey, including fish, frogs, and crustaceans, underwater. They are speedy swimmers, required for chasing fast-moving fishes. (But being so specialized for fast underwater movement has its drawbacks: because of the structure of their legs and feet, grebes cannot walk; when washed ashore, they can only hop and flop their way back into the water.) In a typical dive a grebe disappears below the surface with a

Distribution:
Worldwide

No. of Living
Species: 19

No. of Species
Vulnerable,
Endangered: 2, 1

No. of Species Extinct
Since 1600: 2 or 3

WHITE-TUFTED GREBE
Podiceps chilensis
9.5–14 in (24–36 cm)
South America

GREAT CRESTED GREBE
Podiceps cristatus
18–23.5 (46–60 cm)
Eurasia, Africa, Australia

BRD

NON-BRD

HORNED GREBE
Podiceps auritus
12–15 in (31–38 cm)
North America, Eurasia

BRD

NON-BRD

PIED-BILLED GREBE
Podilymbus podiceps
12–15 in (30–38 cm)
North America, South America

EARED GREBE
Podiceps nigricollis
11–13.5 in (28–34 cm)
North America, Eurasia, Africa

WESTERN GREBE
Aechmophorus occidentalis
20–29 in (51–74 cm)
North America

AUSTRALASIAN GREBE
Tachybaptus novaehollandiae
9–10.5 in (23–27 cm)
Australasia, Indonesia

HOARY-HEADED GREBE
Poliocephalus poliocephalus
12 in (30 cm)
Australia, New Zealand

thrust of its powerful feet and then reappears some distance away, perhaps with a fish held crosswise in its bill. In a bout of foraging, a grebe will make many consecutive dives, usually staying below the surface each time for less than a minute (typically 10 to 40 seconds), often remaining on the surface between dives for as little as 10 seconds. Most dives are to depths of 3 to 12 feet (1 to 3.5 m). Some grebes, such as the Western Grebe, use their sharply pointed bills to spear prey; fish found in the stomachs of these grebes usually have a hole through the middle. Grebes occasionally also take prey from the water's surface, including insects and their larvae, mollusks and crustaceans; some plant materials may also be consumed. Grebes swallow some of their own feathers, an adaptation to deal with a fish diet. The feathers apparently protect the stomach from sharp fish bones until the bones are sufficiently digested to pass through the rest of the gut. Some grebes also have been found to have small pebbles in their stomachs, which they may swallow to assist the stomach's food-grinding action.

Other than during nesting, grebes tend to spend most of their time alone. They breed on freshwater, usually still water areas with plentiful emergent vegetation. But they spend much of the year on saltwater, where feeding conditions are better and the surface remains unfrozen in winter. Relatively weak flyers, most take to the air infrequently, but some species, those at higher latitudes, are migratory. During migrations and in wintering areas, some grebes become more social, forming into flocks.

Grebes are monogamous and usually nest solitarily. After sometimes spectacular courtship displays (mutual swimming, diving, and thrashing about in the water) a pair builds a floating nest of soggy plant stems that form a platform; if the water level changes, the nest rides with it. Eggs are covered by a parent with nest material whenever the nest is left unattended. Young are able to swim soon after they hatch, but parents help feed and protect them.

Most grebe species have large, widespread populations and are secure; they are not heavily hunted, perhaps because their flesh has an unpleasant taste. Two are considered vulnerable, the Madagascar Grebe and New Zealand Dabchick; one is critically endangered, Peru's Junín Flightless Grebe. Three species became extinct in the twentieth century: the Colombian Grebe, last seen in 1977; Guatemala's Atitlan Grebe, last seen in the 1980s; and Madagascar's Alaotra (Rusty) Grebe, last seen in 1982.

WAVED ALBATROSS
Diomedea irrorata
33.5–36.5 in (85–93 cm)
Pacific waters off northern South America

BLACK-FOOTED ALBATROSS
Diomedea nigripes
27–29 in (68–74 cm)
Northern Pacific

SHORT-TAILED ALBATROSS
Diomedea albatrus
33–37 in (84–94 cm)
Northern Pacific

WANDERING ALBATROSS
Diomedea exulans
42–53 in (107–135 cm)
Southern Ocean

BLACK-BROWED ALBATROSS
Diomedea melanophrys
32.5–36.5 in (83–93 cm)
Southern Ocean

BULLER'S ALBATROSS
Diomedea bulleri
30–32 in (76–81 cm)
Southern Ocean (Pacific side)

LIGHT-MANTLED ALBATROSS
Phoebetria palpebrata
31 in (79 cm)
Southern Ocean

Albatrosses

Lords of the ocean skies and the globe's largest seabirds, ALBATROSSES are among the world's most respected birds. Sailors have long venerated these huge soaring birds (historically because they believed albatrosses contained the souls of lost comrades), and many cultures frown on killing them, perhaps in awe of their amazing flight capabilities. Albatrosses are included in the seabird order Procellariiformes along with the other "tubenosed seabirds," the shearwaters, petrels, and storm-petrels. All lead a pelagic existence; except for island nesting, they spend their entire lives at sea. (Most stay far from continents, so seeing them requires an ocean cruise or a visit to remote islands.) They are called tube-nosed because their nostrils emerge through tubes on the top or sides of their distinctly hooked upper bills (see p. 33). Family Diomedeidae contains the fourteen albatross species, distributed over the world's southern oceans and the northern Pacific. They are large, heavy birds with very long wings and long, heavy bills. The largest is the Wandering Albatross, about 3.5 feet (1.1 m) long, with a wingspan to 11.5 feet (3.5 m). Most albatrosses have pale or whitish heads and underparts but dark tails and wing-tops; faces can be gray, white, or yellowish. Two species are mainly a chocolate brown color.

Albatrosses feed, either solitarily or in small groups, and often at night, on fish, squid, and other invertebrates (crabs, krill) near the surface. Larger species sit on the water and seize prey in their bills; smaller, more agile species also seize prey from the surface while flying. Flying involves a type of nonflapping flight known as dynamic soaring, which takes advantage of

Distribution: Southern oceans, northern Pacific

No. of Living Species: 14

No. of Species Vulnerable, Endangered: 8, 1

No. of Species Extinct Since 1600: 0

strong winds blowing across the ocean's surface. This efficient but peculiar flight takes them in huge loops from high above the ocean, where the wind is fastest, down toward the surface, where friction slows the wind, and then up into the faster wind again to give them lift for the next loop—a kind of roller-coaster flight that requires virtually no wing flapping. Albatross wings are so long and narrow that these birds literally need wind to help them fly, and on absolutely calm days (which, luckily, are rare on the open ocean) they must wait out windless hours sitting on the sea's surface. To take flight in windy conditions they need only spread their wings, and the wind provides sufficient lift to make them airborne. But in low winds, they must face into the wind (like an airplane taking off) and make a takeoff run; at island breeding colonies, often there are actual runways, long, clear paths on the islands' windier sides, usually on slopes, along which they make their downhill takeoff runs. Albatrosses are nicknamed "gooney birds" because they are awkward on land and often make untidy takeoffs and landings.

Monogamous breeders, most albatrosses apparently mate for life. On their remote breeding islands they engage in elaborate courtship dances in which male and female face each other, flick their wings, bounce their heads up and down, and clack their bills together. Nests are scrapes on the bare ground sometimes surrounded by vegetation and pebbles. Male and female alternately incubate eggs and brood young. The other adult flies out to sea, sometimes for days, and searches for food. When it returns, it feeds the chick regurgitated fish, squid, and stomach oil. When the chick's demand for food becomes overwhelming, both adults leave it for long periods as they search vast areas of ocean for enough food. A nestling albatross, at times, can actually weigh more than its parents. They take 5 to 7 years to mature, staying at sea during this period before finally returning to their birthplace to breed. Many do not breed until they are 7 to 9 years old; some live more than 40 years.

Because they cannot become airborne readily from land, albatrosses are easy victims for humans and introduced predators. During the 1800s and early 1900s they were widely killed for their feathers, and entire breeding colonies were destroyed. Currently eight species are considered vulnerable, and one, the Amsterdam Albatross, is endangered; it has a population of only about one hundred and breeds on a single Indian Ocean island.

Petrels and Shearwaters

The seventy-two species of PETRELS and SHEARWATERS comprise a major group of seabirds, but one largely unknown, unseen, and unappreciated except by sailors and bird fanciers. The family, Procellariidae, is included in the tube-nosed seabird order, Procellariiformes, which also encompasses albatrosses and storm-petrels, and has a worldwide oceanic distribution. All species are found in marine habitats, and they spend their entire lives at sea except for short periods of nesting on islands. They are called "tube-nosed" seabirds because their nostrils emerge through tubes on the top or sides of their distinctly hooked upper bills. Like many seabirds, they have a large gland between and above their eyes that permits them to drink seawater; it filters salt from the water and concentrates it. This highly concentrated salt solution is excreted in drops from the base of the bill and the nostril tubes then direct the salt drops to the end of the bill, where they drop off.

Petrels comprise a large group of mid- to large-size seabirds with long wings and hooked bills. Shearwaters, small to midsize, are very similar to petrels, but their bills are longer and thinner, and have smaller hooks. Lengths range from 10 to 39 inches (25 to 99 cm); wingspans can be up to 6.5 feet (2 m). Both groups, along with other tubenoses, are usually dark above and lighter below, although some are all dark or all light; coloring is limited to black, brown, gray, and white.

Petrels and shearwaters are often excellent flyers, some using dynamic soaring (see p. 31) like albatrosses, some alternating flapping flight with gliding. They feed at sea by day or night on squid, fish, and crustaceans. Some

Distribution: All oceans

No. of Living Species: 72

No. of Species Vulnerable, Endangered: 20, 11

No. of Species Extinct Since 1600: 0

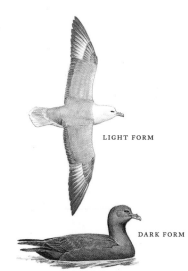

LIGHT FORM

DARK FORM

SOOTY SHEARWATER
Puffinus griseus
15.5–20 in (40–51 cm)
Oceans worldwide

NEWELL'S SHEARWATER
Puffinus newelli
14 in (35 cm)
Hawaii

NORTHERN FULMAR
Fulmarus glacialis
17.5–19.5 in (45–50 cm)
Northern Pacific, Arctic waters

BROAD-BILLED PRION
Pachyptila vittata
10–12 in (25–30 cm)
New Zealand, Southern Atlantic

SOUTHERN GIANT PETREL
Macronectes giganteus
34–39 in (86–99 cm)
Southern Ocean

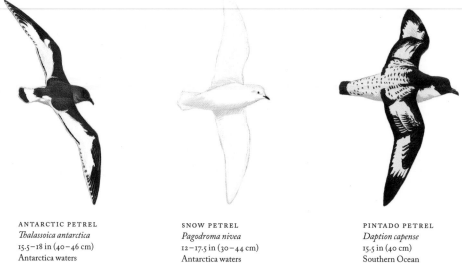

ANTARCTIC PETREL
Thalassoica antarctica
15.5–18 in (40–46 cm)
Antarctica waters

SNOW PETREL
Pagodroma nivea
12–17.5 in (30–44 cm)
Antarctica waters

PINTADO PETREL
Daption capense
15.5 in (40 cm)
Southern Ocean

pluck prey from the surface of the sea using their wings to hover. Petrels and shearwaters are sometimes solitary, but some species occur in small to large flocks; some Short-tailed Shearwater flocks, for instance, number in the millions. Petrels and shearwaters are ocean wanderers, often traveling huge distances; some, for example, breed in the Australian region, and then wander to the northern Pacific.

One of the bird world's more curious spectacles—birds using takeoff runways—occurs daily during shearwater breeding seasons on small islands, such as some of those off Australia's east coast. Here, Short-tailed and Wedge-tailed Shearwaters nest. The birds (locally called muttonbirds because dense breeding colonies in the past provided large amounts of human food in the form of the birds' chicks) feed at sea during the day, land on the islands toward evening, then waddle to their nest burrows to spend the night. At dawn, they waddle to the island's edge (usually a windier shore), mill about in a crowd on a dune above the water, then one by one run down the slope of the dune toward the water, spread their wings, and fly. Given their long, narrow wings, they usually need the takeoff run, like an airplane, to create enough lift to become airborne. Observing shearwater runway launches is a special treat to bird-watchers who visit breeding colonies; the birds generally ignore quiet human spectators.

Petrels and shearwaters breed monogamously, usually in large, dense colonies, often but not always on small oceanic islands. Some nest in the open, with nests simply small depressions in the ground, but most nest either in a burrow they dig themselves or take over, or in a natural cavity, such as rock cavities in cliffs. The burrow or cavity is reused each year by the same pair. Both parents incubate the single egg and feed and brood the chick. These birds probably live an average of 15 to 20 years.

Some species of petrels and shearwaters are incredibly numerous, and the Sooty Shearwater has been nominated as the world's most abundant bird. But because of their often highly restricted nesting sites on small islands and vulnerability during the nesting period, many species in the group are at risk; twenty are considered vulnerable, another eleven, some with extremely small populations, endangered. The Hawaiian Petrel, sometimes considered a subspecies of Dark-rumped Petrel, breeds only in Hawaii; in the past its nests were regularly raided by introduced rats, cats, and mongooses, and it is now endangered.

WHITE-VENTED STORM-PETREL
Oceanites gracilis
6.5 in (16 cm)
Pacific waters off South America

WEDGE-RUMPED STORM-PETREL
Oceanodroma tethys
8 in (20 cm)
Pacific waters off South America

FORK-TAILED STORM-PETREL
Oceanodroma furcata
8.5 in (22 cm)
Northern Pacific

BLACK STORM-PETREL
Oceanodroma melania
9 in (23 cm)
Pacific waters off North and South America

BLACK-BELLIED STORM-PETREL
Fregetta tropica
8 in (20 cm)
South Atlantic, South Pacific, Southern Ocean

WHITE-FACED STORM-PETREL
Pelagodroma marina
8 in (20 cm)
Atlantic and Indian Oceans, Southern Pacific

PERUVIAN-DIVING PETREL
Pelecanoides garnotii
8–9.5 in (20–24 cm)
Pacific waters off northern South America

COMMON DIVING-PETREL
Pelecanoides urinatrix
8–10 in (20–25 cm)
Southern Ocean

Storm-petrels; Diving-petrels

STORM-PETRELS and DIVING-PETRELS are the smallest members of the tube-nosed seabird order, Procellariiformes, which also includes albatrosses, shearwaters, and petrels. All in the order are found only in marine habitats, and they spend most of their lives at sea. They are called tube-nosed because their nostrils emerge through tubes on the top or sides of their hooked upper bills (see p. 33). Twenty species of storm-petrels, with a worldwide oceanic distribution, comprise family Hydrobatidae. These small seabirds, only 5 to 10 inches (13 to 25 cm) long, have proportionately shorter wings and longer legs than the closely related petrels and shearwaters. Like other tubenoses, they are often dark above and lighter below, with colors limited to black, brown, gray, and white.

Although storm-petrels are widespread and often quite abundant, relatively little is known about them. The paradox stems from the difficulty of studying small birds that spend their lives at sea, usually approaching land only to nest on islands. Further, they typically have less contact with people than other seabirds, even following fishing boats much less than do other groups. A consequence of the relative lack of interaction between people and storm-petrels is that a certain mystery surrounds these birds; seafarers in several parts of the world even consider touching a storm-petrel to be bad luck.

Highly agile in the air, storm-petrels fly low and erratically over the sea surface, almost like swallows; they settle on the water only occasionally. They are perhaps best known for their feeding behavior, which occurs by day or night. With their small mouths they take mainly tiny planktonic

STORM-PETRELS

Distribution:
All oceans

No. of Living
Species: 20

No. of Species
Vulnerable,
Endangered: 1, 0

No. of Species Extinct
Since 1600: 1

crustaceans, but also squid and small fish. Although occasionally making shallow dives from the air, the classic storm-petrel feeding method is to pluck prey from the surface of the sea while using their wings to flutter and hover, with their legs dangling, just above the water. Some species actually patter their feet along the water's surface, the tiny splashes or the act of dragging their feet in the water perhaps aiding their feeding in some way. To people, the pattering often makes it look like the birds are trying to walk on the water, and this may point to the origin of the word *petrel*: the Greek word *petros* and the French word *péterelle* refer to the biblical disciple Peter, who tried to walk on water. Storm-petrels, along with other tubenoses, have perhaps the best developed sense of smell of any birds (probably related to the long tubular nostrils). They use this ability to locate nest sites and young when returning from extended foraging trips and probably use it to locate some types of food.

Storm-petrels are hole- or burrow-nesters that breed monogamously in colonies of a few dozen pairs to tens of thousands. The burrow or cavity is reused each year by the same pair. To reduce predation by day-active predators such as gulls, skuas, and raptors, adults of all but one species visit breeding colonies only at night. The single chick is fed a partially digested paste of small fish and crustaceans, eventually growing to outweigh its parents. One night, when it is 7 to 11 weeks old, it leaves the nest burrow, flies out to sea, and probably does not return to land for 2 to 4 years, when first ready to breed.

Because of their often restricted nesting sites on small islands and vulnerability to predators during nesting, some storm-petrels are at risk. One, the Polynesian Storm-petrel, with a small population, is considered vulnerable. Mexico's Guadalupe Storm-petrel, owing mainly to predation by feral cats on its nesting island, recently became extinct.

DIVING-PETRELS, family Pelecanoididae, are four species of small, stocky, tube-nosed seabirds that use their wings like paddles, penguinlike, to feed on planktonic crustaceans underwater. They have a pouch under their tongues in which they store food to eat later or to bring back to their young at the nest. Diving-petrels occur over the southern oceans. One species, the Peruvian Diving-petrel, which breeds on four small islands off South America's western coast, is endangered.

Tropicbirds; Frigatebirds

TROPICBIRDS (family Phaethontidae, with three species) and FRIGATEBIRDS (family Fregatidae, five species) are members of order Pelecaniformes, along with pelicans, boobies, and cormorants; both have mainly tropical ocean distributions. Tropicbirds, considered among the most striking and attractive of tropical seabirds, are midsize white or white and black birds with two very long, thin tail feathers, called streamers, which provide the birds an unmistakable flight silhouette. A pair of tropicbirds engaged in their spectacular aerial courtship display, tail streamers tracing arcs in the sky, is an unforgettable sight. Tropicbirds have long, narrow wings, very short legs unsuited for walking on land (they spend most of their time in the air or on the sea), and webbed feet. They range from 28 to 59 inches long (70 to 105 cm; about half of which is tail streamer) and have wingspans of about 3.3 feet (1 m). Frigatebirds are very large soaring birds (28 to 45 inches, 70 to 114 cm, long), black or black and white, with huge pointed wings that span up to 6.5 feet (2 m) and long forked tails. Males have large red throat pouches that they inflate, balloonlike, during courtship displays.

Tropicbirds are often seen flying alone or in pairs over the ocean or near shore, or inland circling in valleys and canyons. They eat fish, squid, and crustaceans, which they obtain by flying high over the water, spotting food, hovering a bit, then plunging into the water to catch the meal. They rarely feed within sight of land, preferring the open ocean. They travel far and wide on oceanic winds; for instance, individuals banded in Hawaii have been spotted at sea 5,000 miles (8,000 km) away. Tropicbirds do alight on

TROPICBIRDS

Distribution:
Tropical oceans

No. of Living
Species: 3

No. of Species
Vulnerable,
Endangered: 0

No. of Species Extinct
Since 1600: 0

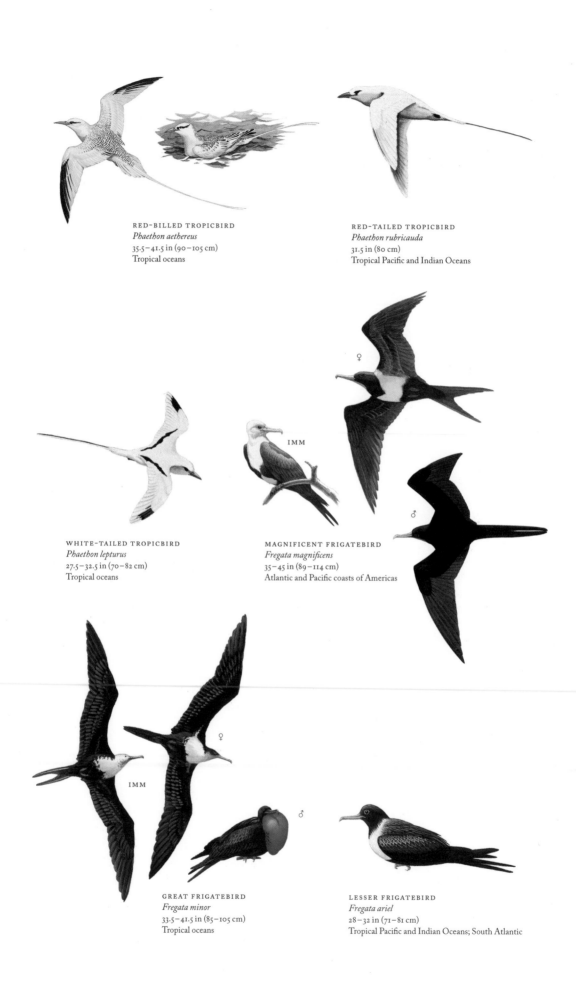

RED-BILLED TROPICBIRD
Phaethon aethereus
35.5–41.5 in (90–105 cm)
Tropical oceans

RED-TAILED TROPICBIRD
Phaethon rubricauda
31.5 in (80 cm)
Tropical Pacific and Indian Oceans

WHITE-TAILED TROPICBIRD
Phaethon lepturus
27.5–32.5 in (70–82 cm)
Tropical oceans

MAGNIFICENT FRIGATEBIRD
Fregata magnificens
35–45 in (89–114 cm)
Atlantic and Pacific coasts of Americas

GREAT FRIGATEBIRD
Fregata minor
33.5–41.5 in (85–105 cm)
Tropical oceans

LESSER FRIGATEBIRD
Fregata ariel
28–32 in (71–81 cm)
Tropical Pacific and Indian Oceans; South Atlantic

the sea surface, but not often; most of their time at sea is spent looking for prey, flying 30 to 65 feet (10 to 20 m) above the water.

Frigatebirds are often seen along tropical seacoasts, especially around fishing boats, from which they may learn to take scraps. They feed on the wing, sometimes soaring effortlessly for hours at a time. They swoop low to catch fish that leap from the water and to pluck squid and jellyfish from the waves. They drink by flying low over the water's surface and sticking their long bill into the water. Although their lives are tied to the sea, frigatebirds cannot swim and rarely, if ever, enter the water voluntarily; with their very long, narrow wings, they have difficulty lifting off. To rest, they land on remote islands, itself a problematic act in the high winds that are common in these places.

Large and beautiful, frigatebirds are a treat to watch as they glide silently along coastal areas, but they have some patterns of behavior that among humans would be indictable offenses. Frigatebirds practice *kleptoparasitism:* they "parasitize" other seabirds, such as boobies, frequently chasing them in the air until they drop recently caught fish. The frigatebird then steals the fish, catching it in midair as it falls. Frigatebirds are also known as Man-o'-war birds, both names referring to warships, and to the birds' *kleptoparasitism*; they also steal nesting materials from other birds, furthering their image as avian pirates.

Frigatebirds and tropicbirds, like other pelican relatives, usually breed in large colonies on small oceanic islands. Tropicbirds breed on cliffs or ledges; frigatebirds, in trees or shrubs. Both nest on bare ground if preferred sites are unavailable, and both are monogamous, mated males and females sharing in nest building, incubation, and feeding young. Frigatebirds exhibit some of the longest-duration parental care among seabirds, with some parents feeding their young at the breeding colony for 15 to 18 months. These seabirds reach sexual maturity slowly (7 or more years in frigatebirds) and live long lives (20 years or more in frigatebirds).

Tropicbirds and some frigatebirds are abundant, but two of the frigatebirds are in jeopardy: the Christmas Island Frigatebird, critically endangered, with a total population of about three thousand, breeds only on part of a single tiny island located between Indonesia and Australia; and the Ascension Frigatebird, considered vulnerable, likewise breeds solely on one tiny island.

FRIGATEBIRDS

Distribution:
Tropical oceans

No. of Living
Species: 5

No. of Species
Vulnerable,
Endangered: 1, 1

No. of Species Extinct
Since 1600: 0

BROWN PELICAN
Pelecanus occidentalis
41.5–60 in (105–152 cm)
Atlantic and Pacific coasts of Americas

AMERICAN WHITE PELICAN
Pelecanus erythrorhynchos
50–70 in (127–178 cm)
Central America, North America

PERUVIAN PELICAN
Pelecanus thagus
41.5–60 in (105–152 cm)
Pacific waters off Peru and Chile

AUSTRALIAN PELICAN
Pelecanus conspicillatus
60–74 in (152–188 cm)
Australia, New Guinea

GREAT WHITE PELICAN
Pelecanus onocrotalus
58.5–69 in (148–175 cm)
Eurasia, Africa

PINK-BACKED PELICAN
Pelecanus rufescens
49–52 in (125–132 cm)
Africa

Pelicans

Highly distinctive in appearance, PELICANS, with their huge, long bills and large bill sacs, or pouches, are known and identifiable even to people who have never seen them in nature. The pelican pouch is among the most caricatured feature of any bird. These very large fishing birds are also familiar to some because they can become quite tame, to the point of perching on piers and taking fish from people. And beach-goers know that a flight of pelicans, passing low and slow overhead, in perfect V-formation or in a single line, is an awesome sight.

Pelicans (family Pelecanidae, with eight species) are distributed essentially worldwide, on fresh- and saltwater. They are members of order Pelecaniformes, the seabird group that includes cormorants, boobies, frigatebirds, and tropicbirds. Ranging in length from 3.3 to 5.9 feet (1 to 1.8 m; females a bit smaller than males), they have compact bodies and short tails but long and fairly broad wings for gliding and soaring (with wingspans to 11.5 feet [3.5 m]). Among the heaviest of all flying birds, pelicans weigh up to 33 pounds (15 kg). Bill length ranges up to an incredible 18 inches (46 cm) or more. Although heavy and ungainly looking, pelicans are excellent flyers. They can use air updrafts to soar high above in circles for hours, and sometimes undertake lengthy daily round-trips between nesting areas and feeding grounds. Other than the Brown Pelican, pelicans are mainly white, with patches of black, brown, or gray.

Most pelicans feed as they swim along the water's surface, using the bill sac as a net to scoop up fish (to 12 inches [30 cm] long). The sac hangs

Distribution:
All continents except
Antarctica

No. of Living
Species: 8

No. of Species
Vulnerable,
Endangered: 1, 0

No. of Species Extinct
Since 1600: 0

from the lower mandible, and the upper mandible acts as a lid. Captured fish are quickly swallowed because the water in the sac with the fish usually weighs enough to prohibit the bird's lifting off from the water. Often, groups of swimming pelicans cooperate in herding fish, trapping them between birds, and then lunging. Brown Pelicans, common in the United States, are the only ones that also plunge from the air, sometimes from a considerable altitude (to 65 feet [20 m]) to dive for meals. Although pelicans eat fish almost exclusively, some crustaceans are also taken.

Pelicans eat a lot, in some species up to 10 percent of their body weight per day; older nestling pelicans take fish up to 1.3 pounds (600 g) in weight. As a necessary result of such intake, pelicans produce a lot of droppings, or guano, which accumulates at some traditional breeding colonies into huge deposits. Guano, rich in phosphates, is a valuable fertilizer and has been long mined in various parts of the world. The Peruvian Pelican, a close relative of the Brown Pelican that is limited to Peru and Chile, is one of a triumvirate of famed "guano birds" (along with the Peruvian Booby and Guanay Cormorant). Over the centuries, these birds laid down thick reserves of guano on their small breeding islands, reserves that supported a very profitable guano industry.

Pelicans usually breed in large colonies on small islands or in places on the mainland that have few mammal predators. Nests are on the ground or in trees or bushes. Monogamy is the rule, mated males and females sharing in nest building, incubation, and feeding the young; at least in some species, a new mate is found each year. Pelicans lay up to five eggs, but often only one young survives to fledging. Typical life span in the wild is probably 15 to 25 years; in captivity, some pelicans live more than 50 years.

Some fishermen view pelicans as competitors for small fish, and these large birds are sometimes persecuted for this reason. North America's pelicans (Brown and American White Pelicans), however, feed mostly on fish not associated with commercial fishing, such as carp, chub, and silverside. The only threatened species is the Spot-billed Pelican of India and Southeast Asia, which, with a recently reduced range and relatively small population, is considered vulnerable. The Brown Pelican was listed by the United States as an endangered species over parts of its range, but it is still common in many areas.

Boobies and Gannets

BOOBIES are large seabirds known for their sprawling, densely packed breeding colonies and their spectacular plunges from heights into the ocean to pursue fish. The three largest species are called GANNETS and some people use this name for all in the group. The term booby apparently arose because the nesting and roosting birds seemed so bold and fearless toward people, which was considered stupid. Actually, these birds bred on isolated islands and cliffs, which meant they had few terrestrial predators, so had never developed, or had lost, fear responses to large land mammals, such as people.

The boobies, family Sulidae (nine species with worldwide oceanic distribution), are part of the order Pelecaniformes, along with pelicans, cormorants, frigatebirds, and tropicbirds. They are large seabirds with tapered, torpedo-shaped bodies; long pointed wings; long tails; long, straight, pointed bills; and often, brightly colored feet. In fact, these birds, mostly white and black, are famous for their spots of bright body coloring. Gannets, mostly southern temperate-water birds, are very similar to boobies but have shorter tails.

Boobies plunge-dive from the air (from heights of up to 100 feet [30 m] or more) to catch fish underwater; they also eat squid. Sometimes they dive quite deep (to more then 50 feet [15 m]), and they often take fish unawares from below, as they rise toward the surface. Some boobies and gannets follow fishing trawlers, catching fish brought up by nets. Unlike most other birds, boobies do not have holes or nostrils at the base of the upper bill for

Distribution:
All oceans

No. of Living
Species: 9

No. of Species
Vulnerable,
Endangered: 1, 1

No. of Species Extinct
Since 1600: 0

IMM

MASKED BOOBY
Sula dactylatra
32–36 in (81–92 cm)
Tropical oceans

IMM

BROWN BOOBY
Sula leucogaster
25–29 in (64–74 cm)
Tropical oceans

IMM

RED-FOOTED BOOBY
Sula sula
26–30.5 in (66–77 cm)
Tropical oceans

BLUE-FOOTED BOOBY
Sula nebouxii
30–33 in (76–84 cm)
Tropical Pacific off North and South America

AUSTRALASIAN GANNET
Morus serrator
33–36 in (84–91 cm)
Oceans off Australia, New Zealand

breathing; the holes are closed over to keep seawater from rushing into their lungs as they plunge-dive. Some boobies, in addition to plunge-diving, also surface-dive occasionally to chase fish underwater and are sometimes spotted foraging on foot in shallow water.

Boobies and gannets tend to breed in large or medium-size colonies on small islands where there are no mammal predators or in isolated mainland areas that are relatively free of predators. Some nest on cliffs or in trees or other vegetation, but most do so on the ground. Nests in a densely packed colony can be as close to one another as 1.5 feet (0.5 m). (Before taking flight, an adult nesting in the middle of such a colony has to thread its way past all the neighboring nests, to the colony edge, and find space to make its takeoff run. Birds returning to the colony simply drop straight from the air onto their own nest.) Boobies are monogamous; mated males and females share responsibility for nest building, incubation, and feeding the young. High year-to-year fidelity to mates, to breeding regions, and to particular nest sites is common. Females usually lay one or two eggs, but if two, typically only a single chick survives to fledging age; one chick often pecks the other to death. Young remain in and around the nest, dependent on the parents, for up to 6 months or more. These are long-lived birds, an average life span lasting 10 to 20 years.

Several boobies—especially the Peruvian Booby, Blue-footed Booby, and Cape Gannet—are among the globe's champion guano producers. Guano, a sought-after fertilizer, is produced from the phosphate-laced droppings of seabirds, from deposits that have accumulated for thousands of years. The islands on which these birds nest have accumulated guano that is yards thick, and this valuable resource has been exploited commercially for centuries. Indeed, the extraction of guano from offshore islands by Peruvian fertilizer companies played a significant role in Peru's historical economic development. Guano mining is destructive because it disturbs the nesting of seabirds and destroys nesting habitat.

Abbot's Booby, a tree-nesting species now limited to a single, small breeding population on the Indian Ocean's Christmas Island, is critically endangered. In addition to loss of some of its nesting habitat (some of it by the guano industry), it is now threatened by an ant species introduced to the island, which is spreading to its nesting sites; the ants can weaken or kill nestlings. One other booby, southern Africa's Cape Gannet, is considered vulnerable.

FLIGHTLESS CORMORANT
Nannopterum harrisi
35–39.5 in (89–100 cm)
Galápagos Islands

GUANAY CORMORANT
Phalacrocorax bougainvillii
28–30 in (71–76 cm)
Coastal South America

RED-LEGGED CORMORANT
Phalacrocorax gaimardi
28–30 in (71–76 cm)
Coastal South America

RED-FACED CORMORANT
Phalacrocorax urile
28–35 in (71–89 cm)
Northern Pacific

LITTLE CORMORANT
Phalacrocorax niger
20–22 in (51–56 cm)
Southern Asia

GREAT CORMORANT
Phalacrocorax carbo
31.5–39.5 in (80–100 cm)
Eurasia, Africa, Australia,
North America

LITTLE PIED CORMORANT
Phalacrocorax melanoleucos
21.5–25.5 in (55–65 cm)
Australasia, Indonesia

ROUGH-FACED SHAG
Phalacrocorax carunculatus
30 in (76 cm)
New Zealand

SPOTTED SHAG
Phalacrocorax punctatus
25–29 in (64–74 cm)
New Zealand

ORIENTAL DARTER
Anhinga melanogaster
33.5–38 in (85–97 cm)
Africa, Asia, Australia

Cormorants; Anhingas

CORMORANTS (family Phalacrocoracidae, with thirty-nine species) and ANHINGAS (family Anhingidae, with two species) are members of order Pelecaniformes, the seabird group that includes pelicans, boobies, frigatebirds, and tropicbirds. They are such successful fishing birds that U.S. sport fisherman and fish farmers occasionally complain to Congress that the birds get more fish than they do and demand that legislators do something about it. Cormorants, known as shags in some regions, have a long historical association with people. Early European seafarers gave them the name *Corvus marinus*, or sea raven; the word *cormorant* derives from this term. Cormorants have been used for centuries in Japan, China, and Central Europe as fishing birds. A ring is placed around a cormorant's neck so that it cannot swallow its catch, and then, leashed or free, it is released into the water. When the bird returns or is reeled in, a fish is usually clenched in its bill. Anhingas are also known as darters, because of the way the birds swiftly thrust their necks forward to spear fish. Owing to their long necks, they are also called snakebirds; indeed, they sometimes swim with only head and neck rising above the water, and the effect is of a snake gliding across the water's surface.

Cormorants, with long necks and long bills with hooked tips, inhabit coasts and inland waterways over much of the world. They are large (18 to 39 inches [45 to 100 cm] long; with wingspans of the biggest species ranging up to 5.2 feet [1.6 m]), usually blackish or black and white. Anhingas, closely related to cormorants, are fresh- and brackish-water birds mostly of

CORMORANTS

Distribution:
Worldwide

No. of Living
Species: 39

No. of Species
Vulnerable,
Endangered: 8, 2

No. of Species Extinct
Since 1600: 1

tropical and subtropical regions. They are also large (31 to 38 in [80 to 97 cm]; wingspan to 4 feet [1.2 m]), dark with white or silvery patches, and have long necks with long, sharply pointed bills.

Diving from the surface of lakes, rivers, lagoons, and coastal saltwater areas, cormorants and anhingas pursue fish underwater. To propel them rapidly through the depths, they have powerful feet set well back on the body and webbed toes. Cormorants, which take crustaceans also, catch food in their bills; anhingas, which also take other aquatic animals such as small turtles, baby crocodiles, and snakes, use their sharply pointed bills to spear their prey. Cormorants dive to at least 165 feet (50 m). They are social birds, foraging, roosting, and nesting in groups (sometimes with hundreds of thousands of individuals); anhingas are less social and sometimes solitary. After swimming, cormorants and anhingas are known for standing on logs, trees, or other surfaces and spreading their wings, presumably to dry them (they may also be warming their bodies in the sun following dives into cold water.)

Cormorants usually breed in large colonies on islands or in isolated mainland areas. On islands and in coastal areas, they typically nest on the ground but sometimes on cliffs or ledges; inland, they tend to nest in trees or bushes, often near or surrounded by water. They are monogamous, mated males and females sharing in nest building, incubation, and feeding young. Usually the male chooses the nest site and brings nesting material (seaweed, plant materials) and the female arranges it. Anhingas, also monogamous, usually nest solitarily or in small groups, and sometimes in association with cormorants, herons, storks, or ibises.

Anhingas and most of the cormorants are very abundant birds. Nonetheless, eight cormorant species are considered vulnerable and two are endangered. The major threats are human disturbance at or around breeding colonies and coastal development on breeding islands. New Zealand's Chatham Island Shag, which is endangered, has a small population and is restricted to three small islands. The other endangered species is the Flightless Cormorant, a larger species with tiny, nonfunctional wings; it is endemic to the Galápagos Islands and has a total population of one thousand or less. A similar species, the Spectacled Cormorant, which had limited flying ability, was hunted to extinction by people about 100 years after it was first discovered on islands in the Bering Sea.

ANHINGAS

Distribution:
All continents except
Antarctica

No. of Living
Species: 2

No. of Species
Vulnerable,
Endangered: 0

No. of Species Extinct
Since 1600: 0

Herons, Egrets, and Bitterns

Tall birds standing upright and still in shallow water or along the shore, staring intently into the water, are usually members of the heron family—the HERONS, EGRETS, and BITTERNS. These are beautiful, medium to large wading birds that occur throughout temperate and tropical regions of both hemispheres. The family, Ardeidae, has about sixty species; it is included in order Ciconiiformes, with the ibises, spoonbills, and storks. Herons and egrets are very similar, but egrets tend to be all white and have longer nuptial plumes—special, long feathers—than the darker-colored herons. Bitterns are elusive, mainly heavily streaked birds that most often prowl dense marsh vegetation.

Herons and egrets have slender bodies, long necks (often coiled when perched or still, producing a short-necked, hunched appearance), long, pointed bills, and long legs with long toes; bittern necks are shorter. They range in size from 1 to 4.5 feet (0.3 to 1.4 m), and some have wingspans that approach 6.5 feet (2 m). Most are attired in soft shades of brown, blue, gray, and black and white. Many are exquisitely marked with small colored patches of facial skin or broad areas of spots, bars, or streaks (such as in the bitterns and tiger-herons). Some species during breeding have a few nuptial plumes trailing down from the head, neck, chest, or back. The male and female look alike, or nearly so.

Herons and egrets frequent all sorts of aquatic habitats: along rivers and streams, in marshes and swamps, and along lake and ocean shorelines. Many spend most of their foraging time standing motionless in or adjacent to the

Distribution:
All continents except
Antarctica

No. of Living
Species: 63

No. of Species
Vulnerable,
Endangered: 5, 3

No. of Species Extinct
Since 1600: 2 or 3

BOAT-BILLED HERON
Cochlearius cochlearius
17.5–20 in (45–51 cm)
Central America, South America

SNOWY EGRET
Egretta thula
19–27 in (48–68 cm)
North America, South America

CATTLE EGRET
Bubulcus ibis
18–22 in (46–56 cm)
Worldwide

STRIATED (GREEN) HERON
Butorides striatus
14–19 in (35–48 cm)
North America, South America,
Africa, Asia, Australia

YELLOW-CROWNED NIGHT-HERON
Nyctanassa violacea
20–27.5 in (51–70 cm)
North America, South America

AMERICAN BITTERN
Botaurus lentiginosus
22–33.5 in (56–85 cm)
North America, Central America

GRAY HERON
Ardea cinerea
35.5–38.5 in (90–98 cm)
Africa, Eurasia

LITTLE BITTERN
Ixobrychus minutus
10.5–14 in (27–36 cm)
Africa, Eurasia, Australia

MADAGASCAR POND-HERON
Ardeola idae
18.5 in (47 cm)
Madagascar, Africa

PIED HERON
Egretta picata
17–21.5 in (43–55 cm)
Australia, New Guinea,
Indonesia

water, waiting in ambush for unsuspecting prey to wander within striking distance. Then, in a flash, they shoot their long, pointed bills into the water to grab or spear the prey. Herons will also slowly stalk prey and, occasionally, even actively pursue it. They take anything edible that will fit into their mouths and go down their throats, mostly small vertebrates, including fish, frogs, and turtles, and small invertebrates like crabs. On land, they take mostly insects, but also vertebrates such as small rodents. Herons typically are day-active, but many of the subgroup known as night-herons forage at least partly nocturnally. Most herons are social birds, roosting and breeding in colonies, but some are predominantly solitary.

Among the most intriguing of the family is the Cattle Egret. It has made a specialty of following grazing cattle and other large mammals, walking along and grabbing insects and small vertebrates that are flushed from their hiding places by the moving mammals. A typical pasture scene is a flock of these egrets interspersed among a cattle herd, with several of the white birds perched atop the unconcerned mammals. Until recently the species was confined to the Old World, but within the last 150 years, the Cattle Egret, on its own, invaded the New World. Perhaps blown off course by a storm, the first ones, probably from Africa, landed in South America in about 1877. Cattle Egrets have now colonized much of the region between northern South America and the southern United States.

Most herons breed in monogamous pairs within breeding colonies of various sizes. They are known for their elaborate courtship displays. Nests, built mainly by females with materials collected by males, are usually platforms of sticks placed in trees, in marsh vegetation, or on the ground. The young, born helpless, are fed regurgitated food by parents. Herons often lay more eggs than the number of chicks they can feed. For instance, many lay three eggs when there is sufficient food to feed only two chicks. In such years of food shortage, the smallest chick dies because it cannot compete for food from the parents against its larger siblings, and because the larger siblings usually attack it.

Herons are large birds that nest in colonies, so are certainly vulnerable to threats from people. However, they are, in general, highly successful and plentiful, and some of them are among the globe's most conspicuous and commonly seen waterbirds. Five species are considered vulnerable, and three others, all Asian, are currently endangered.

JABIRU
Jabiru mycteria
51–60 in (129–152 cm)
Mexico, Central America,
South America

WOOD STORK
Mycteria americana
32.5–40 in (83–102 cm)
North America, South America

ASIAN OPENBILL STORK
Anastomus oscitans
32 in (81 cm)
Southern Asia

YELLOW-BILLED STORK
Mycteria ibis
37.5–41.5 in (95–105 cm)
Africa, Madagascar

SADDLE-BILLED STORK
Ephippiorhynchus senegalensis
57–59 in (145–150 cm)
Africa

WOOLLY-NECKED STORK
Ciconia episcopus
34–37.5 in (86–95 cm)
Africa, southern Asia

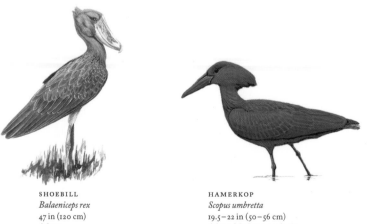

SHOEBILL
Balaeniceps rex
47 in (120 cm)
Africa

HAMERKOP
Scopus umbretta
19.5–22 in (50–56 cm)
Africa, Madagascar, Arabia

Storks; Shoebill; Hamerkop

STORKS are unmistakable; these huge, ungainly-looking wading birds occur worldwide in tropical and temperate regions. Some are nearly as tall as a person. The nineteen species (family Ciconiidae) are usually included in the same order as herons (Ciconiiformes; along with the ibises and spoonbills), which they resemble generally. However, they differ in significant ways, both structurally (being overall heavier birds with more massive bills) and behaviorally (by flying with the neck extended, not retracted as in herons, and often soaring high in the sky on their broad wings). Storks are represented in the United States solely by the Wood Stork, seen mainly in Florida. Nonetheless, storks are well known to Americans, at least by reputation, because of the legends of storks delivering newborn babies. Such stories have existed for 500 years or more and are traceable to northern Europe, where the European White Stork often nests on rooftops and is one of the region's most famous and celebrated birds.

Most storks are white with black wing patches, but some are predominantly dark and many have patches of red or yellow on head, bill, and/or legs. Africa's Marabou, the largest, stands up to 5 feet (1.5 m) tall and has a wingspan up to 9.5 feet (2.9 m), placing it among the largest flying birds. Another African species, Abdim's Stork, is the smallest, at only 30 inches (75 cm). Male and female storks look similar, but males are often a bit larger.

Storks feed by walking slowly through shallow water or damp fields and marshy areas looking for suitable prey, essentially anything that moves: small rodents, young birds, frogs, reptiles, fish, earthworms, mollusks, crustaceans, and insects. Food is grabbed with the tip of the bill and swallowed quickly.

STORKS

*Distribution:
All continents except
Antarctica*

*No. of Living
Species: 19*

*No. of Species
Vulnerable,
Endangered: 2, 3*

*No. of Species Extinct
Since 1600: 0*

SHOEBILL

*Distribution:
Central Africa*

*No. of Living
Species: 1*

*No. of Species
Vulnerable,
Endangered: 0, 0*

*No. of Species Extinct
Since 1600: 0*

HAMERKOP

*Distribution:
Africa, Madagascar,
Arabia*

*No. of Living
Species: 1*

*No. of Species
Vulnerable,
Endangered: 0, 0*

*No. of Species Extinct
Since 1600: 0*

Some use their long pointed bills as daggers to impale fish. African and Asian Openbill Storks have specially shaped bills they use to open the shells of freshwater snails and mussels. Many storks are strongly opportunistic: they patrol at the periphery of bush fires, catching insects and small rodents fleeing flames, and visit garbage dumps. Storks can be found either alone or, if food is plentiful in an area, in groups. They are excellent flyers, often soaring high overhead for hours during hot afternoons. Some species are known to fly 50 miles (80 km) or more daily between feeding areas and roosting or nesting sites.

Storks are monogamous, most nesting colonially in trees or in reeds near water. They are known for their vigorous courtship displays, which often include male and female throwing their heads up and down and clattering their bills together. Nests are made of sticks and often become enormous because pairs, instead of building new nests, simply add new material each year to old ones. Both sexes incubate eggs; chicks take regurgitated food from the adults' throats.

Main threats to storks are habitat loss and hunting, because of their large size and conspicuousness during nesting. Fifteen of the nineteen species are threatened in at least some parts of their ranges. For instance, owing to the draining of marshes and swampy areas for development, several stork species have become rare or virtually extinct in Thailand. Globally, however, only two species are considered vulnerable, and three others, all Asian, are endangered.

The SHOEBILL and HAMERKOP are unique, storklike, mainly African birds that are different enough to be placed in their own single-species families, but both are included in the stork order, Ciconiiformes. Hamerkops (family Scopidae; also called Anvil-headed Storks) are largish (22 inches, 55 cm) brown birds with long, backward pointing crests and long, slightly hooked bills that produce a distinctive hammer-headed silhouette. They are birds of wetlands, feeding on amphibians, fish, crustaceans, worms, and insects. Hamerkops are often considered magical, and numerous local taboos have developed around them; consequently, most people leave them alone and the birds sometimes even become semitame. Shoebills (family Balaenicipitidae) are large (3.9 feet, 1.2 m), gray, storklike birds with massive bills. They live in swamps, marshes, and flooded grasslands, eating fish, frogs, snakes, lizards, turtles, and even young crocodiles. Little is known of their populations, but the species is not globally threatened.

Ibises and Spoonbills

IBISES and SPOONBILLS are largish, strange-looking, long-legged wading birds that, although a bit alien to most Americans, are quite common and conspicuous in many parts of the world. There are a total of twenty-six ibis species and six spoonbills (together comprising family Threskiornithidae; in order Ciconiiformes, along with herons and storks). They are globally distributed, but only four species occur in the United States, all but one of them mainly in eastern coastal regions. The anatomical feature that renders the appearance of these birds strange, and allows one to rapidly distinguish them from other large waders, is the bill: very long, thin, and down-curved in ibises; long and expanded at the end like a flattened spoon in spoonbills.

Distribution:
All continents except
Antarctica

No. of Living
Species: 33

No. of Species
Vulnerable,
Endangered: 1, 6

No. of Species Extinct
Since 1600: 0

These are medium to large birds, 20 to 43 inches (50 to 110 cm) long; the largest have wingspans of up to 4.4 feet (1.35 m). Most ibises are white, brown, or blackish, often with a glossy patch on the wings, and have bare heads. Most spoonbills are all or mainly white. As with many other birds adapted to walk and feed in swamps, marshes, wet fields, and along waterways, ibises and spoonbills have long legs and very long toes that distribute the birds' weight, allowing them to walk among marsh plants and across floating vegetation without sinking. Within a species male and female ibises and spoonbills look alike. Although the two groups look different because of their bills, they are very closely related; there have even been instances of an ibis mating with a spoonbill and producing hybrid young.

Ibises and spoonbills are day-active gregarious birds that feed in marshes, swamps, coastal lagoons, shallow bays, lakes, and mangroves. Ibises insert

ROSEATE SPOONBILL
Ajaia ajaja
27–34 in (69–86 cm)
North America, South America

WHITE IBIS
Eudocimus albus
22–27.5 in (56–70 cm)
North America, South America

PLUMBEOUS IBIS
Theristicus caerulescens
28–30 in (71–76 cm)
South America

BUFF-NECKED IBIS
Theristicus caudatus
28–32 in (71–81 cm)
South America

ROYAL SPOONBILL
Platalea regia
29–30 in (74–76 cm)
Australasia, Indonesia

STRAW-NECKED IBIS
Threskiornis spinicollis
23–35 in (59–89 cm)
Australia, New Guinea

SACRED IBIS
Threskiornis aethiopicus
25.5–30 in (65–76 cm)
Africa, Madagascar, Iraq

HADADA IBIS
Bostrychia hagedash
25.5–57 in (65–145 cm)
Africa

their long bills into soft mud and poke about for insects, snails, crabs, frogs, or tadpoles. Apparently they feed by touch, not vision; whatever the bill contacts that feels like food is grabbed and swallowed. Spoonbills also feed by touch, lowering their bills into the water and sweeping them from side to side, stirring up the mud, then grabbing fish, frogs, snails, or crustaceans that they contact. Ibises and spoonbills are monogamous breeders, nesting in colonies or solitarily. They make stick nests, sometimes mixed with green vegetation, in bushes, trees, or reed beds.

Among the oddest of America's birds is the Roseate Spoonbill, which in the United States occurs mainly along the Gulf Coast. The only spoonbill in the New World, it is pink; this trait, combined with its large size and conspicuous spatula-like bill, make it one of the easiest birds in the hemisphere to identify. The species is also distinguished for the significant, if not widely acknowledged, role it played in the initiation of North America's conservation movement. Many of the ibises and spoonbills historically have been systematically killed for their beautiful long feathers; the Roseate Spoonbill was intensively hunted in the United States for its wings, which were used in their entirety to make decorative pink fans. The shooting of the birds in the 1800s and early 1900s, to supply the fan industry, drastically reduced the bird's populations in Florida and Texas. Publicity about the wholesale killing helped bring about the start of the wildlife conservation movement. The spoonbill first obtained federal protection in the 1940s and has now rebounded, even recolonizing parts of the Gulf Coast from which it had been exterminated. That a member of the ibis/spoonbill family helped stimulate contemporary wildlife conservation is especially fitting because one of the group, Africa's Sacred Ibis, could be perhaps the first bird protected by humans. It inhabited the Nile River, considered the source of life by the ancient Egyptians, so that ancient civilization treated it with great respect. The bird was believed to be the incarnation of Thoth, the god of knowledge and wisdom, and it is likely there would have been proscriptions about harming the species.

One ibis species is considered vulnerable and five others are endangered, four of them critically so: the Dwarf Olive and Northern Bald Ibises of Africa and the White-shouldered and Giant Ibises of Asia, each of which have total populations of three hundred or fewer individuals.

IMM

GREATER FLAMINGO
Phoenicopterus ruber
47 in (120 cm)
Eurasia, Africa, Central America, South America,
West Indies

CHILEAN FLAMINGO
Phoenicopterus chilensis
41.5 in (105 cm)
South America

IMM

LESSER FLAMINGO
Phoeniconaias minor
31.5–35.5 in (80–90 cm)
Africa, southern Asia

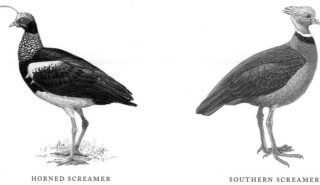

HORNED SCREAMER
Anhima cornuta
33–37 in (84–94 cm)
South America

SOUTHERN SCREAMER
Chauna torquata
32.5–37.5 in (83–95 cm)
South America

Flamingos;
Screamers

Approaching remote lakes in some parts of the world, one may be bewildered when, from afar, it appears as if the water is covered by a pink froth. As the lake is neared, bewilderment turns to wonder as the froth is revealed to be dense concentrations of large, slender, pink wading birds, FLAMINGOS. The five species of flamingos (family Phoenicopteridae) have a broad but highly specialized distribution, occurring in parts of Mexico, the Caribbean, South America, and the Old World, but only in shallow lakes, lagoons, or estuaries with very high salt concentrations or those that are strongly alkaline. The "American Flamingo" occurs naturally in various parts of the United States but does not breed here; it is actually a subspecies of the Greater Flamingo, the world's most widely distributed species (which occurs in both the Old and New Worlds).

Flamingos have long necks, long legs, and heavy, black-tipped, conspicuously down-curved bills. The sexes look the same, but males are often a bit larger. They stand 33 to 59 inches (85 to 150 cm) tall and are pink, pink and crimson, or pink and white. The Greater Flamingo, a shocking pink, is the largest and pinkest of the group.

Aside from their striking coloration and long-legged beauty, flamingos are best known for their filter feeding. A flamingo eats by lowering its bill into the water, resting it upside down on the bottom, and sucking in water, mud, and bottom debris. The materials are pushed through comblike bill filters, which catch the flamingo's meal: tiny crustaceans such as brine shrimp. The flamingos' food explains their choice of habitats, as salty or brackish waters

FLAMINGOS

Distribution:
All continents except
Australia, Antarctica

No. of Living
Species: 5

No. of Species
Vulnerable,
Endangered: 1, 0

No. of Species Extinct
Since 1600: 0

usually have high densities of tiny invertebrate animals. Flamingos are highly social, occurring sometimes, where they have healthy populations, in groups of thousands—which again, given their coloring, makes for quite a sight. They nest in colonies, each pair building a mud-mound nest on top of which the single egg is placed. Both male and female incubate the egg and feed the chick with a milky fluid produced by glands in their digestive tracts.

Two flamingos whose distributions are restricted to high altitude salt lakes in the Andes Mountains, the Andean and Puna Flamingos, have seriously declining populations, and the former species is considered vulnerable. The flamingo feeding method, dredging the bottom sediment from shallow water, makes them vulnerable to toxic chemicals in the mud, such as lead from the lead shot used in shotguns. The Lesser Flamingo, with its main abundance in Africa, is the most numerous species, with a total population of perhaps 4 to 6 million.

SCREAMERS

*Distribution:
Neotropics*

*No. of Living
Species: 3*

*No. of Species
Vulnerable,
Endangered: 0, 0*

*No. of Species Extinct
Since 1600: 0*

SCREAMERS are large, stocky birds that on the ground look like plump blackish geese or chickenlike birds, but while flying resemble eagles with long legs. The three species belong to a uniquely South American family, Anhimidae, which is related to the duck family. They range in length from 30 to 37 inches (75 to 95 cm), weigh up to 11 pounds (5 kg), and have smallish chickenlike heads.

The Northern Screamer is limited to northern Colombia and Venezuela, but the Horned and Southern Screamers have broad South American distributions. Screamers are essentially waterbirds but they spend much of their time out of lakes and marshes, feeding in other open habitats such as meadows and flooded savanna. They graze mainly on aquatic vegetation, consuming roots, leaves, flowers, and seeds. Surprisingly for large waterbirds, they often perch in treetops to give their loud calls. Screamers are strong fliers and frequently soar to great heights. Unlike many ducks, and more in keeping with geese and swans, they may mate for life. Their nests, made of sticks, reeds, and other vegetation, are in or near shallow water.

None of the screamer species are currently threatened, but the Northern Screamer is heading in that direction. Its total population may now be under ten thousand, and it suffers from habitat loss (draining of its wetlands for agriculture), hunting, egg collection, and even from capture as pets. Horned Screamer populations in the Amazon region fell dramatically in the 1980s and 1990s, probably owing to increased hunting.

Ducks, Geese, and Swans

DUCKS, GEESE, and SWANS are among the world's most familiar and recognized birds. They have been objects of people's attention since ancient times, chiefly as food sources. They typically have tasty flesh, are fairly large and therefore economical to hunt, and are easier and less dangerous to catch than many other animals. They also domesticate easily and breed in captivity. Several species have been domesticated for thousands of years. The Muscovy Duck, native to South America, in its domesticated form is a common farmyard inhabitant in many parts of the world, including Africa, and white "domestic ducks" (there are other color forms as well) are descendants of the Mallard, the world's most abundant wild duck. Wild ducks adjust well to the proximity of people, to the point of taking food from them, a behavior that surviving artworks show has been occurring for at least 2,000 years. Hunting ducks and geese for sport is also a long-practiced tradition. As a consequence of these long interactions, and the research on these animals engendered by their use in agriculture and sport, substantial scientific information has been collected on the group; some ducks and geese are among the most well-known of birds. Many species have beautiful markings, and the group as a whole is a favorite with bird-watchers.

The 150 or so species of ducks, geese, and swans (family Anatidae), known collectively as waterfowl, are distributed throughout the globe in habitats ranging from open seas to high mountain lakes. Although an abundant, diverse group throughout most temperate regions, representation is relatively limited in tropical areas. These birds vary quite a bit in coloring and size

Distribution:
Worldwide

No. of Living
Species: 157

No. of Species
Vulnerable,
Endangered: 14, 12

No. of Species Extinct
Since 1600: 4

BLACK-BELLIED WHISTLING DUCK
Dendrocygna autumnalis
17–21 in (43–53 cm)
North America, South America

MUSCOVY DUCK
Cairina moschata
26–33 in (66–84 cm)
Central America, South America

ORINOCO GOOSE
Neochen jubata
24–26 in (61–66 cm)
South America

TORRENT DUCK
Merganetta armata
17.5 in (45 cm)
South America

PUNA TEAL
Anas puna
16.5 in (42 cm)
South America

EGYPTIAN GOOSE
Alopochen aegyptiacus
28.5 in (72 cm)
Africa

SPUR-WINGED GOOSE
Plectropterus gambensis
29.5–39.5 in (75–100 cm)
Africa

COMB DUCK
Sarkidiornis melanotos
22–29.5 in (56–75 cm)
South America, Africa, Southern Asia

TRUMPETER SWAN
Cygnus buccinator
59−71 in (150−180 cm)
North America

GREATER WHITE-FRONTED GOOSE
Anser albifrons
25.5−34 in (65−86 cm)
North America, Eurasia

EMPEROR GOOSE
Chen canagicus
26−35 in (66−89 cm)
Alaska, Siberia

NENE (HAWAIIAN GOOSE)
Branta sandvicensis
22−28 in (56−71 cm)
Hawaii

COMMON MERGANSER
Mergus merganser
23−26 in (58−66 cm)
North America, Eurasia

WOOD DUCK
Aix sponsa
17−20 in (43−51 cm)
North America

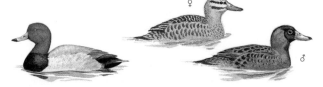

TUFTED DUCK
Aythya fuligula
15.5−18.5 in (40−47 cm)
Eurasia, Africa

REDHEAD
Aythya americana
15.5−22 in (40−56 cm)
North America

MASKED DUCK
Oxyura dominica
12−14 in (30−36 cm)
Central America, South America,
West Indies

MAGPIE GOOSE
Anseranas semipalmata
27.5–35.5 in (70–90 cm)
Australia, New Guinea

BLACK SWAN
Cygnus atratus
43.5–55 in (110–140 cm)
Australia, New Zealand

CAPE BARREN GOOSE
Cereopsis novaehollandiae
29.5–39.5 in (75–100 cm)
Australia

HARDHEAD
Aythya australis
19 in (48 cm)
Australia

RADJAH SHELDUCK
Tadorna radjah
20–24 in (51–61 cm)
Australia, New Guinea

GREEN PYGMY-GOOSE
Nettapus pulchellus
12–14 in (30–36 cm)
Australia, New Guinea

PINK-EARED DUCK
Malacorhynchus membranaceus
14–17.5 in (36–45 cm)
Australia

CAPE SHOVELER
Anas smithii
20.5 in (52 cm)
Africa

(from ducks only 12 inches [30 cm] long to swans 5.5 feet [1.6 m] long that have wingspans of 7.5 feet [2.3 m]), but all share the same major traits: duck bills, long, slim necks, pointed wings, webbed toes, and short tails. There is a preponderance within the group of grays and browns, and black and white, but many have at least small patches of bright color. In some species male and female look alike, but in others there are differences; in the ducks, males are brighter than females and often quite gaudy.

Ducks, geese, and swans are birds of wetlands, spending most of their time in or near the water. They eat aquatic plants or small fish, but some forage on land for seeds and other plant materials. Many of the typical ducks are divided into divers and dabblers. Diving ducks, such as mergansers, scoters, scaups, and goldeneye, plunge underwater for their food; dabblers, such as pintails, shovelers, wigeon, teal, and the Mallard, take food from the surface of the water or put their heads down into the water to reach food at shallow depths. When not breeding, almost all waterfowl gather into flocks, usually of their own species but sometimes mixed. Small ducks fly in clumps, larger ones in lines, and geese and swans in V-formations.

Ducks place their nests on the ground in thick vegetation, in tree holes or rock crevices, or on the water surface. Typically nests are lined with downy feathers that the female plucks from her own breast. In many of the ducks, females perform most of the breeding duties, including incubation and shepherding and protecting the ducklings. Some of these birds, particularly among the geese and swans, have lifelong bonds, and male and female share equally in breeding duties. Young are able to run, swim, and feed themselves soon after they hatch.

Waterfowl populations are much affected by the draining of wetlands (for agriculture or development) and water pollution. Most species are abundant, their populations secure, but a few are currently rare or endangered. The only critically endangered duck in the New World is the Brazilian Merganser, with a population of only several hundred. Two species endemic to Hawaii, the Nene (Hawaiian Goose) and Hawaiian Duck, have been severely reduced through predation by introduced mongooses, cats, and rats, along with hunting and habitat loss; the duck, very closely related to Mallards, is endangered, and the goose, even after intensive conservation efforts, is still vulnerable.

TURKEY VULTURE
Cathartes aura
25–32 in (64–81 cm)
North America, South America

BLACK VULTURE
Coragyps atratus
22–27 in (56–68 cm)
North America, South America

KING VULTURE
Sarcoramphus papa
28–32 in (71–81 cm)
Central America, South America

GREATER YELLOW-HEADED VULTURE
Cathartes melambrotus
29–32 in (74–81 cm)
South America

ANDEAN CONDOR
Vultur gryphus
39.5–51 in (100–130 cm)
South America

New World Vultures

Birds at the very pinnacle of their profession, eating dead animals, VULTURES are highly conspicuous and among the most frequently seen birds in many regions of South and Central America, Mexico, and parts of the United States and southern Canada. That they feast on rotting flesh does not reduce the majesty of these large, soaring birds as they circle for hours high over field and forest. The family of American vultures, Cathartidae, has only seven species; several are abundant animals with broad, continent-wide distributions. The two largest are known as condors, the California and Andean Condors. Traditional classifications place the family within order Falconiformes, with the hawks, eagles, and falcons, but some biologists believe the group is more closely related to storks. (Birds called vultures in Eurasia and Africa, members of the hawk and eagle family, are only distantly related to New World vultures.)

Vultures (excluding the huge condors) range from 22 to 32 inches (56 to 81 cm) long, with wingspans to 6.5 ft (2 m). The two condors range up to 51 inches (130 cm), and the largest, the Andean, has a wingspan up to 10.5 feet (3.2 m). Generally black or brown, vultures have hooked bills and curious, featherless heads, usually with the bare skin colored red, yellow, or orange, or some combination of these. The Turkey Vulture, the United States' most widespread species, is named for its red head, which reminds people of a turkey head. The vultures' bare heads probably aid them in preventing gore from accumulating on feathers there, which might interfere with seeing, hearing, and breathing. Male and female vultures look alike; males are slightly larger than females.

Distribution: New World

No. of Living Species: 7

No. of Species Vulnerable, Endangered: 0, 1

No. of Species Extinct Since 1600: 0

Although vultures survive entirely or almost entirely on carrion, a few, such as Black and King Vultures, occasionally kill and eat small animals, usually newborn or those otherwise defenseless; some, especially the Black Vulture, also take fruit. The latter species, one of the most frequently encountered birds of tropical America, is also very common at trash dumps. Indeed, vultures are important scavengers around South and Central American towns and villages, helping to clean up garbage and remove dead animals from highways. (Aside from esthetic and safety considerations, such work is important ecologically, quickly recycling back into food webs energy and nutrients trapped in animal carcasses.) Most vultures soar during the day in groups, looking for and, in at least some species, perhaps sniffing for, food. Turkey Vultures can find carcasses in deep forest and also buried carcasses, strongly implicating smell, as opposed to vision, as the method of discovery. Vultures can cover huge areas and survey great expanses of habitat each day in their search for dead animals.

Vultures are monogamous breeders. Both sexes incubate eggs, which are placed on the ground in protected places or on the floor of a cave or tree cavity, and both sexes feed young regurgitated carrion for 2 to 5 months until they can fly. Young vultures at the nest very rarely become food for other animals, and it may be that the odor of the birds and the site, awash as it is in decaying animal flesh, keeps predators away. Also, when threatened, vultures may spit up partially digested carrion, which, we can assume, is a strong defense against harassment.

Most New World vultures are somewhat common or very common birds. The only species in dire trouble is the California Condor, now critically endangered; for a time it was extinct in the wild. The main causes of the decline of these condors were hunting (they were persecuted especially because ranchers believed they ate newborn cattle and other domesticated animals); their ingestion of poisonous lead shot from the carcasses they fed on; and the thinning of their eggshells owing to the accumulation of pesticides (DDT) in their bodies. The last few free-ranging individuals were caught during the mid-1980s in southern Californian for use in captive breeding programs. The total captive population is now more than one hundred, and several have been released back into the wild in California and Arizona, with varying success. The Andean Condor has undergone considerable population declines (for similar reasons) and is now rare in its high Andean habitats.

Hawks, Eagles, and Kites; Osprey; Secretarybird

HAWKS, EAGLES, and KITES are raptors, or birds of prey, birds that make their living hunting, killing, and eating other animals. When one hears the term raptor, one usually thinks of soaring hawks and eagles that swoop to catch unsuspecting rodents, but the feeding behavior of these birds is quite diverse. The two main raptor families (both included in order Falconiformes) are the Accipitridae, containing the hawks, kites, and eagles (accipitrids), and the Falconidae, containing the falcons. In addition to birds called hawks, kites, and eagles, accipitrids include buzzards, bazas, griffons, and Old World vultures. The approximately 235 species are common and conspicuous animals over most of their range, which is global except for Antarctica.

Accipitrids vary considerably in size and coloring, but all are similar in form—fierce-looking birds with hooked, pointed bills, and powerful feet with hooked, sharp claws (talons). They vary in size from South America's tiny Pearl Kite, at about 10 inches (15 cm) and with a wingspan of 21 inches (53 cm), to large eagles and vultures up to 4.9 feet (1.5 m) long that weigh up to 13 pounds (6 kg) and 26 pounds (12 kg), respectively; the Himalayan Griffon has the longest wingspan, about 10 feet (3 m). Female accipitrids are usually larger than males. Most species are gray, brown, black, white, or a combination of those colors, usually with brown or black spots, streaks, or bars on various parts. The plumage of these birds is quite beautiful when viewed close-up, which, unfortunately, is difficult to do. Males and females are usually alike in color pattern. A number of species have conspicuous crests.

Many accipitrids are birds of open areas, above which they soar during the

HAWKS, EAGLES, AND KITES

Distribution:
Worldwide

No. of Living
Species: 236

No. of Species
Vulnerable,
Endangered: 22, 12

No. of Species Extinct
Since 1600: 0

EGYPTIAN VULTURE
Neophron percnopterus
23–27.5 in (58–70 cm)
Africa, Eurasia

CAPE VULTURE
Gyps coprotheres
43.5 in (110 cm)
Southern Africa

AFRICAN MARSH-HARRIER
Circus ranivorus
17.5–19.5 in (44–49 cm)
Africa

IMM

AUGUR BUZZARD
Buteo augur
21.5–23.5 in (55–60 cm)
Africa

AFRICAN FISH-EAGLE
Haliaeetus vocifer
25–28.5 in (63–73 cm)
Africa

AFRICAN CROWNED EAGLE
Stephanoaetus coronatus
31.5–39 in (80–99 cm)
Africa

DARK CHANTING-GOSHAWK
Melierax metabates
17.5 in (45 cm)
Africa

SECRETARYBIRD
Sagittarius serpentarius
49–59 in (125–150 cm)
Africa

OSPREY
Pandion haliaetus
23 in (58 cm)
Worldwide

GOLDEN EAGLE
Aquila chrysaetos
29.5–35.5 in (75–90 cm)
North America, Eurasia, Africa

NORTHERN GOSHAWK
Accipiter gentilis
19–27 in (48–68 cm)
North America, Eurasia

DARK FORM

LIGHT FORM

SWAINSON'S HAWK
Buteo swainsoni
19–22 in (48–56 cm)
North America, South America

SHORT-TAILED HAWK
Buteo brachyurus
14.5–18 in (37–46 cm)
North America, South America

BROAD-WINGED HAWK
Buteo platypterus
13.5–17.5 in (34–44 cm)
North America, South America

♂ ♀

♀

WHITE-TAILED KITE
Elanus leucurus
15–17 in (38–43 cm)
North America, South America

SNAIL KITE
Rostrhamus sociabilis
15.5–17.5 in (40–45 cm)
South America, Central America,
Florida, Cuba

WHITE HAWK
Leucopternis albicollis
18.5–20 in (47–51 cm)
Central America, South America

HARPY EAGLE
Harpia harpyja
35–41.5 in (89–105 cm)
Central America, South America

CRESTED GOSHAWK
Accipter trivirgatus
14.5–18 in (37–46 cm)
Southern Asia

BRAHMINY KITE
Haliastur indus
17.5–20 in (45–51 cm)
Southern Asia, Australia, New Guinea

BAT HAWK
Macheiramphus alcinus
17.5 in (45 cm)
Africa, Southeast Asia

WHISTLING KITE
Haliastur sphenurus
20–23 in (51–59 cm)
Australia, New Guinea

WEDGE-TAILED EAGLE
Aquila audax
32–41 in (81–104 cm)
Australia, New Guinea

SPOTTED HARRIER
Circus assimilis
19.5–24 in (50–61 cm)
Australia, Indonesia

day, using the currents of heated air that rise from the sun-warmed ground to support and propel them as they search for meals. But they are found in all types of habitats, including woodlands, forests, and rainforests. They are meat-eaters. Most hunt and eat live prey, but many will also eat carrion. They usually hunt alone, although, when mated, the mate is often close by. Most take mainly vertebrate animals, including some larger items such as large rodents, rabbits, monkeys, and, in Australia, even small kangaroos. A number of species specialize on reptiles, including snakes. Some, such as the Bald Eagle, eat fish. About fifty species exist solely or largely on insects, and some at least occasionally eat fruit. Prey is snatched with talons first, then killed and ripped apart with the bill. Many raptors are territorial, solitary individuals or breeding pairs defending an area for feeding and, during the breeding season, for reproduction. Displays that advertise a territory and may be used in courtship consist of spectacular aerial twists, loops, and other acrobatic maneuvers.

Nests are constructed of sticks that both sexes place in a tree or on a rock ledge. Usually only the female incubates and gives food to the nestlings. The male hunts, bringing food to the nest for the female and for her to provide to nestlings. Both sexes feed the young when they get a bit older; they can first fly at 4 to more than 15 weeks of age, depending on species size. After fledging, young remain with the parents for several more weeks or months until they can hunt on their own.

Although many raptors are common birds, they tend to occur at low densities. Most species are secure, but some are threatened by habitat destruction, and large raptors such as eagles are persecuted by ranchers for allegedly killing livestock. Conservation is difficult because raptors roam very large areas, some breeding and wintering on different continents, and for many, insufficient information exists about their ecology and behavior. Globally, twenty-two accipitrids are vulnerable; four are endangered; and eight are critically endangered.

Two accipitrid relatives are the OSPREY (family Pandionidae, with one species), a large (22 inches [56 cm]) fish-eating raptor of coastal areas and some inland lakes and rivers that occurs worldwide; and the long-legged SECRETARYBIRD (family Sagittariidae, with one species), a terrestrial African eagle (to 59 inches [150 cm]) that strides about open grassland and savannas, foraging for its prey of rodents, insects, and small reptiles.

OSPREY

Distribution:
All continents except
Antarctica

No. of Living
Species: 1

No. of Species
Vulnerable,
Endangered: 0, 0

No. of Species Extinct
Since 1600: 0

SECRETARYBIRD

Distribution:
Sub-Saharan Africa

No. of Living
Species: 1

No. of Species
Vulnerable,
Endangered: 0, 0

No. of Species Extinct
Since 1600: 0

CRESTED CARACARA
Polyborus plancus
19.5–23 in (49–59 cm)
North America, South America

BLACK CARACARA
Daptrius ater
16–18.5 in (41–47 cm)
South America

LAUGHING FALCON
Herpetotheres cachinnans
17.5–21 in (45–53 cm)
Mexico, Central America, South America

APLOMADO FALCON
Falco femoralis
14.5–17.5 in (37–45 cm)
Mexico, Central America, South America

PEREGRINE FALCON
Falco peregrinus
13.5–19.5 in (34–50 cm)
Worldwide

GYRFALCON
Falco rusticolus
19–23.5 in (48–60 cm)
North America, Eurasia

COLLARED FALCONET
Microhierax caerulescens
6.5 in (17 cm)
Southern Asia

FOX KESTREL
Falco alopex
14–15.5 in (35–39 cm)
Africa

NANKEEN KESTREL
Falco cenchroides
12–14 in (30–35 cm)
Australia, New Guinea

Falcons

FALCONS, speedy, streamlined, predatory birds that often feed by snatching smaller birds out of the air, have long had close relationships with people. Falconry, in which captive falcons are trained to hunt and kill game at a person's command, is considered one of the oldest sports; it is known to have been practiced in Asia 4,000 years ago. More recently, the American Kestrel, a small falcon, has associated itself with human-altered landscapes, becoming one of the Americas' most common telephone-wire birds. The family, Falconidae, which occurs worldwide, has sixty-one species, including the forest-falcons and caracaras, which are limited to the New World, as well as the "true" falcons. Some have very broad distributions; at the extreme, the Peregrine Falcon is found globally, enjoying the most extensive natural distribution of any bird. The falcon family is included in order Falconiformes, along with the other birds of prey, the hawks, kites, and eagles.

Falcons are usually distinguishable from hawks by their long, pointed wings, which allow the rapid, acrobatic flight for which these birds are justifiably famous. They share with hawks strongly hooked bills and sharply curved talons. They range in length from 6 to 23 inches (15 to 59 cm), the largest being the Gyrfalcon, a fearsome predator of the far north that has a wingspan to 4.3 feet (1.3 m). Falcons come in combinations of gray, brown, black, and white, usually with patches of dark spots, streaks, or bars. Females are usually larger than males, in some species noticeably so.

Occurring in most habitats, from remote natural areas to city centers, falcons have remarkable eyesight and swift flight capabilities, which are

Distribution:
All continents except
Antarctica

No. of Living
Species: 61

No. of Species
Vulnerable,
Endangered: 4, 0

No. of Species Extinct
Since 1600: 1

useful for their fast, aerial pursuit and capture of flying birds: they are birdhawks. Most people are familiar with stories of Peregrines diving through the atmosphere (called stooping) at speeds above 100 mph (160 kph) to stun, grab, or knock from the sky an unsuspecting bird. But some falcons eat more rodents than birds, and some even take insects. For example, the small kestrels perch on trees, rocks, or wires, scanning the ground for large insects or small mammals, birds, lizards, or snakes. Kestrels also have the ability to hover over a site where prey has been sighted. Some falcons specialize on particular prey. Latin America's Laughing Falcon, a bird of open fields and forest edge, specializes on snakes. After grabbing its dinner, it immediately bites off the head, a smart move because it even takes highly venomous snakes. Another species specializes on catching bats on the wing at dawn and dusk. Caracaras, slow flyers, have long legs and often forage by walking on the ground; some forest dwellers consume mainly wasps and fruit. Forest-falcons perch motionless for long periods on tree branches, waiting to ambush prey such as birds and lizards. Falcons usually live and hunt alone or in solitary pairs.

Falcons nest on cliff edges, in rock cavities, in tree hollows, or in old stick nests of other birds; some make a stick nest, others apparently make no construction. In most, the male hunts for and feeds the female while she lays eggs and incubates them; both sexes feed nestlings. The parents continue to feed the youngsters for several weeks after fledging until they are proficient hunters.

Four falcon species are considered vulnerable, three of them Old World kestrels; no falcons are currently endangered. Only one New World species may be at risk, southern South America's Striated Caracara. Conservation of falcons and other raptors is difficult because the birds are often persecuted for a number of reasons (hunting, pet trade, ranchers protecting livestock), and they roam very large areas. Peregrine Falcons disappeared from the eastern United States by the mid-1970s, a consequence mainly of decades of exposure to pesticides like DDT, which caused their eggshells to thin and break. After DDT was banned, Peregrines eventually returned to many of their old haunts. These beautiful falcons have been introduced to cities such as Chicago by people seeking to reestablish populations in areas from which they disappeared. They roost on skyscrapers, breed on high building ledges or bridges, and maintain themselves, and keep the cities cleaner, by eating pigeons.

Megapodes

The MEGAPODES, or mound-builders, of the Australia/New Guinea region are some of the world's most intriguing birds and a staple subject of nature documentaries. The plain but amazing facts are these: some of the species construct enormous mounds of soil and vegetation (up to 40 feet [12 m] across and 15 feet [4.5 m] high) and lay their eggs in tunnels in the mounds. The heat emitted by the decaying, fermenting plant material is the main source of warmth for incubation (supplemented by the sun). These are the only birds to use for incubation a heat source other than their own bodies. Males attending the mounds apparently regulate the temperature toward one best for their eggs' development by scratching more matter onto the mounds or taking some of it off. They do this by thrusting their bill, and sometimes their whole head, into the mound, to sense its temperature. When eggs hatch, it usually takes chicks 2 days or so to dig themselves out of the mound. They are strongly precocial, able to run and feed themselves, and are on their own as soon as they emerge. Some megapode species do not build mounds but lay eggs in sandy soil or in holes dug on beaches and let the sun incubate them.

Megapodes (meaning "big foot"; family Megapodiidae) are included in order Galliformes with the pheasants, partridges, grouse, quail, and turkeys. Their classification is controversial, but there seem to be about twenty-one species, in Australia, New Guinea, eastern Indonesia, the Philippines, and on some Pacific islands. They are mostly dull brownish birds with small crests, about 20 inches (50 cm) long. The ones known as brush-turkeys are larger, to 28 inches (70 cm), black with bare heads and necks and large tails.

Distribution:
Australasia,
Philippines, Pacific
islands

No. of Living
Species: 21

No. of Species
Vulnerable,
Endangered: 7, 2

No. of Species Extinct
Since 1600: possibly 1
or 2

MALLEEFOWL
Leipoa ocellata
23.5 in (60 cm)
Australia

ORANGE-FOOTED SCRUBFOWL
Megapodius reinwardt
14–18.5 in (35–47 cm)
Australia, New Guinea, Indonesia

AUSTRALIAN BRUSH-TURKEY
Alectura lathami
23.5–27.5 in (60–70 cm)
Australia

MALEO
Macrocephalon maleo
21.5 in (55 cm)
Sulawesi

RED-BILLED BRUSH-TURKEY
Talegalla cuvieri
17.5–22 in (45–56 cm)
New Guinea

Also known generally as mound-birds, thermometer birds, incubator birds, and in Australia, as scrubfowl and brush-turkeys, megapodes are mostly omnivores. Australia's three species, the Orange-footed Scrubfowl, Australian Brush-turkey, and Malleefowl, are the best known of the group. They eat a lot of plant material such as seeds, shoots, roots, buds of herbs, fruit, and berries, but also small invertebrates such as insects and worms. They are ground-dwelling birds, seldom flying unless given no other choice. They are usually shy and inconspicuous, but individuals in parks and other public areas, familiar with people, show themselves readily. Scrubfowl are usually seen in pairs (they may mate for life), which usually appear to have territories that are defended from other scrubfowl. Brush-turkeys are often solitary in most of their daily activities, but when food is plentiful (such as around picnic areas), groups of up to twenty or more may gather. Male brush-turkeys are strongly territorial, defending the area around their breeding mounds with aggressive displays, chasing, and deep, booming vocalizations. They are sedentary, the same males, for example, owning the same territories for several years. Malleefowl are quiet, inconspicuous birds that, in pairs, occupy large territories that may contain several nest mounds (but only one is used each breeding season).

One major problem for mound-building megapodes is that huge mounds of dirt and vegetation are easy to detect. In Australia, for instance, monitor lizards and foxes take advantage of this, locating mounds and excavating eggs. In New Guinea, people also get involved. Traditionally, a mound belongs to the person who finds it. He may decide to dig up all the eggs at once, or gather them slowly, over a period of time to keep them fresh longer.

Nine megapode species or subspecies are vulnerable or already endangered. The Australian Brush-turkey and Orange-footed Scrubfowl are secure, but Australia's endemic Malleefowl is considered vulnerable, its population declining, mainly from habitat loss and predation on adults and eggs by foxes. The Polynesian Megapode, endemic to Tonga, and the Micronesian Megapode, of Palau and the United States' Northern Mariana Islands, are both endangered, the Polynesian species critically so. Both are restricted to small islands and have tiny populations. Adults of both species are sometimes still hunted by people, and adults and eggs are preyed upon by a variety of introduced predators including rats, cats, dogs, and pigs.

SCALY-BREASTED PARTRIDGE
Arborophila chloropus
12 in (30 cm)
Southeast Asia

CRIMSON-HEADED PARTRIDGE
Haematortyx sanguiniceps
10 in (25 cm)
Borneo

SILVER PHEASANT
Lophura nycthemera
24–47 in (61–120 cm)
Southeast Asia

RED JUNGLEFOWL
Gallus gallus
16.5–29.5 in (42–75 cm)
Southern Asia

GREAT ARGUS
Argusianus argus
28.5–78.5 in (72–200 cm)
Southeast Asia

CRESTED WOOD-PARTRIDGE
Rollulus rouloul
10 in (26 cm)
Southeast Asia

CRESTED FIREBACK
Lophura ignita
22–27.5 in (56–70 cm)
Southeast Asia

Pheasants, Partridges, and Grouse; Buttonquail

America's PHEASANTS, PARTRIDGES, GROUSE, and PTARMIGAN are not generally considered real beauties, being known more as drab brown game birds. But the main family of these chickenlike birds, Phasianidae, with a natural Old World distribution, contains some of the globe's most visually striking larger birds, chiefly among the pheasants, like the Silver Pheasant, Crested Fireback, and Common Peafowl illustrated here. The most historically (and gastronomically) significant, if usually unheralded, member of the group is Asia's Red Junglefowl, the wild ancestor of domestic chickens.

All chickenlike birds (except buttonquail) are contained in order Galliformes. In the past, most (excluding the megapodes and curassows) were included in family Phasianidae, but more recently, the grouse (treated here), which occur over North America and northern Eurasia, have been separated into their own family of 18 species, Tetraonidae, and the New World quail into their own family (treated on p. 87). Phasianidae itself now contains 155 species, including partridges, francolins, junglefowl, Old World quail, and pheasants. Several Old World species, such as Chukar, Gray Partridge, and Ring-necked Pheasant, were introduced to North America as game birds and are now common here.

Birds in these groups are stocky, with short, broad, rounded wings; long, heavy toes with claws adapted for ground-scratching; short, thick, chickenlike bills; and short or long tails, some of the pheasants having tails to 5 feet (1.5 m) long. Some small quails, such as the Harlequin Quail, are only about 6 inches (15 cm) long. Many species, particularly among the pheasants, are exquisitely marked with bright colors and intricate patterns,

PHEASANTS AND PARTRIDGES

Distribution:
Old World

No. of Living
Species: 155

No. of Species
Vulnerable,
Endangered: 32, 9

No. of Species Extinct
Since 1600: 3

RUFFED GROUSE
Bonasa umbellus
17–19 in (43–48 cm)
North America

SPRUCE GROUSE
Falcipennis canadensis
15–17 in (38–43 cm)
North America

ROCK PTARMIGAN
Lagopus mutus
13–15 in (33–38 cm)
North America, Eurasia

BLUE GROUSE
Dendragapus obscurus
17.5–22.5 in (44–57 cm)
North America

WHITE-TAILED PTARMIGAN
Lagopus leucurus
12.5 in (32 cm)
North America

CHUKAR
Alectoris chukar
12.5–1V5.5 in (32–39 cm)
Southern Asia

KALIJ PHEASANT
Lophura leucomelanos
19.5–29 in (50–74 cm)
Southern Asia

COMMON PEAFOWL
Pavo cristatus
35.5–90.5 in (90–230 cm)
Southern Asia

BLACK FRANCOLIN
Francolinus francolinus
14 in (35 cm)
Southern Asia

RED-BILLED FRANCOLIN
Francolinus adspersus
13–15 in (33–38 cm)
Africa

GRAY-WINGED FRANCOLIN
Francolinus africanus
12.5 in (32 cm)
Africa

YELLOW-NECKED SPURFOWL
Francolinus leucoscepus
14 in (35 cm)
Africa

HARLEQUIN QUAIL
Coturnix delegorguei
7 in (18 cm)
Africa, Madagascar, Arabia

BROWN QUAIL
Coturnix ypsilophora
8 in (20 cm)
Australia, New Guinea, Indonesia

BARRED BUTTONQUAIL
Turnix suscitator
6.5 in (16 cm)
Southern Asia

GROUSE

Distribution:
North America,
Eurasia

No. of Living
Species: 18

No. of Species
Vulnerable,
Endangered: 1, 1

No. of Species Extinct
Since 1600: 0

BUTTONQUAIL

Distribution:
Africa, southern Asia,
Australia

No. of Living
Species: 16

No. of Species
Vulnerable,
Endangered: 2, 1

No. of Species Extinct
Since 1600: 0

and conspicuous crests are common. The long tails of male peafowl (the peacock) are among nature's most ornate and colorful constructions. Other species are dully colored. The sexes can look alike or different; males are often a bit larger than females.

Quails, francolins, partridges, and pheasants feed and nest on the ground. Pheasants are forest birds; partridge and some quail tend to be in forest undergrowth or in nearby open areas; francolins and quail prefer open grasslands and scrub. All these birds eat seeds, fruit, and insects, often scratching with their large feet and bills to expose food under the soil surface or in leaf litter. Most species travel in small family or multifamily groups of perhaps four to fifteen individuals, but some are more solitary. With their short, rounded wings, these birds are not built for sustained flight; most fly only short distances, such as when making an escape from predators.

Mating in this group is highly variable, some species being monogamous, but others, such as many pheasants, being promiscuous. In the latter species, a male's contribution to breeding is limited to mating with several females. The females incubate eggs and rear young by themselves. Nests are placed on the ground, either in a bare scrape or perhaps one lined with leaves or grass. Young are born covered with downy feathers and are soon ready to leave the nest, follow parents, and feed themselves.

In the grouse family, one species is vulnerable and one, the Gunnison Sage Grouse (restricted to Colorado and Utah), is endangered. Within the pheasants and partridges, more than thirty species are vulnerable and nine are endangered. About half of these are pheasants, which all occur in Asia. Because they are forest birds at a time when Asian forests are increasingly being cleared, and because they are desirable game birds, the large, showy pheasants will remain a conservation problem for some time.

BUTTONQUAIL look like small quail but differ in foot structure, in having more pointed wings, and because males are considerably smaller than females. The group is actually more closely related to rails and cranes than to any of the quails, and, indeed, the small buttonquail family (Turnicidae, with sixteen species) is included in the rail/crane order, Gruiformes. Buttonquail, which occur in Africa, southern Asia, and Australia, inhabit grassy areas, eating grass seeds and small insects from the ground. Breeding behavior is peculiar in that many of the roles of the sexes are reversed, and one female mates with several males (polyandry). Two buttonquail species are vulnerable and one, in Australia, is endangered.

New World Quail

NEW WORLD QUAIL, fairly small chickenlike game birds of North, Central, and South America, are included in order Galliformes with other, similar birds—pheasants, partridges, grouse, guineafowl, and turkeys. The family, Odontophoridae, has thirty-one species, variously called quail, tree-quail, wood-quail, wood-partridges, and bobwhites. Many are drably colored—various shades of brown and gray with black and white spots and streaks, colors and patterns that serve them well as camouflage in their on-the-ground lifestyles. But some have exquisitely marked faces and heads, the former detailed with fine dark and light stripes, the latter equipped with conspicuous crests. It is for these often perky crests (along with their reputation as important game birds), in such common species as California Quail, Gambel's Quail, and Mountain Quail, that this group is probably best known in North America.

New World quail have stocky, compact bodies with short necks; short, broad wings; short, strong legs; and short, thick, slightly down-curved bills. Most have short tails. They range in size from 6.5 to 14.5 inches (17 to 37 cm), the largest being Mexico's Long-tailed Tree-quail. The sexes can look alike or different.

Quail are terrestrial birds that feed and nest on the ground. Most flights are very brief, but over short distances, such as when making a quick escape from a predator or a hunter, these birds are powerful, swift flyers. In fact, one of their main defenses is the speed with which they can take flight. When approached, they initially hide and remain motionless, but then make

Distribution:
New World

No. of Living
Species: 31

No. of Species
Vulnerable,
Endangered: 4, 1

No. of Species Extinct
Since 1600: 0

SPOTTED-BELLIED BOBWHITE
Colinus leucopogon
8.5 in (22 cm)
Central America

NORTHERN BOBWHITE
Colinus virginianus
9.5 in (24 cm)
North America

SCALED QUAIL
Callipepla squamata
10 in (25 cm)
North America

GAMBEL'S QUAIL
Callipepla gambelii
10 in (25 cm)
North America

CALIFORNIA QUAIL
Callipepla californica
10 in (25 cm)
North America

SINGING QUAIL
Dactylortyx thoracicus
8.5 in (22 cm)
Mexico, Central America

MONTEZUMA QUAIL
Cyrtonyx montezumae
8.5 in (22 cm)
North America

amazingly quick takeoffs and accelerations. The explosive launch combined with the loud whirring sound of their wings produces a startle effect that often prevents them from becoming a predator's dinner. In most situations, however, they prefer walking or running to flying.

New World quail occur in diverse habitats, from forest and forest edge to savanna and open agricultural lands. They spend most of their time on the ground, although some in Central and South America roost at night in trees. They are mainly vegetarians, eating seeds and other plant matter such as shoots, some leaves, buds, berries, roots, and tubers, but also fairly opportunistic, taking insects and other small invertebrate animals, especially during the breeding season. All peck at food but some also scratch the ground with their large feet, clearing away vegetation and soil, looking for buried delicacies. Most species are gregarious, traveling in small family groups, called coveys, of up to thirty individuals, but some, such as the California Quail, sometimes congregate in groups of a thousand or more.

In contrast to many other of the world's chickenlike birds, a good many of which are promiscuous breeders, with no long-term pair-bonds between males and females, most of the New World quail are apparently monogamous, males and females forming pair-bonds and sharing parental care. Nests are on the ground, simple shallow depressions lined with vegetation. Both sexes incubate eggs. One of the most intriguing things about quail and other chickenlike birds is the rapid development of their young, which are amazingly precocial, able to run after their parents and begin pecking at small food items just hours after hatching. The young take 3 to 4 weeks to reach a good size and independence but, because baby quail are such tasty prey for so many carnivorous birds and mammals, they can fly at less than 2 weeks, giving them a better chance of survival.

New World quail have been food sources for people for many centuries and they are still widely hunted. This, combined with habitat destruction, has led to sharp reductions in some species. Four, in Mexico and South America, are vulnerable. One species, Colombia's Gorgeted Wood-Quail, is critically endangered, with a tiny range and a total population below a thousand. North American species are generally in good shape and many are common targets of hunters. The Northern Bobwhite, for example, known throughout the eastern United States for the male's sharp "bob-white" call, has a huge population; an estimated 20 million per year are taken by hunters.

CHACO CHACHALACA
Ortalis canicollis
19.5–22 in (50–56 cm)
South America

HIGHLAND GUAN
Penelopina nigra
23–25.5 in (59–65 cm)
Central America

SPECKLED CHACHALACA
Ortalis guttata
17.5–23.5 in (45–60 cm)
South America

SPIX'S GUAN
Penelope jacquacu
26–30 in (66–76 cm)
South America

BLUE-THROATED PIPING GUAN
Pipile cumanensis
23.5–27 in (60–69 cm)
South America

SALVIN'S CURASSOW
Mitu salvini
29.5–35 in (75–89 cm)
South America

BARE-FACED CURASSOW
Crax fasciolata
29.5–33.5 in (75–85 cm)
South America

Curassows, Guans, and Chachalacas

CURASSOWS, GUANS, and the entertainingly named CHACHALACAS are medium-size to large chickenlike birds that occur over warmer regions of the New World. Their habits are little known because, being mostly forest birds of Central and South America that often inhabit relatively inaccessible areas, studying them is difficult. The group is noted for the striking handsomeness and odd head ornaments of some of the curassows and guans; the loud morning and evening calls (sound like "cha-cha-LAW-ka") of chachalaca males, which are some of the most characteristic background sounds of tropical American forests; and the fact that, owing to forest destruction and hunting, fully 25 percent of the fifty species in the family are currently on conservation watch lists. The family, Cracidae, is included in order Galliformes, along with other chickenlike birds such as pheasants, partridges, quail, grouse, and turkeys.

Members of the family range in length from about 16 to 36 inches (40 to 92 cm); the largest of the group, the Great Curassow, the size of a small turkey, weighs up to 9.5 pounds (4.3 kg). Chachalacas are the smallest; curassows and some of the guans, the largest. All have long legs and long, heavy toes. Many have conspicuous crests. The colors of their bodies are generally drab, with gray, brown, olive, or black and white; some appear glossy in the right light. They typically have small patches of bright coloring such as yellow, red, or orange on parts of their bills, their cheeks, or on a hanging throat sac. Male Great Curassows, for instance, all black above and white below, have a bright yellow knob on the top of the bill; and Horned Guans, also black

Distribution: South Texas, Mexico, Central and South America

No. of Living Species: 50

No. of Species Vulnerable, Endangered: 7, 7

No. of Species Extinct Since 1600: 1 (extinct in the wild)

above and light below, have a conspicuous reddish "horn" arising from the forehead. Within the group, males are larger than females; the sexes are generally similar in coloring, except for the curassows, in which females are drabber than males.

Birds in this family tend to inhabit moist forests at low and middle elevations. Guans and curassows are birds of deep forest, but chachalacas prefer more open areas such as forest edges and even clearings. Guans and chachalacas are mainly arboreal, staying high in the treetops as they pursue their diet of fruit, young leaves, and tree buds, and the occasional frog or large insect. For such large birds in trees, they move with surprising grace, running quickly and carefully along branches. One's attention is sometimes drawn to them when they jump and flutter upward from branch to branch until they are sufficiently in the clear to take flight. While guans and chachalacas will occasionally come down to the forest floor to feed on fallen fruit, curassows (and larger guans) are terrestrial birds, more in the tradition of turkeys and pheasants. They stalk about on the forest floor, seeking fruit, seeds, and insects. Paired off during the breeding season, members of this family typically are found at other times of the year in small flocks of ten to twenty.

Curassows, guans, and chachalacas are monogamous breeders, the sexes sharing nesting duties. Several of the guans are known for producing loud whirring sounds with their wings during the breeding season, presumably during courtship displays. A simple, open nest of twigs and leaves is placed in vegetation or in a tree. Young leave the nest soon after hatching to hide in the surrounding vegetation, where they are fed by the parents (in contrast to most of the chickenlike birds, which feed themselves after hatching). Several days later, the fledglings can fly short distances. The family group remains together for a time, the male leading the family around the forest.

Significant conservation problems exist for members of this group, especially the larger guans and curassows. They are chiefly forest birds at a time when neotropical forests are increasingly being cleared. They are also desirable game birds, hunted by local people for food. As soon as new roads penetrate virgin forests in Central and South America, one of the first chores of workers is to shoot curassows for their dinner. Unfortunately, curassows reproduce slowly, raising only small broods each year. Seven species are considered vulnerable, and seven others, endangered (three of them critically so).

Guineafowl;
Turkeys

GUINEAFOWL and TURKEYS are medium-size to large chickenlike game birds that sometimes are included in a huge, catch-all family with pheasants, quail, grouse, and partridges, but they are also sometimes separated into their own families, and some recent research supports this view. All are contained in order Galliformes. The six guineafowl species (family Numididae) are native to Africa. However, owing to their use as domesticated food sources and ornamental birds, guineafowl have been spread far and wide; the Helmeted Guineafowl now occurs in feral populations in such far-flung sites as Madagascar, Yemen, the West Indies, and Florida.

Guineafowl are heavy-bodied with strong, rather short legs and short, arched bills. Their heads and necks are mostly featherless, the bare skin often red or blue. Some have bushy crests, and the Helmeted Guineafowl sports a bony knob on its head. Plumage is highly distinctive—black or dark gray heavily marked with white spots. The spotting is unseen at a distance, but up close, the effect is quite beautiful. The spotted plumage gives rise to the Latin name of the Helmeted Guineafowl, *Numida meleagris* (and also to the family and genus names of turkeys): Meleager was a Greek hero whose death so distressed his relatives that they turned into birds, the tears they shed forming white spots on their dark mourning clothes. Guineafowl range in length from 16 to 28 inches (40 to 72 cm); the largest, the Vulturine Guineafowl, weighs up to 3.5 pounds (1.6 kg).

Favoring lightly wooded and savanna habitats, guineafowl feed on the ground, snapping up seeds, leaves, fruit, insects, spiders, and even small

GUINEAFOWL

Distribution:
Africa

No. of Living
Species: 6

No. of Species
Vulnerable,
Endangered: 1, 0

No. of Species Extinct
Since 1600: 0

HELMETED GUINEAFOWL
Numida meleagris
21–25 in (53–63 cm)
Africa

VULTURINE GUINEAFOWL
Acryllium vulturinum
23.5–28.5 in (60–72 cm)
Africa

CRESTED GUINEAFOWL
Guttera pucherani
18–22 in (46–56 cm)
Africa

OCELLATED TURKEY
Meleagris ocellata
28–36 in (71–91 cm)
Central America

WILD TURKEY
Meleagris gallopavo
35.5–43.5 in (90–110 cm)
North America

frogs. They use their powerful legs and feet to scratch the ground in search of food and to dig for roots and bulbs. Although reluctant to take to the air, they invariably fly up into tree branches to sleep at night; they will also fly to escape predators.

Guineafowl are apparently monogamous. Nests consist of simple shallow scrapes in soil under dense foliage or long grass, perhaps lined with leaves or a little grass. The female incubates alone while the male stays protectively nearby. Young leave the nest soon after hatching and can feed themselves but stay close to their parents for protection. Most guineafowl species are secure; one, West Africa's White-breasted Guineafowl, is considered vulnerable, owing to a rapidly declining population and deforestation of its rainforest habitat.

There are only two turkeys (family Meleagrididae): the Wild Turkey of the United States and northern Mexico and the Ocellated Turkey of Mexico's Yucatán Peninsula (including northern Guatemala and Belize). Turkeys, however, actually roam more widely because people have introduced the former species to such diverse spots as Hawaii, Europe, Australia, and New Zealand. Ground dwellers, turkeys are unmistakable because of their large size, generally dull (but iridescent) plumage, and bare heads and necks with bright red or blue skin and hanging wattles. They range up to 3.6 feet (1.1 m) tall, and weigh up to 22 pounds (10 kg). The Ocellated's name arises with the eyelike images (ocelli) adorning the birds' plumage.

Ocellated Turkeys are primarily birds of low-elevation wet forests and clearings, but they can also be found in open, brushy areas. Usually in small groups, they feed on seeds, berries, nuts, and insects. Wild Turkeys occupy a broad range of habitats, including forests, shrubland, grassland, and agricultural areas. Like the Ocellated, they are usually found in groups. They eat seeds, leaves, fruit, and tubers. Turkey females place their eggs in a shallow scrape in a hidden spot on the ground. Young are born ready to leave the nest and feed themselves (eating insects for their first few weeks). They remain with the mother for several months for protection. Males do not participate in nesting or raising the young.

Owing to overhunting and destruction of its forest habitats, the Ocellated Turkey is at risk; it has been eliminated from parts of its range and is still hunted, even in nature preserves. By the 1940s hunting had severely reduced Wild Turkey populations. Conservation efforts restored the species to abundance, and limited hunting is now permitted in much of the United States.

TURKEYS

Distribution:
North and Central
America

No. of Living
Species: 2

No. of Species
Vulnerable,
Endangered: 0, 0

No. of Species Extinct
Since 1600: 0

IMM

GRAY-NECKED WOOD-RAIL
Aramides cajanea
13–15.5 in (33–40 cm)
Central America, South America

PURPLE GALLINULE
Porphyrula martinica
10.5–14 in (27–36 cm)
North America, South America

EURASIAN COOT
Fulica atra
15 in (38 cm)
Eurasia, Africa, Australia

IMM

CLAPPER RAIL
Rallus longirostris
12–15.5 in (31–40 cm)
North America, South America

COMMON MOORHEN
Gallinula chloropus
12–15 in (30–38 cm)
Eurasia, Africa, North America,
South America

WHITE-BREASTED WATERHEN
Amaurornis phoenicurus
11–13 in (28–33 cm)
Southern Asia

PURPLE SWAMPHEN
Porphyrio porphyrio
15–19.5 in (38–50 cm)
Eurasia, Africa, Australia

BLACK-TAILED NATIVE-HEN
Gallinula ventralis
12–15 in (30–38 cm)
Australia

BLACK CRAKE
Amaurornis flavirostris
8.5 in (22 cm)
Africa

Rails, Gallinules, and Coots

RAILS are a worldwide group of often secretive small to medium-size birds of wetlands and forest floors. The family, Rallidae (included in order Gruiformes, with the cranes, bustards, and buttonquail), has 134 species; it includes wood-rails, crakes, GALLINULES, moorhens, and COOTS. Rails are known among bird-watchers for the elusiveness of some of their kind, especially marsh-dwelling species that are often heard but hardly ever seen. Ornithologists appreciate rails for a seeming paradox: although apparently weak flyers, some make migrations between continents, and rails have successfully colonized many remote oceanic islands. The family is also distinguished for its conservation status. Fully a quarter of living species are currently vulnerable or already endangered, and about 15 have become extinct during the past 200 years, more than in almost any other avian family. A major factor contributing to this dismal record is that many island species, removed as they were from mammalian predators, evolved flightlessness, which rendered them highly vulnerable when people and their mammalian pets and pests reached their isolated outposts.

Most rails make their living swimming and stalking about marshes and other wetlands (mostly freshwater but also brackish water, mangroves, and salt marshes), seeking plant and animal foods, with the emphasis on animal, including insects, crayfish, frogs, and snakes. Some, such as coots, are vegetarians, feeding at the surface and diving for leaves and stems of aquatic plants. Chief traits permitting the rails' aquatic lifestyle are long legs and very long toes that distribute the birds' weight, allowing them to walk

Distribution: Worldwide except polar regions

No. of Living Species: 134

No. of Species Vulnerable, Endangered: 17, 15

No. of Species Extinct Since 1600: 22

among marsh plants and across floating vegetation without sinking. Species that mainly inhabit forests or grasslands specialize on animal foods, taking little plant material.

Rails, short-bodied with short, broad wings, are 5 to 25 inches (12 to 63 cm) long. Compressed from side to side, they move easily through dense marsh vegetation. Bills, some brightly colored, vary in size and shape. Many of the group, such as coots, moorhens, and gallinules, resemble ducks when swimming, but they do not have ducklike bills and their feet are not webbed. Most species are clad in camouflage colors, the browns, russets, and grays blending into the dense vegetation through which they move. Several species, however, such as the Purple Gallinule, are riotously colorful and can be highly conspicuous. Others are drably colored but still conspicuous: the American Coot, mainly gray, is one of North America's most frequently spotted waterbirds. Many rails, including coots, move about shallow water and land with a characteristic head-bobbing walk. Little research has been done on rail behavior, but it is believed that most species are seasonally or permanently territorial, living in pairs or family groups.

Rails tend to be monogamous. They build nests with leaves of aquatic plants, usually just above the water surface in dense vegetation. Females lay large clutches of eggs, and both sexes incubate. Young are able to move about soon after hatching, but the adults feed them for several weeks, then shepherd them about for up to 2 months until they are independent. Young coots and gallinules have brightly colored heads and bills, presumably so adults can find their young hiding in vegetation.

Many rails are very abundant animals and not at risk. Coots, in fact, are among the most common and successful waterbirds in many parts of the world. But many in the family are threatened. This is particularly the case for island-bound, often flightless species where introduced predators such as rats, cats, and dogs find the adults, as well as their eggs and young, easy prey. Few of these rails have survived. The Wake Rail and Laysan Crake (endemic to the mid-Pacific's Wake and Laysan islands, respectively), both flightless, became extinct within the past 75 years; as did the Hawaii Crake (endemic to Hawaii), within the past 150 years. Habitat loss and introduced wetland plants that render some wetlands unsuitable for rail foraging or breeding also take their toll on these birds. Currently, seventeen rails are considered vulnerable, and fifteen are endangered; the Guam Rail (flightless; endemic to Guam), is extinct in the wild but still exists in captivity.

Cranes

CRANES, large, long-necked, long-legged wading birds, have been admired by people for millennia. Ancient Egyptians, Greeks, and Romans had strong ties to these birds, as evidenced by surviving artwork and literature. Eastern civilizations such as China and Japan have long associated cranes with positive attributes and emotions: longevity, happiness, luck, and more recently, peace. Crane characteristics that attract human attention are the size of the birds (some approach the height of people), their graceful movement and elegant appearance (Africa's crowned cranes being particularly beautiful), their soaring flight capabilities, and their renowned courtship dances. In North America, cranes are also recognized for their role in the emergence of the wildlife conservation movement and remain symbols of conservation.

Cranes (fifteen species, family Gruidae; in order Gruiformes, with rails and bustards) occur on all continents but South America and Antarctica. Two, Sandhill and Whooping Cranes, occur in North America. Cranes range in length from 3 to 5.7 feet (0.9 to 1.75 m); the largest, the Sarus Crane of Asia and Australia, has a wingspan up to 9.2 feet (2.8 m). Colored mainly gray, white, and black, most cranes have a patch of red, naked skin on top of their head. Bills generally are longish but relatively delicate, especially when compared with those of the similar-looking storks. Cranes have special ornamental inner wing feathers that hang down over a short tail, often giving the appearance of a bustle. Within a species, male and female cranes look alike.

Primarily wetland and open-country birds, cranes forage by walking slowly and steadily in water or on swampy ground, searching for food; they

Distribution:
Old World and North America

No. of Living Species: 15

No. of Species Vulnerable, Endangered: 6, 3

No. of Species Extinct Since 1600: 0

SANDHILL CRANE
Grus canadensis
39.5–47 in (100–120 cm)
North America, Cuba

WHOOPING CRANE
Grus americana
51–63 in (130–160 cm)
North America

BROLGA
Grus rubicunda
31.5–51 in (80–130 cm)
Australia, New Guinea

GRAY CROWNED CRANE
Balearica regulorum
39.5–43.5 in (100–110 cm)
Africa

BLUE CRANE
Anthropoides paradisea
43.5–47 in (110–120 cm)
Southern Africa

eat plant materials (leaves, buds, seeds, grasses) and small animals—just about anything they can swallow. They are particularly attracted to grain-growing areas after harvesting, feeding in fields on grain spillage. Their sharp bill is sometimes used as a probe, and their feet scrape out roots and tubers of marsh plants. In some areas, following breeding, cranes gather in great numbers. They fly with the neck fully extended, often in flocks, and sometimes soar to great heights. Northern Hemisphere species are, in general, highly migratory, moving to distant wintering grounds. They use the rising warm air of thermals to circle higher and higher on set wings, then glide off in the appropriate migration direction to find another thermal and repeat the maneuver. Some migrations are impressive: the Eurasian Crane flies over the Himalayas at altitudes near 33,000 feet (10,000 m).

Cranes are monogamous and the same partners often remain together for many years, returning annually to the same nesting area. Pairs defend territories while they breed, and they maintain their long-term pair-bond with haunting calls and exuberant dancing. Courtship dances can include jumping in the air, giving elaborate bows and wingspreads, and pulling up and tossing bits of plants and soil. Pairs share breeding duties, building a bulky mound nest of vegetation. The young often leave the nest soon after hatching and follow the parents; they are fed by the adults until they gradually learn how to feed themselves. They remain with their parents for their first year, and, where migratory, accompany the parents in their first migrations.

Owing to their large size and conspicuousness to hunters, as well as to habitat loss, many cranes are now at risk. Six species are considered vulnerable; three are endangered (Siberian and Red-crowned Cranes of Asia and North America's Whooping Crane). Efforts to save the Whooper, one of America's most spectacular birds, have been ongoing for half a century, and the near-extinction of this species stimulated aspects of America's conservation movement. Subject both to hunting and disturbance on its nesting grounds, by 1941 it was thought that no more than two dozen Whoopers remained. With complete protection and public education, numbers gradually increased to the present total of several hundred birds. There is a large captive flock and also a substantial population that nests in Wood Buffalo National Park in Canada and winters on the Texas coast. Beginning in 1993 captive-bred birds were released annually in an attempt to start a second, nonmigratory, flock in Florida.

LIMPKIN
Aramus guarauna
22–28 in (56–71 cm)
Central America, South America,
West Indies, Florida

GRAY-WINGED TRUMPETER
Psophia crepitans
17.5–20.5 in (45–52 cm)
South America

PALE-WINGED TRUMPETER
Psophia leucoptera
17.5–20.5 in (45–52 cm)
South America

RED-LEGGED SERIEMA
Cariama cristata
29.5–35.5 in (75–90 cm)
South America

BLACK-LEGGED SERIEMA
Chunga burmeisteri
27.5–33.5 in (70–85 cm)
South America

Limpkin; Trumpeters; Seriemas

LIMPKINS, TRUMPETERS, and SERIEMAS are large relatives of rails and cranes with mainly tropical and subtropical distributions in the Americas; they are included, along with the rails and cranes, in order Gruiformes.

Limpkins, family Aramidae, are large wading birds, smaller than cranes but having much the same shape, with long legs, neck, and bill. Camouflage coloring—brown with whitish markings—renders them highly suited for their lives among tall marsh vegetation. The single species, which ranges in length from 22 to 28 inches (56 to 71 cm), occurs from Florida to northern Argentina. Limpkin sexes look alike.

Inhabiting marshes, ponds, swamps, riversides, and mangroves, Limpkins eat mainly snails, but also take freshwater mussels. They forage by poking around in shallow water, finding snails by touch, then cracking open their shells to devour the inhabitants. Limpkins, usually monogamous, build a nest of leaves and twigs on the ground, in a marsh, or in trees. Parental duties are shared. Soon after young hatch they are mobile, able to follow parents and run from predators. By 2 months of age, young can fly and can open snails themselves. The Limpkin is common in many parts of its large range. However, following decades of intensive hunting as a game bird, Limpkins in some areas, such as Florida and the West Indies, were at risk by the early twentieth century, but with conservation measures, their populations have rebounded.

A little smaller and bulkier than Limpkins, trumpeters are terrestrial

LIMPKIN

Distribution:
Florida, West Indies,
Central and South
America

No. of Living
Species: 1

No. of Species
Vulnerable,
Endangered: 0, 0

No. of Species Extinct
Since 1600: 0

TRUMPETERS

*Distribution:
South America*

*No. of Living
Species: 3*

*No. of Species
Vulnerable,
Endangered: 0, 0*

*No. of Species Extinct
Since 1600: 0*

SERIEMAS

*Distribution:
South America*

*No. of Living
Species: 2*

*No. of Species
Vulnerable,
Endangered: 0, 0*

*No. of Species Extinct
Since 1600: 0*

birds inhabiting dense tropical forests. The family, Psophiidae, has three species, restricted to northern and central regions of South America. They most closely resemble guans (which are in a different order), but have a humpbacked appearance. They are chicken-size, 18 to 20 inches (45 to 52 cm) long, and have short, robust bills, fairly long necks, and longish legs; the sexes look alike. True to their name, trumpeters all have in their vocal repertoires loud deep calls that sound like a trumpet. Where they are hunted, they are very shy, and so are rarely seen in their forest habitats. In some regions trumpeters are kept as pets by local people and are often spotted running around villages, bullying domestic chickens and sometimes acting like guard dogs.

Trumpeters feed by walking along the forest floor, looking for fallen fruits and scratching with their feet to stir up insects. Their flying is generally limited to flights into trees to escape danger and to roost at night. Trumpeters are almost always in flocks of five to twenty or more individuals. The breeding of trumpeters is poorly understood. Some appear to breed in cooperative groups, with only the dominant male and female actually nesting, the other individuals helping. Nests are often in tree cavities. The young follow parents away from the nest soon after hatching, being fed by adults for several weeks. None of the trumpeters are currently at risk; they apparently maintain healthy populations in undisturbed areas of rainforest.

The two Seriemas species (family Cariamidae) occur in central and eastern South America. Tall, long-legged birds of open habitats, they rarely fly, even to escape predators; instead they run or try to hide. They will fly, however, to a high night roost or if a predator gets too close. Running for these large birds is not an inconsiderable defense: the Red-legged Seriema is known to speed along at more than 30 mph (50 kph). Seriemas are grayish and brown, with hawklike heads and long necks and tails. They range in length from 28 to 35 inches (70 to 90 cm), the Red-legged Seriema being larger and more brightly colored than the Black-legged Seriema.

Birds of semiarid savanna, grasslands, and open woodlands, seriemas stalk about, usually alone or in pairs, foraging for small animals. They take mainly insects but also frogs, lizards, snakes, and rodents. Small prey is grabbed and swallowed; larger items are held with the feet and ripped apart with the bill. Breeding in seriemas is not well studied. Apparently monogamous, a pair builds a stick nest in a tree and lines it with leaves and perhaps cattle dung. Both parents feed the young. Neither seriema is currently at risk.

Sungrebes;
Sunbittern; Kagu;
Mesites

The best known members of order Gruiformes are the cranes and rails, both
with near-global distributions; but the order encompasses a wide variety of
other birds, including a number of small families that each have only one
to three species, including the sungrebes (finfoots), Sunbittern, Kagu, and
mesites (treated here) and the Limpkin, trumpeters, and seriemas (treated
in another account). These groups consist mainly of ground-living birds
that prefer walking (or swimming) to flying (with one, the Kagu, being
flightless), and most are shy and retiring midsize or large birds with relatively
limited distributions.

The SUNGREBE and two species of FINFOOTS comprise family
Heliornithidae. Ranging in length from 10 to 23 inches (26 to 59 cm), they
are dark above, light below, with long necks, slender bodies, longish tails,
and sharply pointed yellow or orange bills. The Sungrebe occurs in South
and Central America, while one finfoot occurs in Africa and the other in
Southeast Asia. Despite the Sungrebe's name, the group is not closely related
to grebes, but are ducklike and largely aquatic. Sungrebes and finfoots swim
among the roots and overhanging leaves of thick vegetation along lakes,
rivers, and streams, where they feed on insects, snails, worms, crustaceans,
and the occasional frog. They do not dive, but instead take their food from
riverbanks and vegetation. They take cover in vegetation at the slightest
hint of danger and will "run" across the surface of the water and fly low to
escape predators. The Sungrebe and finfoots make flat nests of sticks and
reeds, often on branches of dead trees low over the water. Parents share

SUNGREBES

*Distribution:
Africa, southern Asia,
Central and South
America*

*No. of Living
Species: 3*

*No. of Species
Vulnerable,
Endangered: 1, 0*

*No. of Species Extinct
Since 1600: 0*

SUNGREBE
Heliornis fulica
10–13 in (26–33 cm)
Central America, South America

MASKED FINFOOT
Heliopais personata
17–21.5 in (43–55 cm)
Southeast Asia

SUNBITTERN
Eurypyga helias
17–17.5 in (43–45 cm)
Central America, South America

KAGU
Rhynochetos jubatus
21.5 in (55 cm)
New Caledonia

WHITE-BREASTED MESITE
Mesitornis variegata
12 in (31 cm)
Madagascar

some nesting duties. Males have a unique adaptation to protect their fledged young. In case of danger, the small chicks climb into pouches in the skin under a father's wings and he then dives underwater or flies to escape. Asia's Masked Finfoot has a small population and is considered vulnerable.

The SUNBITTERN, occurring from southern Mexico to southern Bolivia, is the sole member of family Eurypigidae. About 18 inches (46 cm) long, it has a beautiful gray-, brown-, and black-patterned plumage, black head with white stripes, long, straight bill, slender neck, and long tail. Sunbitterns live along tropical forested streams and flooded forest, where they walk along searching for insects, spiders, crabs, frogs, crayfish, and small fish. They are known for their spectacular courtship and threat display in which the tail and wings are spread to reveal a striking sunburst pattern of yellow, brown, and black. Sunbitterns make a rounded cuplike nest of leaves and mud in a tree or bush, and apparently both mates incubate and feed the young.

The KAGU is a largish (22 inches [55 cm]) white and gray bird endemic to the island of New Caledonia, west of Australia. It looks like a cross between a heron and a rail, and is placed in its own family, Rhynochetidae. Its plumage color, long reddish legs and bill, and large red eyes make it unmistakable. Although flightless, the Kagu often uses its wings to help move through its forest and shrubland habitats, flapping to move faster when, for example, running from predators. The Kagu eats only animal food, including insects, spiders, centipedes, snails, worms, and lizards. It is monogamous; the sexes share breeding duties. Nests, on the ground, are simple affairs of leaves. The Kagu, with a small, fragmented population, is endangered.

Three species of MESITES comprise family Mesitornithidae. These are midsize (12 inches [30 cm]) ground birds of forests, woodlands, and thickets, all restricted to Madagascar. They have long, thick tails, short, rounded wings, and are brownish with lighter underparts. They eat seeds, insects, and fruit. In the past, some of the mesites were thought to be flightless, but all can fly, albeit weakly. Little is known about mesite breeding. A stick nest is built in a bush, tree, or clump of vegetation. Depending on species, the female only or both parents incubate eggs and care for the young. Madagascar's mesites are considered one of the world's most threatened bird families; the three species are all vulnerable, suffering from habitat loss and degradation.

SUNBITTERN

Distribution:
Central and South
America

No. of Living
Species: 1

No. of Species
Vulnerable,
Endangered: 0, 0

No. of Species Extinct
Since 1600: 0

KAGU

Distribution:
New Caledonia

No. of Living
Species: 1

No. of Species
Vulnerable,
Endangered: 0, 1

No. of Species Extinct
Since 1600: 0

MESITES

Distribution:
Madagascar

No. of Living
Species: 3

No. of Species
Vulnerable,
Endangered: 3, 0

No. of Species Extinct
Since 1600: 0

AUSTRALIAN BUSTARD
Ardeotis australis
35.5–47 in (90–120 cm)
Australia, New Guinea

KORI BUSTARD
Ardeotis kori
35.5–47 in (90–120 cm)
Africa

WHITE-BELLIED BUSTARD
Eupodotis senegalensis
19.5–23.5 in (50–60 cm)
Africa

BLACK-BELLIED BUSTARD
Lissotis melanogaster
23.5 in (60 cm)
Africa

BLUE KORHAAN
Eupodotis caerulescens
21.5 in (55 cm)
Southern Africa

WHITE-QUILLED BUSTARD
Eupodotis afraoides
19.5 in (50 cm)
Southern Africa

Bustards

BUSTARDS are medium-size to large ground birds of open landscapes of Europe, Asia, Australia, and, especially, Africa, where about twenty of the twenty-five species occur (seventeen being endemic there). The family, Otididae, is included in the order Gruiformes with the cranes and rails. Despite the size and frequent conspicuousness of some of the species, bustards are not a well known or particularly appreciated group. Their coloring (patterns mainly of brown, gray, black, and white), especially from afar, makes them seem rather drab, and bustards historically have not had close associations with people, other than as game birds. But they are recognized by the ecologically knowledgeable for their highly meritorious effect on many agricultural and other pests; they consume enormous numbers of harmful insects (locusts, grasshoppers, beetles, termites) and small rodents. Also, bustards are known among animal behaviorists for their courtship displays and nonmonogamous breeding. Two species from southern Asia are called floricans and, in southern Africa, where eleven species occur, many are called korhaans (crowing hens), for their croaking, clattering calls.

Like their cousins the cranes, bustards are long-necked and long-legged, but they have stocky bodies and short, sharp bills. They range in length from 16 to 47 inches (40 to 120 cm), and the largest, Africa's Kori Bustard, weighs up to 42 pounds (19 kg), thus making it one of the world's heaviest flying birds. Males tend to be larger and more boldly marked than females.

Bustards prefer dry, open habitats. The Australian Bustard, for instance, that continent's only species, inhabits grassland and savanna areas. Bustards

Distribution:
Old World

No. of Living
Species: 25

No. of Species
Vulnerable,
Endangered: 1, 3

No. of Species Extinct
Since 1600: 0

spend almost all their time on the ground even though, with their long, broad wings, they are fairly strong fliers. To avoid detection, they often lie flat on the ground, their cryptic coloration providing superb camouflage. When disturbed, they stalk furtively away, only reluctantly taking flight when pressed. Smaller bustards especially are shy and usually very hard to spot or approach. Some bustards lead relatively solitary lives; others remain in year-round pairs or small family groups; in several, flocks of up to fifty birds occasionally gather. Bustards forage by walking methodically and stopping to peck at food items on the ground and in foliage. They take plant materials including leaves, seeds, flowers, berries, and roots, and animals such as insects, lizards, and mice, as well as eggs and young of small ground birds. At brush fires, they wait for prey fleeing flames; after a fire, they stalk about looking for heat-killed bugs.

Bustards may breed monogamously or, more commonly, polygamously. In the latter, males during the breeding season occupy traditional lek sites, where they gather to advertise themselves to females. After a female selects a male and mates, she departs and nests on her own; no pair-bonds are formed between the sexes. The advertising displays of male bustards can be spectacular. In southern Africa's Red-crested Bustard, a male utters a series of increasingly loud whistles interspersed with tongue clicks, then dramatically launches himself into the air to a height of perhaps 50 to 65 feet (15 to 20 m), closes his wings, and falls back to the ground. Males of some species grow elongated chest feathers that are puffed out during displays, briefly turning themselves into seeming "bird balls." Males also inflate throat sacs during displays, furthering their rounded images. After mating, a female lays eggs in a simple depression or scrape in the ground and incubates them. Young bustards leave the nest within a few hours but stay with the mother for 4 to 6 weeks. In monogamous species, such as the Blue Bustard, pairs (sometimes with offspring from previous years) live in permanently defended territories; only females incubate, but both sexes care for the young.

Although bustards are generally in decline because of hunting and habitat loss to agriculture, only a few are currently threatened: one species is considered vulnerable and three others, restricted to India and its environs, are endangered. The Australian Bustard has been eliminated from some parts of its range by human settlement and by use of its grassland habitats by sheep. Also, some bustards are in demand for training falcons to kill, notably in the Middle East.

Jacanas

JACANAS are small to medium-size waterbirds that occur in freshwater wetlands throughout tropical and many subtropical regions of the world. They are usually fairly abundant near shore in marshes, swamps, shallow parts of lakes, and even in artificial wetlands in and around human settlements. Owing to this, to their aquatic-plant-hopping ways, and to their striking coloration, they are conspicuous, often-watched birds. Jacanas are best known among biologists as one of the few bird groups that routinely practice polyandry, a mating system in which one female mates with several males. The family, Jacanidae, with eight species, many with broad distributions, is included in order Charadriiformes, with shorebirds and gulls. The name *jacana* is derived from a native Brazilian name for the Wattled Jacana, South America's sole species.

Outstanding physical characteristics of jacanas are their long legs and incredibly long toes and claws, which distribute the birds' weight, permitting them to walk among marsh plants and across floating vegetation without sinking (they are called lily-trotters in various parts of the world). They range in length from 6 to 23 inches (15 to 58 cm); a good part of the length of the largest species, southern Asia's Pheasant-tailed Jacana, resides in its elongate tail. Jacana bodies are relatively narrow, making it easier for them to push through dense vegetation, and they have short tails (all but the Pheasant-tailed) and rounded wings. Most have brown or blackish bodies but come equipped with bright yellow or blue bills and red or blue foreheads; some have striking white or yellow markings. Some species spread their wings

Distribution:
Worldwide tropics and subtropics

No. of Living Species: 8

No. of Species Vulnerable, Endangered: 0

No. of Species Extinct Since 1600: 0

NORTHERN JACANA
Jacana spinosa
6.5–9 in (17–23 cm)
Mexico, Central America, West Indies

WATTLED JACANA
Jacana jacana
9.5 in (24 cm)
South America

BRONZE-WINGED JACANA
Metopidius indicus
12 in (30 cm)
Southern Asia

PHEASANT-TAILED JACANA
Hydrophasianus chirurgus
15.5–23 in (39–58 cm)
Southern Asia

COMB-CRESTED JACANA
Irediparra gallinacea
9 in (23 cm)
Australia, New Guinea, Indonesia

AFRICAN JACANA
Actophilornis africanus
9–12 in (23–31 cm)
Africa

during flight or displays, exposing large patches of bright yellow, which render the birds instantly visible. Male and female jacanas look alike, but females are larger.

Jacanas are common birds of ponds, lakes, marshes, and wet fields. They walk along, often on top of lily pads and other floating plants, picking up insects, snails, small frogs and fish, and some vegetable matter, such as seeds. Often they turn over leaves in search of prey. Judging by the care with which they place each foot, carefully distributing their weight on their long outstretched toes and testing each leaf in turn, lily-trotting is a tricky business. In Africa they sometimes walk close to feeding elephants and onto basking hippopotamuses in their search for insects flushed by the large mammals. They do not rely heavily on flight but rather swim in the water or walk on and through vegetation near the water. Several species have a sharp spur on each wing that is used for fighting other jacanas and predators.

Africa's Lesser Jacana, the smallest in the family, is a monogamous breeder, but the other species employ polyandry, the rarest type of mating system among birds. In a breeding season, a female mates with several males, and the males then carry out most of the breeding chores. Males each defend small territories from other males; each female has a larger territory that encompasses several male territories. Males build nests of floating, compacted aquatic vegetation. Following mating, the female lays eggs in the nest, after which the male incubates them and then leads and protects the chicks; the young feed themselves. Females sometimes remain near nests and attack predators that approach chicks. Jacana fathers are able to move their young chicks in case of flooding or danger by holding them under their closed wings and running to safety. Young are dependent on the father for up to 3 months. Meanwhile, the female has mated with other males in her territory and provided each with a clutch of eggs to attend.

None of the jacanas are currently considered at risk (but little is known about populations of the small and secretive Lesser Jacana). The main threat to these waterbirds is the destruction or degradation of their wetland habitats, which occurs with increasing frequency throughout their range. Marshes are drained for agricultural development, and sometimes floating aquatic vegetation in lakes is removed for navigation or aesthetic purposes, rendering habitat unsuitable for jacanas. On the plus side, some jacanas do well in agricultural areas, especially wet fields and rice paddies.

SPOTTED SANDPIPER
Actitis macularia
7.5 in (19 cm)
North America, South America

SANDERLING
Calidris alba
8 in (20 cm)
Worldwide

RUDDY TURNSTONE
Arenaria interpres
8.5–10 in (21–26 cm)
Worldwide

ANDEAN SNIPE
Gallinago jamesoni
12 in (30 cm)
South America

WANDERING TATTLER
Heteroscelus incanus
11 in (28 cm)
North America, Siberia, Australia,
New Guinea, tropical Pacific Islands

SHORT-BILLED DOWITCHER
Limnodromus griseus
10–11.5 in (25–29 cm)
North America, South America

BAR-TAILED GODWIT
Limosa lapponica
14.5–16 in (37–41 cm)
Eurasia, Africa, Australasia

Sandpipers, Phalaropes, and Snipes; Painted-snipes

Shorebirds are the small to medium-size, leggy birds with slim bills that one sees along coasts and around wetlands just about everywhere. Their main food, tiny invertebrate animals, is superabundant on the world's beaches and mudflats, and owing to this, shorebirds are highly successful, often occurring in huge numbers. Because they are plain-looking and common, shorebirds usually are of intense interest only to serious bird-watchers, many of whom enjoy the challenge of correctly identifying the myriad species with similar drab appearance. But shorebirds are fascinating ecologically, with intriguing migrations and breeding systems. Also, although most are not endowed with striking plumage, shorebirds such as sandpipers nonetheless provide some of nature's most compelling sights as their flocks rise from sandbar or mudflat to fly fast and low over the ocean surf, wheeling quickly in the air as if they were a single organism.

More than ten families can be considered to contain shorebirds, all included in order Charadriiformes with gulls and puffins. The largest, with a worldwide distribution, is the SANDPIPER group (family Scolopacidae), which contains about eighty-five species of sandpipers, snipes, woodcocks, godwits, dowitchers, curlews, knots, turnstones, stints, and phalaropes. All have a characteristic look, being mainly dully colored birds (especially during the nonbreeding months), darker above, lighter below, slender, with long, thin legs for wading through wet meadows, mud, sand, or surf. Depending on feeding habits, bills are straight or curved and vary in length from short to very long. Overall length ranges from 5 to 26 inches (12 to 66 cm), the largest

SANDPIPERS, PHALAROPES, AND SNIPES

Distribution:
All continents except
Antarctica

No. of Living
Species: 87

No. of Species
Vulnerable,
Endangered: 6, 4

No. of Species Extinct
Since 1600: 2

ROCK SANDPIPER
Calidris ptilocnemis
8.5 in (22 cm)
North America, eastern Asia

SURFBIRD
Aphriza virgata
10 in (25 cm)
North America, South America

DUNLIN
Calidris alpina
6.5–8.5 in (16–22 cm)
North America, Eurasia, Africa

WILSON'S PHALAROPE
Phalaropus tricolor
9 in (23 cm)
North America, South America

UPLAND SANDPIPER
Bartramia longicauda
10–12.5 in (26–32 cm)
North America, South America

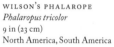

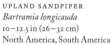

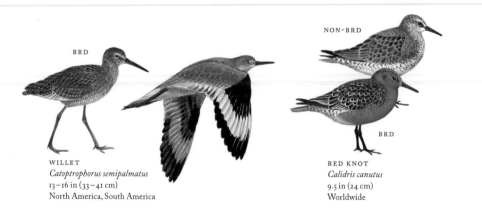

WILLET
Catoptrophorus semipalmatus
13–16 in (33–41 cm)
North America, South America

RED KNOT
Calidris canutus
9.5 in (24 cm)
Worldwide

NON-BRD

NON-BRD

COMMON REDSHANK
Tringa totanus
11 in (28 cm)
Eurasia, Africa

COMMON SANDPIPER
Actitis hypoleucos
8 in (20 cm)
Eurasia, Africa, Australia

TEREK SANDPIPER
Xenus cinereus
9.5 in (24 cm)
Eurasia, Africa, Australia

RUFF
Philomachus pugnax
8–12.5 in (20–32 cm)
Eurasia, Africa

EASTERN CURLEW
Numenius madagascariensis
21–26 in (53–66 cm)
Asia, Australasia

COMMON GREENSHANK
Tringa nebularia
13.5 in (34 cm)
Eurasia, Africa, Australia

NON-BRD

CURLEW SANDPIPER
Calidris ferruginea
8.5 in (21 cm)
Asia, Africa, Australia

SOUTH AMERICAN PAINTED-SNIPE
Rostratula semicollaris
8.5 in (21 cm)
South America

PAINTED-SNIPES

Distribution:
South America. Africa,
Asia, Australia

No. of Living
Species: 2

No. of Species
Vulnerable,
Endangered: 0, 0

No. of Species Extinct
Since 1600: 0

being the curlews, which have extremely long down-curved bills. The sexes look alike or nearly so in most species.

Sandpipers typically are open-country birds, associated with coastlines and inland wetlands, grasslands, and pastures. They are excellent flyers but they spend a lot of time on the ground, foraging and resting; when chased, they often prefer running to flying away. They pick their food up off the ground or use their bills to probe for it in mud or sand, taking insects and other small invertebrates, particularly crustaceans. Species with curved bills tend to forage more by probing into holes and crevices. Some also snatch bugs from the air as they walk and from the water's surface as they wade or swim. Larger, more land-dwelling species, such as curlews, also eat small amphibians, reptiles, and rodents. Most shorebirds breed in the Northern Hemisphere, many on the arctic tundra, and make long migrations to distant wintering sites, often in the Southern Hemisphere.

Many shorebirds breed in monogamous pairs that defend small breeding territories. Others, such as phalaropes, practice polyandry, in which some females have more than one mate. In these species, the normal sex roles of breeding birds are reversed: the female establishes a territory on a lakeshore that she defends against other females. More than one male settles within the territory, either at the same time or sequentially during a breeding season. After mating, the female lays a clutch of eggs for each male. The males incubate their clutches and care for the young. Females may help care for some of the broods. Most shorebird nests are simply small depressions in the ground in which eggs are placed; some of these scrapes are lined with shells, pebbles, or vegetation. Shorebird young are able to run from predators and feed themselves soon after they hatch. Parents usually stay with the young to guard them at least until they can fly, perhaps 3 to 6 weeks after hatching.

Owing mainly to habitat loss and, in the case of island-dwelling species, to the invasion of breeding or wintering sites by introduced predators (such as cats and rats), six sandpipers are vulnerable and four are endangered, two of them critically.

PAINTED-SNIPES, family Rostratulidae, are midsize (8 to 11 inches [20 to 28 cm]), long-billed wading birds of swamps and wet grasslands. They eat seeds and small invertebrates such as insects, worms, and snails. There are only two, the Greater Painted-snipe, broadly distributed in the Old World, and the South American Painted-snipe, restricted to southern South America; neither is threatened.

Plovers and Lapwings

PLOVERS and LAPWINGS comprise a family of small to medium-size birds that, with a total of about sixty-seven species, is the second most diverse group of shorebirds, after the sandpipers. The family, Charadriidae (included in order Charadriiformes with gulls and puffins), is distributed in open habitats essentially worldwide. Plovers and lapwings in the past were widely hunted and were celebrated primarily for the meals they and their eggs could provide. Today they are known mainly as ever-vigilant denizens of shorelines and other open habitats; some, such as North America's Killdeer (a plover), Eurasia's Northern Lapwing, southern Asia's Red-wattled Lapwing, and Australia's Masked Lapwing, are common, conspicuous inhabitants of suburban parks, grassy fields, and agricultural sites.

Plovers and lapwings, 5 to 15 inches (12 to 38 cm) long, are similar in form, having relatively large eyes, large rounded heads, short thick necks, and fairly short, usually straight bills. Most are brownish above and light below. Many have dark eye stripes and dark bands on chest or neck. The twenty-five species of lapwings, which typically are larger than plovers and have longer legs, are also more boldly marked; many have head decorations such as a crest or wattles. Some species have wing spurs (sharp bony protrusions on the shoulders, used in fighting). Male and female look alike or nearly so. Perhaps the oddest species in the family is the Wrybill, a small, gray plover endemic to New Zealand, which has a long dark bill that curves to the right; the sideways-curving bill aids the birds in probing for insects under small rocks. The Magellanic Plover, a small (8 inches [20 cm]), pale gray species

Distribution:
All continents except
Antarctica

No. of Living
Species: 67 (includes
Magellanic Plover)

No. of Species
Vulnerable,
Endangered: 5, 3

No. of Species Extinct
Since 1600: 1

PIED LAPWING
Vanellus cayanus
9 in (23 cm)
South America

DIADEMED SANDPIPER-PLOVER
Phegornis mitchellii
7 in (18 cm)
South America

RED-WATTLED LAPWING
Vanellus indicus
13.5 in (34 cm)
Southern Asia

LITTLE RINGED PLOVER
Charadrius dubius
6.5 in (16 cm)
Eurasia, Africa

MASKED LAPWING
Vanellus miles
12–14.5 in (30–37 cm)
Australasia

BLACK-FRONTED DOTTEREL
Elseyornis melanops
6.5 in (17 cm)
Australia, New Zealand

AFRICAN WATTLED LAPWING
Vanellus senegallus
13.5 in (34 cm)
Africa

BLACK-HEADED LAPWING
Vanellus tectus
10 in (25 cm)
Africa

MAGELLANIC PLOVER
Pluvianellus socialis
8.5 in (21 cm)
Southern South America

with short legs, which is restricted to southern South America, is different enough from other plovers in form and behavior that some authorities place it in a separate single-species family, Pluvianellidae.

Plovers and lapwings occupy an array of open habitats, being common on beaches and around many kinds of wetlands; they are also found in grasslands, tundra, and semidesert areas. Feeding day or night on insects, spiders, worms, small crustaceans, and the odd berry or seed, plovers and lapwings are known for their "run, stop, and peck" foraging behavior: they typically run or walk a few steps, then stop with head held high, presumably scanning for prey and predators, then move again or bend to peck at food spotted on the ground. Many species maintain feeding territories during nonbreeding periods, small areas that they defend from others of their species for a few days up to a few months. In the air, lapwings, on broad wings, fly somewhat slowly; plovers, with pointed wings, fly fast and straight. Some of the plovers are among the globe's champion long-distance migrants. Many make long journeys over vast expanses of open ocean, a good example being the Pacific Golden Plover, which flies apparently nonstop from wintering sites in Hawaii to its arctic breeding grounds in Alaska, and from wintering sites in Australia and southern Asia to breeding grounds in northern Russia.

Most lapwings and plovers are territorial during breeding, pairs defending an area around the nest; territories, however, are sometimes small and grouped together. Monogamous, these birds have nests that are simply small depressions in the ground; some of these scrapes are lined with pebbles or shell fragments. Lapwings that breed in marshes pull plant materials together to form a nest. Both sexes incubate eggs. After hatching, chicks can feed themselves, but parents lead, brood, and protect them. When potential predators approach, plovers and lapwings engage in mobbing behavior (flying over the predator and trying to drive it off by diving at it) and distraction displays (walking away from the nest or young and behaving as if injured—for instance, by keeping a wing outstretched, as if broken—to draw the predator away from the nest).

Owing mainly to loss or disturbance of breeding habitat, and for island-dwelling species, to the introduction of mammal predators such as cats and rats, several plovers and lapwings are in trouble. Three species are currently endangered, and five, including two in North America (Piping and Mountain Plovers), are vulnerable.

BEACH THICK-KNEE
Burhinus magnirostris
21–22.5 in (53–57 cm)
Southeast Asia, Australia

BUSH STONE-CURLEW
Burhinus grallarius
21.5–23 in (54–59 cm)
Australia, New Guinea

SPOTTED THICK-KNEE
Burhinus capensis
14.5–17.5 in (37–44 cm)
Africa

RUFOUS-BELLIED SEEDSNIPE
Attagis gayi
11.5 in (29 cm)
South America

LEAST SEEDSNIPE
Thinocorus rumicivorus
7 in (18 cm)
South America

PLAINS-WANDERER
Pedionomus torquatus
6–7.5 in (15–19 cm)
Australia

Thick-knees;
Seedsnipes;
Plains-wanderer

THICK-KNEES, also known as stone-curlews, are largish, long-legged shorebirds that lead mostly terrestrial lives. Some of the nine species (family Burhinidae, included in order Charadriiformes; nearly globally distributed) pursue typical shorebird existences, frequenting beaches, mangroves, riverbanks, estuaries, and the margins of lakes, but others shun water, inhabiting grasslands, savanna, or even woodlands. Many people know these birds for their wailing, nocturnal calls.

Thick-knees have unusually large heads (*dikkop*, a name for these birds in Africa, means "thick-head" in Afrikaans), big, pale yellow eyes (for night vision), long yellowish legs, and knobby, thick "knees" (which actually correspond, anatomically, more to the human ankle joint). Their bills are mostly short and stout, but two species that feed mainly on crabs have massive bills. Thick-knees range from 12 to 23 inches (32 to 59 cm) long; Australia's Bush Stone-curlew is the largest. They are cryptically colored, brown with darker streaking and mottling above, lighter below.

Essentially "land waders," thick-knees are strongly tied to the ground. They fly well, but mostly walk or run, even preferring running from a predator over flying away (if surprised, however, they take flight). They are usually gregarious, often occurring in small groups but sometimes gathering into flocks of hundreds. Unusual among shorebirds, most thick-knees spend days quietly in shady, sheltered spots, such as under bushes, then become active at twilight, foraging nocturnally for insects, worms, small mammals, and reptiles.

Thick-knees tend to be monogamous. A male courts a female with brief running and leaping displays. Eggs, placed in a scrape on the ground, are

THICK-KNEES

Distribution:
All continents except
Antarctica

No. of Living
Species: 9

No. of Species
Vulnerable,
Endangered: 0, 0

No. of Species Extinct
Since 1600: 0

SEEDSNIPES

Distribution:
South America

No. of Living
Species: 4

No. of Species
Vulnerable,
Endangered: 0, 0

No. of Species Extinct
Since 1600: 0

PLAINS-WANDERER

Distribution:
Australia

No. of Living
Species: 1

No. of Species
Vulnerable,
Endangered: 0, 1

No. of Species Extinct
Since 1600: 0

incubated by both sexes. Within a day of hatching, young can move from the nest; parents feed and guard them. Young fly at 2 to 3 months of age; some are not independent until older than 6 months. Africa's Water Thick-knee is known to nest near brooding crocodiles—presumably because the crocodiles keep (other) predators away. Thick-knees are fairly secretive, so not much is known about the sizes of their populations; two species, in Asia and Australia, are considered at risk.

SEEDSNIPES, with their short bills and short legs, look like crosses between pigeons and quails and are considered to be the most bizarre and unusual of the shorebirds due to their physical forms, habitat choices, and vegetarian diets. There are four species (family Thinocoridae, in the order Charadriiformes), which are restricted to western and southern South America. Some of them, such as the Rufous-bellied Seedsnipe, occur on rocky slopes and ridges high in the Andes Mountains (to just below snow line). Others, like the Least Seedsnipe, are found in semidesert areas or grasslands.

Seedsnipes (6 to 12 inches [16 to 30 cm] long) are mottled brown above and whitish or reddish brown below, coloring that makes them highly camouflaged in their natural habitats. They are browsers, taking small leaves, buds, and seeds from low vegetation as they walk along. Seedsnipes are apparently monogamous and are often seen in pairs or small family groups. The nest is a scrape in the ground, sometimes lined with moss or other vegetation. In some species, only the female incubates, while the male stands guard nearby. Young are led from the nest by parents soon after they hatch. None of the seedsnipes are threatened, but some species are poorly understood.

Closely related to seedsnipes is the PLAINS-WANDERER, an odd bird endemic to southeastern Australia. The only member of its family, Pedionomidae (in the order Charadriiformes), the Plains-wanderer is a drab, mainly light brown bird, about 7 inches (18 cm) long, with a short yellow bill and long yellow legs. Like the seedsnipes, it differs from most of its shorebird cousins by its habitat of sparse, open grasslands (where it feeds on seeds and insects). Plains-wanderers also employ an uncommon mating system among birds, polyandry. In a breeding season, a female mates with several males sequentially, and the males carry out the majority of the breeding chores, including most egg incubation and leading and protecting the young. Associated with the reversal of sex roles, female Plains-wanderers are heavier and more boldly marked than males. Loss of habitat to agriculture has endangered the Plains-wanderer.

Oystercatchers;
Crab-plover

OYSTERCATCHERS are striking medium-size gull-like birds, blackish or blackish and white, with long orange-red bills. The daggerlike bills are flattened side to side and used like shucking knives to open clams and other mollusks. The family, Haematopodidae (in the order Charadriiformes with shorebirds, terns, and gulls), occurs almost worldwide, mainly along seacoasts, but is not represented in polar regions or in some parts of Africa or Asia. There seem to be ten species, but the classification of the group is very controversial, especially because some of the species are nearly identical in appearance. Other than being appreciated by bird-watchers for their bright, contrasting coloring, oystercatchers' relationships with people are mainly competitive; they take oysters and other shellfish that *Homo sapiens* would rather consume themselves, but their economic impact is probably minor.

Sixteen to 20 inches (40 to 50 cm) long, oystercatchers, in addition to their striking plumage and bills, are identifiable by their pinkish legs and red eye-rings (one species has a yellow eye-ring). Old World oystercatchers all have black backs, but three of the four New World species have brown backs. The conspicuous bills of these birds average about 3 inches (8 cm) in length, but range in some species up to about 4 inches (10 cm).

Along seacoasts, oystercatchers use their strong, flattened bills to pry open oysters, clams, and other bivalve mollusks, and also to pry various other marine invertebrates (gastropods, crustaceans such as crabs, some echinoderms) from rocky shorelines and dismember them; they occasionally take small fish. Where they occur inland, such as in some grasslands and freshwater wetlands, they feed chiefly on insects and worms. During nonbreeding

OYSTERCATCHERS

Distribution:
Worldwide

No. of Living
Species: 10

No. of Species
Vulnerable,
Endangered: 0, 1

No. of Species Extinct
Since 1600: 1

AMERICAN OYSTERCATCHER
Haematopus palliatus
16.5 in (42 cm)
North America, South America

BLACK OYSTERCATCHER
Haematopus bachmani
17.5 in (44 cm)
North America

PIED OYSTERCATCHER
Haematopus longirostris
19.5 in (50 cm)
Australia, New Guinea

IMM

AFRICAN BLACK OYSTERCATCHER
Haematopus moquini
17.5 in (44 cm)
Southern Africa

CRAB-PLOVER
Dromas ardeola
15.5 in (40 cm)
Southern Asia, Africa, Madagascar

months, oystercatchers usually spend their time in small foraging flocks of fifty or fewer individuals; during breeding, pairs are territorial.

Oystercatchers are monogamous, and they tend to remain faithful between years to mate and to breeding site; some are known to have had the same breeding territory for 20 years. Various species breed, in mainland areas or on islands, on beaches, dunes, rocky shores, or in salt marshes. Females tend to select the nest site; males do most of the nest preparation. The nest is a simple scrape in the ground, sometimes lined with shell fragments and pebbles. Both sexes incubate eggs and feed young. Chicks are mobile soon after hatching and usually leave the nest within 24 hours but are dependent on their parents for food for about 2 months, and some remain with their parents for more than 6 months. They reach sexual maturity slowly, first breeding at 3, 4, or more years of age.

Most oystercatchers are abundant. However, several species are in trouble, mostly owing to disturbance at their breeding sites, for instance by coastal development and off-road recreational vehicles on the birds' nesting beaches, and in some regions, by reduction in their food availability by people who gather shellfish. Three current at-risk species are southern Africa's African Black Oystercatcher, New Zealand's Variable Oystercatcher, and Australia's Sooty Oystercatcher. In addition, the Chatham Oystercatcher, endemic to some tiny islands off New Zealand, with a population of only a few hundred, is endangered. The Canary Islands Oystercatcher became extinct about 1940.

The CRAB-PLOVER is a medium-size shorebird that shares with oystercatchers a long, robust, daggerlike bill. The single species (family Dromadidae, in order Charadriiformes) occurs only in Africa and Asia, breeding along the Persian Gulf, Red Sea, and parts of the Indian Ocean and wintering mainly along western Indian and eastern African coasts. The birds, about 16 inches (40 cm) long, are distinctively black and white with long grayish legs. Their most conspicuous trait is the powerful black bill, which they use to stab and dismember their main food, crabs. They also eat other crustaceans and some worms and mollusks. They feed in flocks on mudflats, sandbanks, rocky shorelines, and coral reefs, being most active at dusk and nocturnally. Breeding is not well studied, but they appear to be monogamous, pairs nesting in colonies. Unique among shorebirds, Crab-plovers nest in burrows, which are excavated in sandy areas. The species, although of limited range and abundance, is not considered at risk.

CRAB-PLOVER

Distribution:
Indian Ocean coasts

No. of Living
Species: 1

No. of Species
Vulnerable,
Endangered: 0, 0

No. of Species Extinct
Since 1600: 0

HAWAIIAN STILT
Himantopus knudseni
14–15.5 in (35–40 cm)
Hawaiian Islands

BLACK-NECKED STILT
Himantopus mexicanus
14–15.5 in (35–40 cm)
North America, South America

BLACK-WINGED STILT
Himantopus himantopus
14–15.5 in (35–40 cm)
Eurasia, Africa

BANDED STILT
Cladorhynchus leucocephalus
14–17 in (35–43 cm)
Australia

RED-NECKED AVOCET
Recurvirostra novaehollandiae
15.5–19 in (40–48 cm)
Australia

PIED AVOCET
Recurvirostra avosetta
17.5 in (44 cm)
Eurasia, Africa

IBISBILL
Ibidorhyncha struthersii
15.5 in (40 cm)
Asia

Stilts and Avocets; Ibisbill

STILTS and AVOCETS comprise a broadly distributed group of midsize, particularly beautiful shorebirds with very long legs and long, fine bills (straight in stilts, up-turned in avocets). They are among the most easily recognized and well known of the various shorebirds because of their distinctive looks, conspicuousness in shallow wetlands, and noisiness. There are about ten species in the family, Recurvirostridae (in order Charadriiformes, with sandpipers and gulls). Uncertainty over the exact number of species arises mainly from the controversial classification of one, the Black-winged Stilt (*Himantopus himantopus*), which occurs on six continents; some authorities consider it a single species but others divide it into from two to five separate species. For instance, the stilt that occurs in Hawaii is one of the Black-winged, sometimes called Black-necked, Stilts, but is often now considered a distinct species, the Hawaiian Stilt (*Himantopus knudseni*); and the Black-winged Stilt in Australasia is often separated out as the Pied, or White-headed, Stilt (*Himantopus leucocephalus*).

Fourteen to 19 inches (35 to 48 cm) long, stilts and avocets are striking with their pied plumage, mostly involving patches of black and white, although three species have areas of reddish brown on head or chest. The widely ranging Black-winged Stilt differs in form geographically, from white with black back and wings to having varying amounts of additional black on the head and neck (three of the birds shown, labeled Black-necked Stilt, Black-winged Stilt, and Hawaiian Stilt, illustrate some of this variation).

Stilts, usually in small groups but sometimes in aggregations in the

STILTS AND
AVOCETS

Distribution:
All continents except
Antarctica

No. of Living
Species: 11

No. of Species
Vulnerable,
Endangered: 0, 1

No. of Species Extinct
Since 1600: 0

thousands, wade about in shallow fresh- and saltwater, using their bills to probe mud for small insects, snails, and crustaceans. In some, worms, small fish, and aquatic plant materials are also taken. All in the family also swim to take food in deeper water. Food is detected visually or by touch; in the latter, birds swish their bills through shallow water or soft mud while comblike parts of the bill filter out small prey, or they probe the bill down into mud and food items contacted are grabbed and swallowed. Stilts and avocets occur in a variety of coastal and other wetlands, but they prefer shallow, salty inland lakes in open areas. They are known for their variety of sharp, brief calls—given frequently when foraging, while flying, and especially when alarmed, such as when potential predators approach nests.

Avocets and stilts are monogamous and tend to breed in small colonies near water. The nest scrape, into which eggs are placed, is often lined with vegetation. Both sexes incubate. Soon after they hatch, young are able to run from predators and feed themselves. Parents usually guard young at least until they can fly, perhaps 3 to 6 weeks after hatching. These birds first breed at 1 to 3 years of age; some are known to live more than 20 years.

Most of the stilts and avocets are abundant. One, New Zealand's Black Stilt, is critically endangered, with a total population under one hundred. Loss of breeding habitat and predation on eggs, young, and adults by introduced mammals such as cats, rats, and weasels have rendered it one of the world's most threatened shorebirds. The Hawaiian Stilt, endemic to the Hawaiian Islands, is considered endangered by U.S. authorities. The Pied Avocet, occurring widely in the Old World and once threatened in Western Europe, has recovered after successful conservation efforts; it is the long-time symbol of the UK's Royal Society for the Protection of Birds.

The IBISBILL, a midsize (16 inches, 40 cm) gray, black, and white shorebird confined to Central Asia, is the sole species of the family Ibidorhynchidae. It is closely related to stilts and avocets and, indeed, has sometimes been included in their family. The Ibisbill's outstanding physical trait is its bill, which is very long, down-curved, and, strikingly, dark red. This species occurs along rivers in mountainous areas, at elevations up to about 13,000 feet (4,000 m). It eats aquatic insects, crustaceans, and small fish. The Ibisbill is not considered threatened.

IBISBILL

Distribution:
Central Asia

No. of Living
Species: 1

No. of Species
Vulnerable,
Endangered: 0, 0

No. of Species Extinct
Since 1600: 0

Pratincoles
and Coursers

PRATINCOLES and COURSERS are a somewhat bizarre group of smallish to midsize Old World shorebirds. Pratincoles especially are puzzling curiosities to North American bird-watchers who view them for the first time when visiting Africa, for instance because, when initially spotted, often on a river beach, they seem to resemble terns. But when they run, they move fast along the ground, like plovers, and when they fly, their forked tail and long pointed wings give them the look of large swallows (they used to be called swallow-plovers). The family, Glareolidae, contains seventeen species, twelve of which are distributed wholly or partly in Africa; others occur in Eurasia and two occur in Australia. The family is included in the order Charadriiformes, with the sandpipers and gulls.

    Mainly brownish and grayish, many pratincoles and coursers are handsomely marked with black and/or white eye stripes, and some have dark neck or chest bands. Even with their bold markings, however, most species are highly cryptic on the ground, fading quickly into the brown soils of their native habitats. Bills are arched and fairly short, and either all dark (coursers; two have partly yellowish bills) or reddish at the base with dark tips (pratincoles). Length varies from 6.5 to 11.5 inches (17 to 29 cm). The sexes look alike or nearly so.

    Pratincoles are usually found near and along inland waterways and sometimes around estuaries. They may chase insect prey on sand and mud beaches by running, but more commonly they pursue flying insects on the wing, usually at dusk and typically in flocks. Their short bills open to

Distribution: Old World

No. of Living Species: 17

No. of Species Vulnerable, Endangered: 0, 1

No. of Species Extinct Since 1600: 0

ORIENTAL PRATINCOLE
Glareola maldivarum
9.5 in (24 cm)
Asia, Australia

AUSTRALIAN PRATINCOLE
Stiltia isabella
9 in (23 cm)
Australia, New Guinea, Indonesia

TEMMINCK'S COURSER
Cursorius temminckii
8 in (20 cm)
Africa

DOUBLE-BANDED COURSER
Smutsornis africanus
8.5 in (22 cm)
Africa

BRONZE-WINGED COURSER
Rhinoptilus chalcopterus
11 in (28 cm)
Africa

THREE-BANDED COURSER
Rhinoptilus cinctus
10.5 in (27 cm)
Africa

reveal a very large gape, which helps funnel food into the mouth. Aerial pursuit of insects is, to put it mildly, an unusual habit among shorebirds. Coursers, birds chiefly of hot, dry, open habitats, with much longer legs than pratincoles, are mainly terrestrial runners (hence, "coursers"), taking all their food (insects plus some snails and seeds) on the ground. They have squarish tails, not requiring the forked tails that give extra flight agility to pratincoles for their aerial insect eating. Some of the coursers have relatively long bills that they use to dig for bugs in loose soil. And some, such as the Double-banded Courser, are unusual among shorebirds because they prefer to spend their days hiding in the shade and only become active at twilight. Many pratincoles also rest during daytime heat, being most active at dawn and dusk. Pratincoles are highly social, almost always in groups; some of the coursers, however, are seen only singly or in small family parties.

Breeding habits are well documented for only a few pratincoles and coursers. Pratincoles nest in colonies; coursers, in single pairs. Apparently all monogamous, they tend to follow general shorebird breeding practices. Eggs are placed on the bare ground or in a shallow scrape. Both male and female incubate. When temperatures are very high, the parent on the nest is thought actually to be cooling rather than heating the eggs. They do this by panting while incubating, thus radiating from their mouths excess heat that they are drawing away from the eggs, and, in at least several species, by wetting their chest feathers just before taking over incubating duty, putting the cool water into contact with the eggs. Young are able to run from the nest and feed themselves soon after hatching. Posthatching parental care is well developed in the ways that parents protect young. If a predator is spotted nearby, they may feign injury, flaring out their wings and flopping about on the ground, all the while moving away from the nest or young to lure away the predator, and in some species, parents have been seen covering their chicks with sand to hide them from predators.

Only one species among the pratincoles and coursers is threatened: the Jerdon's Courser of southern India is critically endangered; it was thought extinct until rediscovered in 1986. Many other species, although currently safe, are experiencing declining populations, mainly owing to loss of their natural habitats as wild lands are altered for agricultural use and water levels are lowered as the result of dams and human water demands.

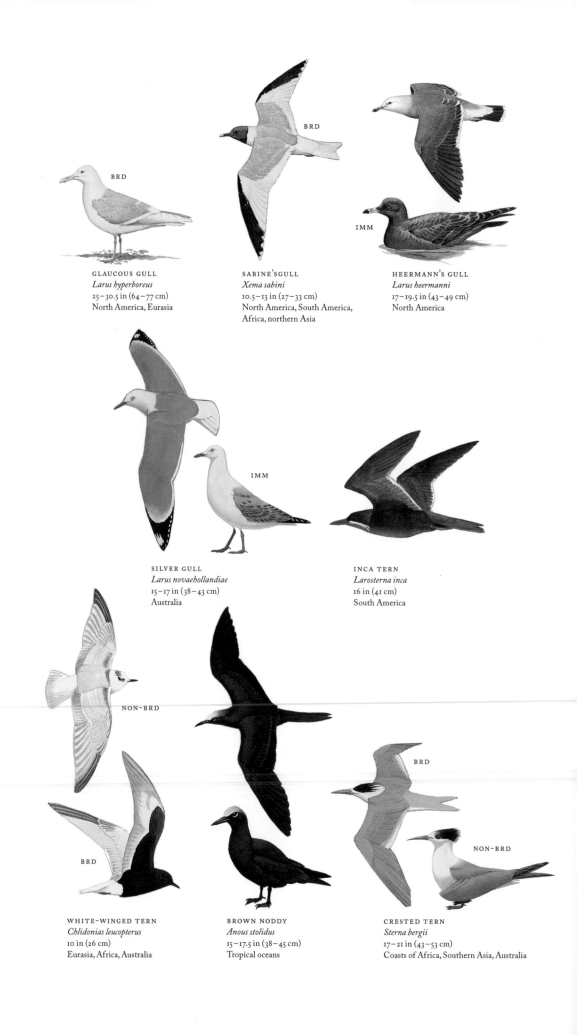

BRD

BRD

IMM

GLAUCOUS GULL
Larus hyperboreus
25–30.5 in (64–77 cm)
North America, Eurasia

SABINE'S GULL
Xema sabini
10.5–13 in (27–33 cm)
North America, South America,
Africa, northern Asia

HEERMANN'S GULL
Larus heermanni
17–19.5 in (43–49 cm)
North America

IMM

SILVER GULL
Larus novaehollandiae
15–17 in (38–43 cm)
Australia

INCA TERN
Larosterna inca
16 in (41 cm)
South America

NON-BRD

BRD

BRD

NON-BRD

WHITE-WINGED TERN
Chlidonias leucopterus
10 in (26 cm)
Eurasia, Africa, Australia

BROWN NODDY
Anous stolidus
15–17.5 in (38–45 cm)
Tropical oceans

CRESTED TERN
Sterna bergii
17–21 in (43–53 cm)
Coasts of Africa, Southern Asia, Australia

Gulls and Terns

GULLS and TERNS are the most common and conspicuous birds over mainland seacoasts and near-shore and offshore islands. These highly gregarious seabirds—they feed, roost, and breed in large groups—are also common in offshore waters and even, in certain species, inland. Gulls, of course, are known among the uninitiated as seagulls, a generic designation that evokes shudders and protests from most bird-watchers. Family Laridae, which is allied with the shorebirds in order Charadriiformes, includes the gulls (a few of which are known as kittiwakes) and terns (some of which are called noddies). The fifty-one species of gulls are distributed worldwide, but they are mainly birds of cooler ocean waters, and even of inland continental areas. Few occur in the tropics or around isolated, oceanic islands. Gulls (10 to 31 inches [25 to 79 cm], long) generally are large white and gray seabirds with fairly long, narrow wings, squarish tails, and sturdy, slightly hooked bills. Many have a blackish head, or hood, during breeding seasons. The forty-four tern species are distributed throughout the world's oceans. Terns (8 to 22 inches [20 to 56 cm] long) are often smaller and more delicate-looking than gulls. They have a slender light build, long pointed wings, deeply forked tail, slender tapered bill, and webbed feet. They are often gray above and white below, with a blackish head during breeding. Noddies are dark birds with lighter crowns and have broader wings and tails than other terns.

Gulls feed on fish and other sea life snatched from shallow water and on crabs and other invertebrates found on mudflats and beaches. Also, they are not above visiting garbage dumps or following fishing boats to grab whatever

Distribution: Worldwide

No. of Living Species: 95

No. of Species Vulnerable, Endangered: 6, 2

No. of Species Extinct Since 1600: 0

goodies fall or are thrown overboard. They also scavenge, taking bird eggs and nests from seabird breeding colonies when parents are gone or inattentive. Gulls also follow tractors during plowing, grabbing whatever insects and other small animals are flushed from cover when the big machines go by. Many larger gulls also chase smaller gulls and terns in the air to steal food the smaller birds have caught, an act termed *kleptoparasitism*. Terns, which eat mainly fish, squid, and crustaceans, feed during the day but also sometimes at night. Their main food-gathering technique is a bit messy: they spot prey near the water's surface while flying or hovering, then plunge-dive into the water to grab the prey, then rapidly take off again, the prey held tightly in the bill. Some terns also feed frequently by simply flying low and slow over the water and picking prey, including some insects, from the water's surface. Most terns seem to fly continuously, rarely setting down on the water to rest. Many in the family are long-distance flyers; the Arctic Tern has the longest migration route of any bird.

These seabirds usually breed in large colonies, often in the tens of thousands, on small islands, where there are no mammal predators, or in isolated mainland areas that are relatively free of predators. Gulls and terns usually breed on flat, open ground near the water (but some breed on cliffs, or inland, in marsh areas, and noddies breed in trees, bushes, or cliffs). Most are monogamous, mated males and females sharing in nest building, incubation, and feeding young. Young are fed when they push their bill into a parent's throat, in effect forcing the parent to regurgitate food stored in its crop, an enlargement of the top portion of the esophagus. These seabirds reach sexual maturity slowly, often not achieving their full adult, or breeding, plumage until they are 3 to 4 years old (terns) or 4 to 5 years old (gulls).

Most gulls and terns are abundant seabirds. Some species have even been able to capitalize on peoples' activities (for instance, using as feeding opportunities garbage dumps and agricultural fields; or, like some noddies, establishing new breeding colonies in trees on islands that, previous to human plantings, were treeless) and during the past few hundred years have succeeded in expanding their numbers and breeding areas. Worldwide, six species are considered vulnerable; two, New Zealand's Black-fronted Tern and eastern Asia's Chinese Crested-tern, are endangered.

Skuas; Skimmers; Sheathbills

SKUAS, sometimes called jaegers, are medium to large seabirds that, owing to their brown plumage, heavy, hooked bills, and predatory natures, might be mistaken for hawks, but are actually closely related to gulls. Skuas are known chiefly for their feeding, being predators on small rodents and, especially, on seabird eggs and young, and for their *kleptoparasitism*— stealing prey from other birds that have already captured it. In nature films, skuas are often seen dragging squirming penguin chicks from their defenseless parents. Sometimes skuas are included in family Laridae with the gulls, but many recent classifications place the seven skuas in a separate family, Stercorariidae (within order Charadriiformes). Long-winged, long-distance fliers, skuas occur in all the world's oceans. They are brown or sooty gray, sometimes with white or yellowish patches, 16 to 25 inches (41 to 64 cm) long.

Skuas occupy all kinds of marine habitats, including far out at sea; they breed on tundra or other short-vegetation habitats in the far north, in Antarctica, or on sub-Antarctic islands. On breeding grounds, skuas prey on lemmings and bird eggs and young. Along migration routes and in wintering areas, they are unsurpassed kleptoparasites. For instance, the Pomarine Skua "parasitizes" gulls and shearwaters at sea, and the Parasitic Jaeger pursues gulls and terns feeding near shore. When a skua sees a bird capture prey, such as a fish, it approaches rapidly, often catching the bird unaware. The skua, usually larger, has an advantage, and the original captor often drops its prey quickly after a brief chase. Small gulls and terns often attempt escape, but it

SKUAS

Distribution:
All oceans

No. of Living
Species: 7

No. of Species
Vulnerable,
Endangered: 0, 0

No. of Species Extinct
Since 1600: 0

LIGHT FORM

DARK FORM

IMM

POMARINE SKUA
Stercorarius pomarinus
18–20 in (46–51 cm)
Arctic tundra; equatorial
and southern oceans

PARASITIC JAEGER
Stercorarius parasiticus
16–18 in (41–46 cm)
Arctic tundra; southern oceans

BROWN SKUA
Catharacta antarctica
20.5–25 in (52–64 cm)
Southern oceans, Antarctica

SOUTH POLAR SKUA
Catharacta maccormicki
19.5–21.5 in (50–55 cm)
Antarctica, northern Atlantic
and Pacific oceans

NON-BRD

BRD

AFRICAN SKIMMER
Rynchops flavirostris
14–16.5 in (36–42 cm)
Africa

SNOWY SHEATHBILL
Chionis alba
13.5–16 in (34–41 cm)
Antarctica, southern South America

seems futile; the skua can fly faster and is as agile in flight as the smaller bird it is chasing. If the prey is not dropped quickly, the skua will grab the other bird by the wing or tail and spin it in the air. The prey is then dropped, and the skua typically catches it before it hits the water.

Mainly monogamous, skuas breed in colonies or in widely scattered vsingle pairs. Nests are open scrapes on the ground. The sexes share breeding duties. Young remain at or near the nest, being fed by the adults, for 2 or more months, until they can fly. Parents are highly aggressive around their nests, flying at and repeatedly striking any animal they regard as a potential predator (including humans and, in northern Europe, grazing sheep). None of the skuas are threatened.

Also closely related to the gulls but placed in their own family, Rynchopidae, are the three species of SKIMMERS, medium-size (14 to 18 inches, 36 to 46 cm) ternlike waterbirds that occur mainly in the tropics and subtropics. Being black and white and having huge red bills and, in flight, long arched wings, they are perhaps the most bizarre-looking of all the gull-like birds. Skimmers use their peculiar bill (the lower part extends beyond the tip of the upper part) to pluck fish and large invertebrates from the water's surface. They fly low over water (mostly large rivers inland, but also some coastal areas) with their bladelike bills open and the lower part skimming the surface (hence, their name). When they strike a fish or small crustacean, they quickly close the bill to catch it. Skimmers are quite social, roosting and breeding monogamously, in groups. One, the Indian Skimmer, is considered vulnerable.

The two SHEATHBILL species (family Chionidae) are medium-size (13 to 16 inches, 34 to 41 cm), white, pigeonlike birds with bill sheaths and fleshy wattles at the base of the bill. They occur in parts of Antarctica and on some sub-Antarctic islands; when not breeding, some wander to southern South America. Sheathbills are terrestrial, inhabiting coastal areas and foraging along shorelines. In flocks of up to fifty, they are omnivorous scavengers, taking, among other foods, penguin eggs and chicks (often their main foods), carrion, animal feces, insects, intertidal invertebrates, and even some algae. They also steal food from other birds, such as penguins and cormorants, by striking them while they are regurgitating food to their young. Sheathbills are monogamous, their nests often placed within or near a penguin breeding colony, a chief food source. Neither sheathbill is threatened.zz

SKIMMERS

Distribution: Tropics and subtropics

No. of Living Species: 3

No. of Species Vulnerable, Endangered: 1, 0

No. of Species Extinct Since 1600: 0

SHEATHBILLS

Distribution: Antarctic and Sub-Antarctic

No. of Living Species: 2

No. of Species Vulnerable, Endangered: 0, 0

No. of Species Extinct Since 1600: 0

COMMON MURRE
Uria aalge
15–17 in (38–43 cm)
Northern oceans and coasts

ANCIENT MURRELET
Synthliboramphus antiquus
10 in (26 cm)
Coasts and islands of northern Pacific

HORNED PUFFIN
Fratercula corniculata
14–16 in (36–41 cm)
Coasts and islands of northern Pacific

TUFTED PUFFIN
Fratercula cirrhata
14–16 in (36–41 cm)
Coasts and islands of
northern Pacific

PIGEON GUILLEMOT
Cepphus columba
12–14.5 in (30–37 cm)
Coasts and islands of northern Pacific

RHINOCEROS AUKLET
Cerorhinca monocerata
14.5 in (37 cm)
Coasts and islands of northern Pacific

LEAST AUKLET
Aethia pusilla
5 in (13 cm)
Coasts and islands of northern Pacific

CRESTED AUKLET
Aethia cristatella
7.5 in (19 cm)
Coasts and islands of northern Pacific

Puffins and Auks

PUFFINS and AUKS are smallish to medium-size diving seabirds that for centuries, owing to their typically pelagic existence, were known mainly only to fishermen and other mariners. Now some of these birds are popularly recognized: puffins, of which there are three species, for their massive, colorful, photogenic bills and overall cute appearance, and in the North American wildlife conservation movement, the Marbled Murrelet, a threatened species that nests in the canopy of old-growth forests and has become a symbol of the fight to protect these forests. Furthermore, puffins and auks, along with other members of the family, Alcidae, including murres, guillemots, and auklets, are now routinely spotted at sea and in their breeding areas by nature-lovers on whale-watching cruises or pelagic trips undertaken especially to view seabirds. The twenty-three species in the family, collectively called alcids, are included in order Charadriiformes with the gulls, terns, and shorebirds. Alcids are restricted to the colder waters of the Northern Hemisphere. Because they pursue fish underwater like penguins, using their wings for propulsion, they are often thought of as the Northern Hemisphere equivalents of penguins (which are restricted to the Southern Hemisphere). Unlike penguins, however, alcids fly.

Six to 16 inches (15 to 41 cm) long, alcids are stocky birds with short wings and a very short tail—in fact, on the water, they are usually identified by their compact bodies, short necks, and tiny tails (in the air, they look like flying footballs). Their feet are placed well back on the body, which aids in swimming. They are mainly dark gray or black above, light below. Bright

Distribution:
Northern oceans

No. of Living
Species: 23

No. of Species
Vulnerable,
Endangered: 4, 0

No. of Species Extinct
Since 1600: 1

coloring is restricted to their bills and feet, which, in some, are red orange and perhaps yellow. These colors are enhanced in the breeding season in puffins and a few others by the growth of an additional hard plate at the base of the bill, which adds more size and color to it. In the Rhinoceros Auklet, that plate includes a "horn" that dramatically alters the shape of the bill during the breeding season. In some species, long feathers that form conspicuous head plumes are also grown for breeding seasons.

Alcids are champion divers, feeding underwater like grebes and cormorants. They prey mostly on fish and crustaceans, but some also eat mollusks and worms, and some specialize on plankton, the ocean's huge population of tiny floating organisms. Unlike loons, grebes, and cormorants, which use their feet for underwater propulsion, alcids use their wings, thus when viewed underwater (possible now for many people in the huge transparent water tanks at some aquariums), they appear to be "flying" underwater. Their wings, although small enough to move through the water rapidly for this wing-propelled diving, are still large enough to hold them in the air if beat fast enough. Alcids swim rapidly underwater, quick enough to chase down the fish upon which they feed. Not only are they fast, they dive quite deeply; some murres are known to have reached depths of about 600 feet (180 m), and even small species such as some auklets dive below 100 feet (30 m).

Alcids are monogamous and probably mate for life; both sexes incubate eggs and feed young. They tend to breed in groups, sometimes in vast colonies. Nesting sites, on small islands or coastal cliffs inaccessible to mammal predators, are varied: murres nest on cliff ledges, many species nest in rock crevices, and puffins and guillemots nest in crevices on rocky islands or dig their own burrows. Eggs are placed on the ground or in a burrow. Marbled Murrelets are an exception. When not nesting on tundra, they nest in forests, where their nest is a depression on a wide, mossy branch high in an old-growth conifer.

Alcids in the past were widely hunted for meat, oil, and feathers. All or most species experienced seriously reduced populations through the first half of the twentieth century. One, the Great Auk, the largest alcid, which was flightless, became extinct around 1850. With hunting now much reduced, many alcids enjoy healthy populations. Currently four species are considered vulnerable: North America's Marbled, Xantus's and Craveri's Murrelets, and the Japanese Murrelet.

Sandgrouse

SANDGROUSE look a lot like pigeons, and anyone not steeped in ornithological knowledge could be forgiven for mistaking these medium-size, cryptically colored land birds for types of pigeons or doves. Their precise classification has befuddled avian biologists for generations. Historically, given their short bills, short, feathered legs, and terrestrial lifestyles, sandgrouse were thought to be allied with chickenlike game birds such as partridges and grouse, and so perhaps to be included in the game bird order (Galliformes), or given their pigeonlike long, pointed wings, plumage, and some pigeonlike elements of internal anatomy, with the pigeons (order Columbiformes). However, sandgrouse differ in significant ways from game birds and from pigeons (for instance, they fly, drink, and call in a manner different from pigeons, and their nests, eggs, and chicks are very different from pigeons'). Some molecular research suggests sandgrouse are perhaps related to shorebirds. Several recent classifications place the sandgrouse, family Pteroclidae, in their own order, Pterocliformes—and that system is followed here. Sandgrouse are not well known or appreciated; their main interactions with people are as agricultural pests and dinner fare (although reports are that they are far from delicious), and for these reasons they are hunted in some areas.

Restricted in their distributions to the Old World, sandgrouse are mainly African, with twelve of the world's sixteen species occurring on that continent (another is endemic to Madagascar, off Africa's coast). Some occur in southern Europe and southern Asia. From 9.5 to 16 inches (24 to 40 cm) long, they are mostly beautifully patterned in shades of sandy brown, with

Distribution:
Old World

No. of Living
Species: 16

No. of Species
Vulnerable,
Endangered: 0, 0

No. of Species Extinct
Since 1600: 0

CHESTNUT-BELLIED SANDGROUSE
Pterocles exustus
12.5 in (32 cm)
Africa, southern Asia

BLACK-FACED SANDGROUSE
Pterocles decoratus
11 in (28 cm)
Africa

YELLOW-THROATED SANDGROUSE
Pterocles gutturalis
12 in (30 cm)
Africa

NAMAQUA SANDGROUSE
Pterocles namaqua
11 in (28 cm)
Southern Africa

DOUBLE-BANDED SANDGROUSE
Pterocles bicinctus
10 in (25 cm)
Southern Africa

patches of gray, black, and white. Males usually have more distinct markings than females.

Sandgrouse favor hot, dry environments, such as savannas, semiarid scrublands, and deserts. They spend most of the time on the ground, moving with a shuffling walk. They feed mainly on small seeds picked up from the ground, and, like pigeons, collect small pebbles in their stomachs to help grind and digest the seeds. With their strong, long wings, sandgrouse are powerful fliers. They are often seen on the wing, commuting between feeding areas and drinking sites. The daily flight that sandgrouse make to and from water, sometimes involving a round-trip of more than 60 miles (100 km), is one of the most characteristic features of these birds; because they feed entirely on dry seeds and derive no moisture from these, access to open water is crucial. Sandgrouse are sociable birds, sometimes gathering in thousands at their preferred drinking sites.

Although often in flocks outside the breeding season, sandgrouse break up into monogamous pairs to breed. They nest (alone or in loose colonies) in a shallow scrape on open ground or in a natural depression, sometimes lined with vegetation. Both parents incubate the eggs, the female during the day, the male taking the night shift. The young leave the nest almost immediately after hatching. They are not fed by their parents, but are shown suitable food to pick up. In response to their dry habitat, sandgrouse have evolved special breast feathers, modified to maximize their ability to absorb and retain water. Males usually fly to pools, squat in the water, ruffle and soak their feathers (known as belly-wetting) and then fly back to their nestlings. The young suck water from the breast feathers. Females also have some of the special water-carrying feathers, but fewer than males, and the water-carrying behavior is best developed in the males. (Although some shorebirds, such as pratincoles, also carry water in this way, they use it for cooling eggs and chicks, not for providing drinking water.) After about 4 to 5 weeks, the young sandgrouse can fly to water with their parents.

Little is known about population sizes of most sandgrouse species, but almost all appear to be moderately to very common inhabitants of the regions in which they occur. None of the sandgrouse are threatened globally, but some are threatened locally; for example, in North Africa and parts of Europe, Black-bellied and Chestnut-bellied Sandgrouse are disappearing from parts of their ranges, perhaps as a result of changing agricultural practices.

INCA DOVE
Columbina inca
8.5 in (21 cm)
North America, Central America

WHITE-FACED QUAIL-DOVE
Geotrygon albifacies
11–14 in (28–36 cm)
Mexico, Central America

SCALY-NAPED PIGEON
Columba squamosa
13–16 in (33–41 cm)
West Indies

EARED DOVE
Zenaida auriculata
8.5–11 in (22–28 cm)
South America

GALÁPAGOS DOVE
Zenaida galapagoensis
7–9 in (18–23 cm)
Galápagos Islands

PACIFIC DOVE
Zenaida meloda
10–13 in (25–33 cm)
South America

PICAZURO PIGEON
Columba picazuro
13.5 in (34 cm)
South America

WHITE-CROWNED PIGEON
Columba leucocephala
11.5–15.5 in (29–40 cm)
West Indies, Central America, Florida

Pigeons and Doves

PIGEONS and DOVES comprise a highly successful group, represented, often in large numbers, almost everywhere on dry land, except for some oceanic islands. Their ecological success is viewed by biologists as somewhat surprising because they are largely defenseless and quite edible, regarded as a tasty entrée by human and an array of nonhuman predators. The family, Columbidae (placed in its own order, Columbiformes), includes about 308 species. They inhabit almost all kinds of habitats, from arid grasslands to tropical rainforests to higher-elevation mountainsides. Smaller species generally are called doves, larger ones, pigeons, but there is a good amount of overlap in name assignments.

Varying in length from 6 to 30 inches (15 to 76 cm), pigeons and doves are plump-looking birds with compact bodies, short necks, and small heads. Legs are usually fairly short, except in the ground-dwelling species. Bills are small, slender, and straight. Typically there is a conspicuous patch of naked skin, or cere, at the base of the bill, over the nostrils. The soft, dense plumage of most are gray or brown, although some have bold patterns of black lines or spots; many have splotches of iridescence, especially on necks and wings. But some groups, such as the fruit-doves (about 120 species), which mainly occur in the Indonesia/Malaysia and Australia/New Guinea regions, are easily among the most gaily colored birds. Green, the predominant hue of fruit-doves, camouflages them in their arboreal habitats. Some pigeons have conspicuous crests, including the three huge, ground-dwelling crowned-pigeons of New Guinea, which have large, elaborate crests. Male and female pigeons are generally, but not always, alike in coloring.

Distribution:
All continents except
Antarctica

No. of Living
Species: 308

No. of Species
Vulnerable,
Endangered: 34, 26

No. of Species Extinct
Since 1600: About 10

EURASIAN COLLARED-DOVE
Streptopelia decaocto
12 in (31 cm)
Eurasia

THICK-BILLED GREEN-PIGEON
Treron curvirostra
9.5–12 in (24–31 cm)
Southeast Asia

PIED IMPERIAL PIGEON
Ducula bicolor
14–16.5 in (35–42 cm)
Southeast Asia

EMERALD DOVE
Chalcophaps indica
10 in (26 cm)
Southern Asia, Australia

MOUNTAIN IMPERIAL-PIGEON
Ducula badia
17–20 in (43–51 cm)
Southern Asia

NAMAQUA DOVE
Oena capensis
11 in (28 cm)
Africa, Madagascar, Arabia

AFRICAN GREEN-PIGEON
Treron calva
10–12 in (25–30 cm)
Africa

WESTERN BRONZE-NAPED PIGEON
Columba iriditorques
10 in (25 cm)
Africa

BROWN CUCKOO-DOVE
Macropygia amboinensis
15.5–17.5 in (40–45 cm)
Australia

COMMON BRONZEWING
Phaps chalcoptera
11–14 in (28–36 cm)
Australia

IMM

ROSE-CROWNED FRUIT-DOVE
Ptilinopus regina
9 in (23 cm)
Australia, Indonesia

CRESTED PIGEON
Ocyphaps lophotes
12–14 in (31–36 cm)
Australia

WOMPOO FRUIT-DOVE
Ptilinopus magnificus
11.5–18 in (29–46 cm)
Australia, New Guinea

CHESTNUT-QUILLED ROCK-PIGEON
Petrophassa rufipennis
12 in (30 cm)
Australia

PARTRIDGE PIGEON
Geophaps smithii
10.5 in (27 cm)
Australia

Most pigeons are at least partly arboreal, but some spend time in and around cliffs, and others are primarily ground dwellers. They eat seeds, fruit, berries, and the occasional insect, snail, or other small invertebrate. Even those species that spend a lot of time in trees often forage on the ground, moving along with their head-bobbing walk. Owing to their small, weak bills, they eat only what they can swallow whole; "chewing" is accomplished in the gizzard, a muscular portion of the stomach in which food is mashed against small pebbles that are eaten expressly for this purpose. Fruit-doves swallow small fruits whole, and the ridged walls of their gizzards rub the skin and pulp from the fruits. Pigeons typically are strong, rapid flyers, which, in essence, along with their cryptic color patterns, provide their only defenses against predation. Most are gregarious to some degree, staying in groups during the nonbreeding portion of the year; some gather into large flocks.

Monogamous breeders, pigeons nest solitarily or in colonies. Nests are shallow, open affairs of woven twigs, plant stems, and roots, placed on the ground, on rock ledges, or in forks of shrubs or trees. Reproductive duties, such nest building, incubation, and feeding the young, are shared by male and female. All pigeons, male and female, feed their young "pigeon milk," a nutritious fluid produced in the crop, an enlargement of the esophagus used for food storage. During the first few days of life, nestlings receive 100 percent pigeon milk; as they grow older, they are fed an increasing proportion of regurgitated solid food.

Pigeons and doves in most parts of the world are common, abundant animals. Some species benefit from people's alterations of natural habitats, expanding their ranges, for example, where forests are cleared for agriculture. But a number of species are threatened, mostly from a combination of habitat loss (generally forest destruction), reduced reproductive success owing to introduced nest predators, and excessive hunting. Currently, thirty-four species are considered vulnerable and twenty-six endangered, twelve of the latter critically so. Many of these sixty species are restricted to islands, where for various reasons, they are especially at risk; indeed, most of the eight to ten pigeon species known to have become extinct during the past 200 years were island-bound. The Dodo, a large, flightless pigeon from the Indian Ocean island of Mauritius, became extinct in the late 1600s. North America's Passenger Pigeon, once one of the continent's most abundant birds, was extinct by the early 1900s, mainly because of overhunting.

Parrots

PARROTS are among the most recognizable of birds; even in the roughly half of the world where they don't occur naturally, most people can identify one. Parrots have long fascinated people and have been captured and transported widely as pets for thousands of years. The fascination stems from the birds' bright coloring, ability to imitate human speech and other sounds, individualistic personalities, and long life spans (up to 80 years in captivity). The 330 or so parrot species that comprise family Psittacidae (in order Psittaciformes with the cockatoos) are globally distributed across the tropics, with some extending into subtropical and even temperate zones. They have a particularly diverse, abundant presence in the neotropical and Australian regions. Parrots are generally consistent in overall form and share some traits that set them distinctively apart from other birds. They are stocky, with short necks and compact bodies. All possess a short, hooked, bill with a hinge on the upper part that permits great mobility and leverage during feeding. Their legs are short, and their feet are adapted for powerful grasping and a high degree of dexterity, more so than any other bird.

Parrots are sometimes divided by size: "parrotlets" are small birds (as small as 4 inches [10 cm]) with short tails; "parakeets" are also small, with long or short tails; "parrots" are medium size, usually with short tails; and the "macaws" of Central and South America, the world's largest parrots, are large (up to 40 inches [100 cm]) with long tails. The group contains some of the globe's most gaudily plumaged birds. Australia, for instance, has some bedazzling common species, including the red and purple Crimson Rosella

*Distribution:
Neotropics, Africa,
southern Asia,
Australasia*

*No. of Living
Species: 331*

*No. of Species
Vulnerable,
Endangered: 43, 46*

*No. of Species Extinct
Since 1600: 19*

SCARLET MACAW
Ara macao
33–35 in (84–89 cm)
Central America,
South America

BLUE-AND-YELLOW MACAW
Ara ararauna
34 in (86 cm)
South America

CHESTNUT-FRONTED MACAW
Ara severa
18–20 in (46–51 cm)
South America

HYACINTH MACAW
Anodorhynchus hyacinthinus
39.5 in (100 cm)
South America

LILAC-CROWNED PARROT
Amazona finschi
13 in (33 cm)
Mexico

COBALT-WINGED PARAKEET
Brotogeris cyanoptera
7.5 in (19 cm)
South America

WHITE-NECKED PARAKEET
Pyrrhura albipectus
9.5 in (24 cm)
South America

BLUE-HEADED PARROT
Pionus menstruus
10 in (26 cm)
Central America, South America

GOLDEN PARAKEET
Guarouba guarouba
14 in (35 cm)
South America

BLUE-FRONTED PARROT
Amazona aestiva
14.5 in (37 cm)
South America

THICK-BILLED PARROT
Rhynchopsitta pachyrhyncha
15–17 in (38–43 cm)
Mexico

ROSE-THROATED PARROT
Amazona leucocephala
12.5 in (32 cm)
West Indies

ORANGE-WINGED PARROT
Amazona amazonica
12 in (31 cm)
South America

IMPERIAL PARROT
Amazona imperialis
17.5 in (45 cm)
West Indies

VERNAL HANGING-PARROT
Loriculus vernalis
5 in (13 cm)
Southern Asia

LONG-TAILED PARAKEET
Psittacula longicauda
15.5–19 in (40–48 cm)
Southeast Asia

ROSY-FACED LOVEBIRD
Agapornis roseicollis
6.5 in (17 cm)
Southern Africa

ROSE-RINGED PARAKEET
Psittacula krameri
14.5–17 in (37–43 cm)
Africa, southern Asia

REGENT PARROT
Polytelis anthopeplus
15.5 in (40 cm)
Australia

♀   ♂

AUSTRALIAN KING-PARROT
Alisterus scapularis
16.5 in (42 cm)
Australia

IMM

CRIMSON ROSELLA
Platycercus elegans
14 in (36 cm)
Australia

BROWN PARROT
Poicephalus meyeri
9 in (23 cm)
Africa

RAINBOW LORIKEET
Trichoglossus haematodus
10–12 in (25–30 cm)
Australia, New Guinea, Indonesia

PURPLE-CROWNED LORIKEET
Glossopsitta porphyrocephala
6 in (15 cm)
Australia

and the riotously colorful Rainbow Lorikeet. The Scarlet Macaw, red, yellow, and blue, is certainly one of the New World's most striking birds. Green, the predominant parrot color, serves admirably as camouflage; parrots feeding amid a tree's high foliage are very difficult to see. In most species the sexes are similar or identical in appearance.

Highly social seed and fruit-eaters, parrots are usually encountered in flocks of four or more, and groups of more than fifty smaller parrots are common. Flocks are usually groups of mated pairs. They move about seeking food in forests, woodlands, savannas, and agricultural areas. They take mostly fruits, nuts, and seeds, leaf and flower buds, and some flower parts and nectar. Although considered fruit-eaters, they often attack fruit just to get at the seeds within. The powerful bill slices open fruit and crushes seeds. Some parrots are specialized feeders. For example, lorikeets, a brilliantly colored subgroup, confined mainly to Australasia and Indonesia, have long tongues with brushlike tips that they use to gather nectar and pollen. During early mornings and late afternoons, raucous, squawking flocks of parrots characteristically take flight explosively from trees, heading in mornings for feeding areas and later for night roosts. A wide range of parrots often visit "licks," exposed soil clay deposits. They eat the clay, which may help detoxify harmful compounds that are consumed in their seed and fruit diet or may supply essential minerals not provided by a vegetarian diet. Parrots are not considered strong flyers and most do not undertake long-distance flights.

Parrots are monogamous, often pairing for life. Most species breed in tree cavities, often in dead trees or branches, and often high off the ground; nests may be lined with wood chips. Females incubate eggs alone while usually being periodically fed regurgitated food by their mates. The helpless young are fed by both parents.

Although many parrots still enjoy healthy populations, eighty-nine species worldwide are threatened (forty-three are vulnerable, thirty-three are endangered, and thirteen critically endangered). Unfortunately, parrots are subject to three powerful forces that, in combination, take heavy tolls on their numbers: they are primarily forest birds (often nesting only in tree hollows), and forests are increasingly under attack by loggers, agricultural interests, and developers; they are considered agricultural pests by farmers owing to their seed and fruit eating; and they are among the world's most popular cage birds. The United States' only native parrot, the Carolina Parakeet, became extinct during the early 1900s.

RED-TAILED BLACK-COCKATOO
Calyptorhynchus banksii
19.5–25.5 in (50–65 cm)
Australia

SULPHUR-CRESTED COCKATOO
Cacatua galerita
17.5–21.5 in (45–55 cm)
Australia, New Guinea

GANG-GANG COCKATOO
Callocephalon fimbriatum
12.5–14.5 in (32–37 cm)
Australia

GALAH
Eolophus roseicapilla
14 in (36 cm)
Australia

LITTLE CORELLA
Cacatua sanguinea
15 in (38 cm)
Australia

COCKATIEL
Nymphicus hollandicus
13 in (33 cm)
Australia

MAJOR MITCHELL'S COCKATOO
Cacatua leadbeateri
14 in (35 cm)
Australia

Cockatoos

Like other parrots, COCKATOOS, a spectacular group that occurs only in the Australia/New Guinea and Indonesia regions, have long captured people's imaginations and been kept in captivity. With their long life spans, their ability to bond with people, and often, their large size, cockatoos especially have been kept as companion animals. The twenty-one species, family Cacatuidae (placed in the order Psittaciformes with the typical parrots), are medium to large parrots (12 to 25 inches, 30 to 64 cm, long) with crests that they can erect. The most recognized member of the family to many North Americans is probably the Sulphur-crested Cockatoo, which is large, white, with a yellow crest, and is a commonly kept pet.

Sharing their general form with other parrots, cockatoos are bulky birds with short necks and short, hooked bills. They have medium to longish tails and their legs are short, with feet adapted for powerful grasping and a high degree of dexterity. Cockatoos come in two main colors, blackish or white. The blackish or dark gray species include the huge, large-billed Palm Cockatoo of Australia and New Guinea, which, at up to 2.2 pounds (1 kg), is the largest cockatoo, and the smallest cockatoo, Australia's Cockatiel, which is a common cage bird worldwide. Among the white species is a group of smaller cockatoos known as corellas; the Sulphur-crested; and Australia's Major Mitchell's Cockatoo, which has a pinkish head and striking red, yellow, and white crest. The Galah, endemic to Australia and considered quite an agricultural pest, is gray and pink. In most cockatoos, the sexes are very similar or identical in appearance, but some have moderate differences.

Distribution: Australasia, parts of Indonesia

No. of Living Species: 21

No. of Species Vulnerable, Endangered: 2, 3

No. of Species Extinct Since 1600: 0

Cockatoos occur in almost all terrestrial habitats within their ranges, from forests (including the margins of rainforests) to shrublands and even in desert regions—wherever they can find food and places to roost and nest. Like other parrots, cockatoos are highly social, usually foraging and roosting in flocks. They eat fruits, nuts, seeds, flower parts, and some insects; some use their strong bills to extract insect larvae from wood. Using their powerful feet to grasp branches and their bills as, essentially, a third foot, cockatoos clamber methodically through trees in search of food. Just as caged parrots do, they will hang gymnastically at odd angles and even upside down, the better to reach some delicious morsel. Cockatoo feet, with their powerful claws, also function as hands, delicately manipulating food and bringing it to the bill. Cockatoo tongues, as in other parrots, are thick and muscular, used to scoop pulp from fruits and hold seeds and nuts for the bill to crush. Although many species feed primarily in trees, some, pursuing seeds, such as Galahs, forage mainly on the ground. Some of the corellas use specialized bills to dig in the ground for roots.

Monogamous breeders, cockatoos form long-term pairs that generally remain together all year. Most species breed in cavities in live or dead trees; nests are lined with wood chips. The female only or both sexes incubate; in the former case, the female may be fed on the nest by her mate. Young are fed, as nestlings and fledglings, by both parents. In some species, the young stay with the parents until the next breeding season.

Cockatoos, generally, are threatened because they nest in tree cavities, and the large trees they breed in are increasingly scarce owing to such human activities as logging and land clearance. Also, they are persecuted by farmers and orchardists because they eat seeds and fruit crops, and are pursued for the pet trade. In Australia, cockatoos such as the Galah, Cockatiel, Red-tailed Black-Cockatoo, and Western Corella, have long been poisoned or shot to protect crops. Sulphur-crested Cockatoos are considered to be real pests; they damage trees in orchards and, apparently exercising their powerful bills, tear up car windshield wipers and house window moldings. The Galahs, taking advantage of agriculture and artificial water supplies, now occur throughout Australia in large numbers and are as much a part of the landscape as kangaroos. Two cockatoo species are considered vulnerable and three are endangered (one each in Australia, Indonesia, and the Philippines).

Turacos

TURACOS are large, colorful, arboreal birds of sub-Saharan African forests, woodlands, and savannas. They are known for their brilliant plumage and have long been hunted for their feathers; turaco feathers are commonly used in ceremonial headdresses of various African groups, including East Africa's nomadic Masai people. Being large and tasty birds, turacos are also pursued for the dinner table. Visitors to African forests and savannas are made quickly aware of these birds by their raucous, often repetitive calls, some of the most characteristic sounds of these habitats. The twenty-three turaco species are all confined to Africa; they are known as louries in southern Africa. The family, Musophagidae, although its classification is controversial, is usually placed in order Cuculiformes with the cuckoos. (*Musophagidae* refers to banana or plantain eating, but despite being fruit-eaters, turacos rarely eat wild bananas.). Some of the turacos are formally called plantain-eaters and others are known as go-away-birds, for their loud distinctive "g'way, g'way" calls.

Turacos, all with conspicuous, sometimes colorful crests, are 16 to 29 inches (40 to 74 cm) long and have short, strong bills; short, rounded wings; and long, broad tails. Many have bare, brightly colored patches of skin around their eyes. Most species are primarily a striking glossy blue, green, or purplish. Studies of turacos show that their bright coloring at least partially reflects the foods they eat. Some fruits in their diet provide copper, and a red copper-based pigment (turacin) unique to turacos provides the brilliant reds in their plumage. Similarly, the deep greens of some species

Distribution: Sub-Saharan Africa

No. of Living Species: 23

No. of Species Vulnerable, Endangered: 1, 1

No. of Species Extinct Since 1600: 0

ROSS'S TURACO
Musophaga rossae
21 in (53 cm)
Africa

KNYSNA LOURIE
Tauraco corythaix
18 in (46 cm)
Southern Africa

HARTLAUB'S TURACO
Tauraco hartlaubi
17 in (43 cm)
Africa

WHITE-BELLIED GO-AWAY-BIRD
Corythaixoides leucogaster
19.5 in (50 cm)
Africa

GREAT BLUE TURACO
Corythaeola cristata
27.5–29.5 in (70–75 cm)
Africa

PURPLE-CRESTED TURACO
Tauraco porphyreolophus
17.5 in (45 cm)
Africa

EASTERN GRAY PLANTAIN-EATER
Crinifer zonurus
19.5 in (50 cm)
Africa

arise from another unique copper pigment called turacoverdin. The denser and greener the forest habitat a given turaco species occupies, the deeper green its plumage. The go-away-birds (three species) and plantain-eaters (two species) inhabit scrublands, savanna, and open woodlands, instead of forests, and they are clad mostly in soft gray, buff, brown, and white. Male and female turacos look alike.

Fairly weak and reluctant fliers, turacos are extremely good at moving around trees on foot, often running or hopping along branches; they have strong legs, and feet with an outer toe that can point forward or backward, important for their agile arboreal behavior. When they do fly, turacos often climb to high points in tree canopies and then launch themselves with floppy glides and a few quick flaps to other trees nearby. Gregarious birds, often congregating in large groups, they fly from tree to tree in "strings," or single-file style. They are more or less sedentary birds, making only short flights in search of food. Most turacos specialize on eating fruit (particularly figs in the forest species) but also take some leaves, buds, and flowers, and the occasional caterpillar, moth, or beetle. Lacking the powerful, seed-cracking bills of parrots, turacos select smaller, softer fruits, and regurgitate large hard seeds that are swallowed. They often gather in fruiting trees in noisy groups. Birds will return to the same trees day after day until the fruit there is exhausted. Go-away-birds, although also taking fruits, berries, and flowers, concentrate their feeding on Acacia buds and seedpods; the two species of plantain-eaters consume chiefly fruit.

Turacos, monogamous breeders, in pairs defend territories during the breeding season and nest solitarily. The go-away-birds may sometimes have helpers at the nest, other adults that do not breed but help the nesting pair by aiding with territory defense and caring for young; helpers are usually young from previous broods. Turacos build flat, flimsy twig nests in trees or bushes. Both male and female participate in nest building, although one may remain at the nest and accept materials gathered by its mate. Both parents incubate eggs and provide regurgitated food to the chicks.

Where not widely persecuted (by fruit growers or by people who want to eat them or capture them as cage birds), many turaco species are common or even abundant. Two are threatened: Ethiopia's Prince Ruspoli's Turaco (vulnerable) and Cameroon's Bannerman's Turaco (endangered). Both occur over very small ranges and have small and declining populations, mostly owing to continued destruction and degradation of their forest habitats.

SQUIRREL CUCKOO
Piaya cayana
18 in (46 cm)
Mexico, Central America, South America

YELLOW-BILLED CUCKOO
Coccyzus americanus
12 in (30 cm)
North America, South America

BLACK-BILLED CUCKOO
Coccyzus erythropthalmus
12 in (30 cm)
North America, South America

MANGROVE CUCKOO
Coccyzus minor
13.5 in (34 cm)
North America, South America

PLAINTIVE CUCKOO
Cacomantis merulinus
7–9 in (18–23 cm)
Southern Asia

COMMON KOEL
Eudynamys scolopacea
15.5–18 in (39–46 cm)
Southern Asia, Australia

GREEN-BILLED MALKOHA
Phaenicophaeus tristis
19.5 in (50 cm)
Southern Africa

Cuckoos

Distribution:
All continents except
Antarctica

No. of Living
Species: 122

No. of Species
Vulnerable,
Endangered: 6, 3

No. of Species Extinct
Since 1600: 1 or 2

CUCKOOS are physically rather plain but behaviorally extraordinary, employing some of the most bizarre breeding practices known among birds. Many Old World cuckoo species are brood parasites: they build no nests of their own, the females laying their eggs in the nests of other species. These other birds often raise the young cuckoos as their own offspring, to the significant detriment of their own young. Brood parasitism by cuckoos gave rise to the word *cuckold*, meaning that a man's wife was unfaithful and so he raised offspring genetically unrelated to him, which recalls the cuckoos' reproductive habits. The name cuckoo comes from the calls made by a Eurasian species, the Common Cuckoo, which is also the source of the sounds for cuckoo clocks.

Cuckoos (which include coucals, couas, and malkohas, among others) comprise family Cuculidae, included in order Cuculiformes with the turacos. There are about 120 species, distributed worldwide in temperate and tropical areas. Most are small to medium-size birds (6 to 18 inches [16 to 46 cm] long), slender, narrow-winged, and long-tailed; bills curve downward at the end. Male and female mostly look alike, attired in plain grays, browns, and whites, often with streaked or spotted patches. Several have alternating white and black bands on their chests and/or tail undersides. Some are more brightly clad. Many of the nine couas, restricted to Madagascar, are greenish, and all have bare blue skin around their eyes. Bronze-cuckoos, five small species of the Australia/New Guinea region, have metallic green backs and wings, and the twelve long-tailed malkohas, of southern Asia, have red or bluish

RAFFLE'S MALKOHA
Phaenicophaeus chlorophaeus
12.5 in (32 cm)
Southeast Asia

GREATER COUCAL
Centropus sinensis
18.5–20.5 in (47–52 cm)
Southern Africa

BLACK-AND-WHITE CUCKOO
Clamator jacobinus
13.5 in (34 cm)
Africa, Southern Asia

RED-CHESTED CUCKOO
Cuculus solitarius
12 in (31 cm)
Africa

BLACK CUCKOO
Cuculus clamosus
12 in (31 cm)
Africa

AFRICAN EMERALD CUCKOO
Chrysococcyx cupreus
8 in (20 cm)
Africa

SHINING BRONZE-CUCKOO
Chrysococcyx lucidus
6.5 in (17 cm)
Australasia, Indonesia

FAN-TAILED CUCKOO
Cacomantis flabelliformis
10 in (26 cm)
Australia, New Guinea

LITTLE BRONZE-CUCKOO
Chrysococcyx minutillus
6.5 in (16 cm)
Southeast Asia, Australia

BRD

NON-BRD

CHANNEL-BILLED CUCKOO
Scythrops novaehollandiae
23.5 in (60 cm)
Australia, New Guinea, Indonesia

PHEASANT COUCAL
Centropus phasianinus
21–31.5 in (53–80 cm)
Australia, New Guinea

PALLID CUCKOO
Cuculus pallidus
12.5 in (32 cm)
Australia

BRUSH CUCKOO
Cacomantis variolosus
8.5–11 in (21–28 cm)
Southeast Asia, Australia

eye rings and yellow and/or reddish bills. Coucals, a group of twenty-eight mostly ground-feeding species, mainly of Africa and Southeast Asia, include some of the largest family members, ranging up to about 30 inches (76 cm).

Cuckoos are mainly shy, solitary birds of forests, woodlands, and dense thickets. Most are arboreal, and are known for their fast, graceful, undulating flight. They eat insects and insect larvae, apparently having a special fondness for caterpillars. Larger cuckoos will also take fruit, berries, and small vertebrates. A few cuckoos, such as the Pheasant Coucal, are chiefly ground dwellers, eating insects, but also vertebrates such as mice, small lizards, and snakes. These coucals, which live in dense vegetation areas, do fly, albeit not gracefully or for long distances; they prefer to run along the ground to catch food or to escape from predators.

About 50 of the 120-odd cuckoos are brood parasites. (In the New World, only 3 cuckoos are parasitic, but they are in the ground-cuckoo group, covered in the next account.) Coucals, couas, and malkohas are not parasites, building their own nests and breeding monogamously. A given parasitic cuckoo species usually parasitizes many different "host" species (hosts are all passerines, or perching birds), but an individual female usually lays in the nests of the same host species. A female cuckoo lays an egg in a host nest usually when the host couple is absent, and she often removes a host egg before laying her own (presumably so the same number of eggs is in the nest when the hosts return). Many host species are smaller than the cuckoos that parasitize them, so some cuckoos actually lay eggs that are smaller than those of similar-size birds, so that their eggs more closely match the size of the host's eggs. Other cuckoos lay eggs that resemble the host's eggs in shell color and patterning. All these machinations are thought to be adaptations so that the hosts cannot recognize the cuckoo eggs and eject them (as happens in some species). When the cuckoo chick hatches, it tosses other eggs and the host's own chicks out of the nest by pushing them with its back. If any host chicks are left in the nest, the cuckoo chick, usually larger, out-competes them for food brought to the nest by the host parents, and the hosts' own young often starve or are significantly weakened.

Cuckoos, chiefly birds of forests, are threatened mainly by forest destruction and degradation. Currently six species are considered vulnerable and three endangered (two critically).

Ground-cuckoos;
Anis; Hoatzin

GROUND-CUCKOOS, ANIS, and the HOATZIN are New World birds with controversial classifications. The ground-cuckoos and anis are sometimes included with cuckoos in family Cuculidae, but here, along with the Hoatzin, they are treated as separate families.

The ten species of New World Ground-cuckoos (family Neomorphidae) are mainly shy inhabitants of dense forests and scrub areas of Central and South America, but the two called roadrunners are distributed over dry open areas of North and Central America. Medium to large (10 to 22 inches [25 to 56 cm] long), they are predominantly brown and streaked, but some have glossy green or blue backs and wings; many have patches of red or blue skin near their eyes.

Ground-cuckoos walk, run, and leap, but rarely fly. They eat insects, centipedes, spiders, scorpions, small frogs, snakes, and lizards. Several of them forage around army ant swarms, eating insects escaping from the swarms. Roadrunners eat insects, scorpions, centipedes, spiders, toads, lizards, eggs, and small snakes, birds, and mammals, and occasionally seeds and fruit. Three ground-cuckoos (including the Striped Cuckoo) are the New World's only brood parasitic cuckoos: females lay their eggs in the nests of other species, which then raise the young cuckoos as their own offspring. Other ground-cuckoos, including the roadrunners, are traditional monogamous breeders that place their stick nest on or near the ground. Some ground-cuckoos are fairly common; one species is considered vulnerable.

Anis are conspicuous, gregarious birds of savannas, brushy scrubs, fields, and river edges. The family, Crotophagidae, contains three ani species,

GROUND-CUCKOOS

Distribution:
New World

No. of Living
Species: 10

No. of Species
Vulnerable,
Endangered: 1, 0

No. of Species Extinct
Since 1600: 0

STRIPED CUCKOO
Tapera naevia
11 in (28 cm)
Central America, South America

LESSER ROADRUNNER
Geococcyx velox
19 in (48 cm)
Mexico, Central America

LESSER GROUND-CUCKOO
Morococcyx erythropygus
10 in (25 cm)
Mexico, Central America

SMOOTH-BILLED ANI
Crotophaga ani
14 in (35 cm)
South America, Central America,
West Indies, Florida

GROOVE-BILLED ANI
Crotophaga sulcirostris
12.5 in (32 cm)
North America, South America

GUIRA CUCKOO
Guira guira
14 in (36 cm)
South America

HOATZIN
Opisthocomus hoazin
24.5–27.5 in (62–70 cm)
South America

distributed through South/Central America and the Caribbean, and South America's Guira Cuckoo. Anis are medium size (12 to 18 inches [30 to 46 cm] long), glossy black all over, with iridescent sheens. Their bills are exceptionally large and arched, or humped, on top.

Anis live in groups of eight to twenty-five individuals (two to eight adults plus juveniles). Each group defends a feeding/breeding territory from other groups throughout the year. They usually forage on the ground, taking insects, but also a little fruit; frequently they feed around cattle, grabbing insects that are flushed from hiding places by the grazing mammals. Their vocalizations are usually loud and discordant; a flock calling has been compared to the din at a boiler factory.

Communal breeders, all anis within a group often contribute to a single nest; several females may lay eggs in it—up to twenty-nine eggs have been found in one nest. Many individuals help build the stick nest and feed the young. Although this behavior appears to benefit all individuals involved, actually it is a group's dominant male and female that gain most. Their eggs go in the communal nest last, on top of all the others, which often get buried.

The Hoatzin, somewhat resembling a dinosaur with feathers, is one of the most intriguing birds of South America. About the size of a turkey (24 to 28 inches [60 to 70 cm] long) and brown, with prominent crest and bluish face, it is restricted to flooded forests and marshes of the upper Amazon region. There has been more controversy over this strange bird's classification than perhaps any other; at various times it was considered part of about ten different families. Now it is placed in a single-species family, Opisthocomidae, in its own order, Opisthocomiformes.

Hoatzins are usually in loose flocks of two to eight in bushes along the edges of slow-moving streams or forested lakes. They eat only leaves and have a digestive system similar to a cow's that uses fermentation to digest otherwise indigestible plant parts. Breeding is monogamous, but the breeding pair often has helpers, other members of the social group, usually grown young from previous nests of the breeding pair, which help with care of eggs and young. The stick nest is placed in a tree branch over water. When in danger, nestlings jump into the water, swim underwater, and then climb another tree. For climbing, they have an extra, opposable, digit (lost as the chick matures) on their wing that they use like a hand. Hoatzins, not threatened, are not often hunted because, owing to some of the plants they eat, their meat smells and tastes awful.

ANIS

Distribution:
Neotropics

No. of Living
Species: 4

No. of Species
Vulnerable,
Endangered: 0, 0

No. of Species Extinct
Since 1600: 0

HOATZIN

Distribution:
Northern South
America

No. of Living
Species: 1

No. of Species
Vulnerable,
Endangered: 0, 0

No. of Species Extinct
Since 1600: 0

ASIAN BARRED OWLET
Glaucidium cuculoides
9.5 in (24 cm)
Southern Asia

BROWN HAWK-OWL
Ninox scutulata
10.5–13 in (27–33 cm)
Asia

SUNDA SCOPS-OWL
Otus lempiji
8 in (20 cm)
Southeast Asia

PEL'S FISHING-OWL
Scotopelia peli
21.5–25 in (55–63 cm)
Africa

MARSH OWL
Asio capensis
11.5–14 in (29–36 cm)
Africa, Madagascar

AFRICAN WOOD-OWL
Strix woodfordii
12–14 in (30–36 cm)
Africa

NORTHERN WHITE-FACED OWL
Ptilopsis leucotis
10 in (25 cm)
Africa

PEARL-SPOTTED OWLET
Glaucidium perlatum
7.5 in (19 cm)
Africa

Owls

OWLS through history have been regarded as symbols of wisdom (from ancient Greece through present-day advertising) or of bad luck, evil, or death (ancient Rome and many other cultures, including some in contemporary Africa). They are probably often thought wise because of their humanlike physical traits—forward-facing eyes, upright bearing, and, in many, conspicuous "ears." Their reputation for evil undoubtedly traces to their nocturnal habits and loud, haunting vocalizations. Most owls, the "typical" owls, are members of family Strigidae, a worldwide group of 188 species that lacks representation only on some remote oceanic islands; barn owls, with 16 species worldwide (excepting northern North America and northern Eurasia), constitute a separate family, Tytonidae. Owls are particularly diverse in the tropics and subtropics.

Distinctive owl features are large heads; small, hooked bills; plumpish bodies; and sharp, hooked claws. Most have short legs and short tails, and many have feather ear-tufts. Owls, clad mostly in mixtures of gray, brown, and black, are usually highly camouflaged against a variety of backgrounds. Most are medium-size birds, but the group includes species that range in length from 5 to 30 inches (13 to 75 cm). Males and females generally look alike. In many species, females are a bit larger; in a few, however, males are. Barn Owls differ from other owls in having a narrow body, long legs, and heart-shaped facial "disk," the flattened face of feathers characteristic of owls.

Owls occupy a variety of habitats: forests, clearings, fields, grasslands, deserts, mountains, marshes. They are considered to be the nocturnal

TYPICAL OWLS

Distribution:
All continents except
Antarctica

No. of Living
Species: 188

No. of Species
Vulnerable,
Endangered: 10, 13

No. of Species Extinct
Since 1600: 4

BARN OWL
Tyto alba
11.5–17.5 in (29–44 cm)
North America, South America, Europe,
Africa, Australia, southern Asia

FERRUGINOUS PYGMY-OWL
Glaucidium brasilianum
6–7.5 in (15–19 cm)
North America, South America

BLACK-BANDED OWL
Ciccaba huhula
13.5 in (34 cm)
South America

IMM

BURROWING OWL
Athene cunicularia
7.5–10 in (19–25 cm)
North America, South America

SNOWY OWL
Nyctea scandiaca
21.5–27.5 in (55–70 cm)
North America, Eurasia

GREAT HORNED OWL
Bubo virginianus
20–23.5 in (51–60 cm)
North America, South America

GREAT GRAY OWL
Strix nebulosa
23–27 in (59–69 cm)
North America, Eurasia

NORTHERN HAWK-OWL
Surnia ulula
15 in (38 cm)
North America, Eurasia

BOREAL OWL
Aegolius funereus
8.5–11 in (21–28 cm)
North America, Eurasia

WESTERN SCREECH-OWL
Otus kennicottii
9 in (23 cm)
North America

NORTHERN SAW-WHET OWL
Aegolius acadicus
8 in (20 cm)
North America

SPOTTED OWL
Strix occidentalis
16–19 in (41–48 cm)
North America

SOUTHERN BOOBOOK
Ninox boobook
10–14 in (25–36 cm)
Australia

RUFOUS OWL
Ninox rufa
15.5–20.5 in (40–52 cm)
Australia, New Guinea

equivalents of the day-active birds of prey—the hawks, eagles, and falcons. Some owls hunt at twilight and sometimes during the day, but most hunt at night, taking prey such as small mammals, birds (including smaller owls), and reptiles; some take fish from the water's surface. Smaller owls specialize on insects, earthworms, and other small invertebrates. Owls hunt by sight and sound. Their vision is very good in low light, the amount given off by moonlight, for instance. Their hearing is remarkable. They can hear much softer sounds than most other birds, and the ears of many species have different-size openings and are positioned on their heads asymmetrically, the better for localizing sounds in space. In darkness, owls can, for example, actually hear small rodents moving about on the forest floor and quickly locate the source of the sound. Additionally, owing to the owl's soft, loose feathers, its flight is essentially silent, permitting prey little chance of hearing its approach. The forward-facing eyes of owls are a trait shared with only

BARN OWLS

Distribution:
All continents except
Antarctica

No. of Living
Species: 16

No. of Species
Vulnerable,
Endangered: 1, 3

No. of Species Extinct
Since 1600: 0

a few other animals: humans, most other primates, and to a degree, the cats. Eyes arranged in this way allow for almost complete binocular vision (one eye sees the same thing as the other), a prerequisite for good depth perception, which, in turn, is important for quickly judging distances when catching prey. On the other hand, owl eyes cannot move much, so owls swivel their heads to look left or right. Owls swallow small prey whole, then instead of digesting or passing the hard bits, they regurgitate bones, feathers, and fur in compact owl pellets. These oblong pellets are often found beneath trees or rocks where owls perch and, when pulled apart, reveal what an owl has been dining on.

Most owls are monogamous. They do not build nests, but either take over nests abandoned by other birds or nest in cavities such as tree or rock holes. Incubation is usually conducted by the female alone, but she is fed by her mate. Upon hatching, the female broods the young while the male hunts and brings meals; later, the young are fed by both parents.

Because owls are cryptically colored and nocturnal, it can be difficult to determine their population sizes. However, many species are known to be threatened, especially among those restricted to islands. Most owls inhabit tropical forests and, of course, the extent of these forests is being inexorably reduced. Among the typical owls, ten are vulnerable and thirteen are endangered (seven of those critically); one barn owl is vulnerable and three are endangered.

Nightjars

NIGHTJARS are unusual-looking, cryptically colored birds that, although common and diverse over many parts of the planet, are considered to be among the most poorly known of birds. Like their close relatives the owls, they are active chiefly at night and this, combined with their secretive habits and frequent silence, renders them difficult to observe and study. Nightjars (family Caprimulgidae, in the order Caprimulgiformes with frogmouths and potoos) are distributed essentially worldwide outside of polar regions, with the eighty-nine species occupying a broad range of forested and open habitats. Many North American nightjars are called nighthawks (because they were once mistaken for hawks flying at night) and poorwills (after the vocalizations of some species). Nightjars as a group are sometimes also called goatsuckers. At twilight, some species fly low over the ground near grazing mammals, such as goats. The birds fly right next to the mammals to catch insects that flush as they walk. Long ago the assumption was that the birds were trying to suck the goats' milk, so they were called goatsuckers; this fanciful name has stuck. Their order and family name comes from the Latin *caprimulgus*, meaning "goat-milker."

Nightjars have a very characteristic appearance, with big eyes, long wings, and medium or long tails. Their small, stubby bills enclose big, wide mouths. Many species have prominent bristles around the mouth, which act as a food funnel. With their short legs and weak feet, they are poor walkers; flying is their usual mode of locomotion. Nightjar plumage is uniformly cryptic: mottled, spotted, and barred mixtures of browns, grays, tans, and

Distribution:
All continents except
Antarctica

No. of Living
Species: 89

No. of Species
Vulnerable,
Endangered: 3, 4

No. of Species Extinct
Since 1600: 0

LESSER NIGHTHAWK
Chordeiles acutipennis
7.5–9 in (19–23 cm)
North America, South America

SAND-COLORED NIGHTHAWK
Chordeiles rupestris
7.5–9.5 in (19–24 cm)
South America

LYRE-TAILED NIGHTJAR
Uropsalis lyra
10.5 in plus long tail (27 cm plus long tail)
South America

NACUNDA NIGHTHAWK
Podager nacunda
12 in (30 cm)
South America

BUFF-COLLARED NIGHTJAR
Caprimulgus ridgewayi
9 in (23 cm)
North America, Central America

LARGE-TAILED NIGHTJAR
Caprimulgus macrurus
10.5 in (27 cm)
Southern Asia, Australia

GREAT EARED-NIGHTJAR
Eurostopodus macrotis
12–15.5 in (31–40 cm)
Southeast Asia

black. They often have white patches on their wings or tails that can be seen only in flight. Nightjars range in length from 6 to 16 inches (15 to 40 cm); a few in Africa and South America have long tail streamers (modified feathers) that can add another 20 or more inches (50+ cm). Two Southeast Asia species known as eared-nightjars have elongated feathers at the sides of their head.

Most nightjars are night-active, with some becoming active at twilight. Excellent flyers, they feed on flying insects, which they catch mainly on the wing, either by forays out from a perched location on the ground or from tree branches, or with continuous circling flight. The Common Nighthawk is a familiar sight at dusk over much of the United States and Canada, as it circles open habitats catching bugs and giving its pinging flight calls. Some nightjars also take insects and spiders from vegetation or on the ground, and at least a few species also consume small rodents or small birds. Before dawn, nightjars return to their daytime roosts, on the ground, in low vegetation, or on tree branches. Their camouflage coloring makes them extremely difficult to see, even when one is close to them. One of the goatsuckers, North America's Common Poorwill, may be the only bird known to hibernate during very cold weather. During their dormant state, poorwills save energy by reducing their metabolic rate and their body temperature.

Most nightjars breed monogamously, alone or in small colonies. No nest is built; instead females lay eggs on the ground in a small depression, often in leaf litter, usually near rocks, logs, or vegetation. Either the female alone or both sexes incubate the eggs and both parents feed young. Young are able to walk soon after hatching and sometimes move to a new location each night. As do many other ground-nesting birds, nightjars engage in broken-wing displays to lure predators away from the nest and young. They flop around on the ground, often with one or both wings held down as if injured, making gargling or hissing sounds, all the while moving away from the nest.

Owing to their shy ways and nocturnal activity, the health of populations of most species of nightjars is difficult to assess. Habitat loss is generally the gravest threat they face. Some species do well in people-altered environments; Common Nighthawks, for instance, often roost and breed on gravel rooftops in urban settings. Three species are considered vulnerable and four endangered (two critically: the Jamaican Pauraque and Puerto Rican Nightjar).

JAVAN FROGMOUTH
Batrachostomus javensis
7.5–10 in (19–25 cm)
Southeast Asia

TAWNY FROGMOUTH
Podargus strigoides
13.5–21 in (34–53 cm)
Australia

COMMON POTOO
Nyctibius griseus
13–15 in (33–38 cm)
Central America, South America

GREAT POTOO
Nyctibius grandis
17.5–21.5 in (45–55 cm)
Central America, South America

NORTHERN POTOO
Nyctibius jamaicensis
15–17.5 in (38–44 cm)
Mexico, Central America, West Indies

OILBIRD
Steatornis caripensis
15.5–19.5 in (40–49 cm)
South America

AUSTRALIAN OWLET-NIGHTJAR
Aegotheles cristatus
9 in (23 cm)
Australia

Frogmouths; Owlet-Nightjars; Potoos; Oilbird

The nocturnal birds discussed here are close relatives of the better-known and more diverse nightjars and are included with them in order Caprimulgiformes.

The twelve species of FROGMOUTHS (family Podargidae) are distributed in southern Asia and the Australian region. Named for their massive, broad, slightly hooked bills, frogmouths are medium to largish (8 to 24 inches [20 to 60 cm]) gray or brown birds with short legs. They roost by day on exposed tree branches, making use of their extremely cryptic plumage and behavior to avoid detection by predators; they perch motionless with their bills pointing upward, and strongly resemble dead branches. The likeness is so close that bird-watchers sometimes stare and stare through binoculars, trying to decide if the object they see is bird or branch. At dusk, in their forest and woodland habitats, solitary frogmouths begin foraging. They sit on various perches, detect prey—insects and spiders, snails, small vertebrates—and then swoop and grab with their bill. Frogmouth nests are small platforms or cups of twigs, moss, and lichens built fairly high in trees. Eggs are incubated by the female only or by both sexes and young are fed by both parents. No frogmouths are currently threatened, but five are considered near-threatened.

OWLET-NIGHTJARS (family Aegothelidae) are a group of nine species restricted largely to forests of the Australasian region; most occur in New Guinea. Owing to their general shape and upright stance, owlet-nightjars look like owls or like crosses between owls and nightjars. They are

Distribution:
Australasia

No. of Living
Species: 9

No. of Species
Vulnerable,
Endangered: 0, 1

No. of Species Extinct
Since 1600: 0

Distribution:
Neotropics

No. of Living
Species: 7

No. of Species
Vulnerable,
Endangered: 0, 0

No. of Species Extinct
Since 1600: 0

Distribution:
Northern South
America

No. of Living
Species: 1

No. of Species
Vulnerable,
Endangered: 0, 0

No. of Species Extinct
Since 1600: 0

small to medium-size (7 to 12 inches [18 to 30 cm]) with small, wide bills, brownish or gray with white or black markings (or sometimes both). They catch flying insects by making short, sallying flights from a fixed perch. They also seize prey, including ants and other insects, spiders, and millipedes, while on the ground. Breeding is not well documented. The Australian Owlet-Nightjar, Australia's sole species, nests usually in tree holes. Male and female help line the nest with leaves or bark. Both sexes may incubate eggs, and both feed the young. One species, the New Caledonian Owlet-Nightjar, is critically endangered.

The seven species of forest-dwelling POTOOS (family Nyctibiidae) occur from Mexico south into Argentina. They are medium-size (8 to 23 inches [21 to 58 cm]), brown or gray-brown with black and white markings or mottling, and, with their upright posture, owl-like. They have large heads, huge eyes, wide but short bills, and short legs. During the day, potoos sit in trees, and with their camouflage coloring and their bills pointed into the air, like frogmouths, they look like dead branches. At night, they are solitary hunters for large insects, small birds, lizards, and occasionally small mammals. They sit on a perch and make repeated sallies to catch flying insects. Not often seen, potoos are known mainly by their mournful vocalizations. Monogamous breeders, potoos build no nest but lay a single egg in a crevice of a large branch or stump, usually high in a tree. Both sexes incubate eggs and feed the young. None of the potoos are threatened.

The OILBIRD (single-species family Steatornithidae) is an unusual species occurring only in parts of northwestern South America. Related to both nightjars and owls, Oilbirds have a batlike existence. Brown with whitish spots, approximately 18 inches (46 cm) long, they have wingspans of about 3.3 feet (1 m) and hawklike bills. They live in immense colonies associated with large caves. The Oilbird is one of the few bird species to use batlike echolocation to help navigate the night and cave darkness. Birds emerge from their cave entrance every evening to spend the night searching for fruits; before dawn they return to the cave. They stay in pairs and roost year-round at their nest sites on ledges within the cave. The nest is made of droppings and partially digested fruit pulp. Both sexes incubate and feed young. Fed only fruits (a low-protein diet), young develop slowly. Oilbirds are not threatened, but young oilbirds have long been harvested from caves for their fat deposits, used to provide oil for lamps and cooking. Many oilbird caves are officially protected or their locations kept secret.

Swifts; Treeswifts

SWIFTS

Distribution:
All continents except
Antarctica

No. of Living
Species: About 95

No. of Species
Vulnerable,
Endangered: 5, 1

No. of Species Extinct
Since 1600: 0

SWIFTS, more adapted to an aerial lifestyle than any other birds, represent the pinnacle of avian flying prowess. Perpetual flight in the past was so much the popular impression that it was actually thought that swifts never landed but remained flying throughout most of their lives. (Indeed, they were long ago believed to lack feet; hence the family name, Apodidae, meaning "without feet.") Swifts are also known for their superb ability to catch insects on the wing and, in some parts of the world, for their nests: the breeding constructions of some species are fashioned wholly or partly from the swifts' (edible; some say delicious) saliva; their nests are harvested as the primary ingredient of bird's nest soup, an Asian delicacy.

The ninety-five or so swift species are distributed worldwide in temperate and tropical areas; their closest relatives are the hummingbirds (swifts and hummingbirds comprise order Apodiformes). Swifts are slender and streamlined, with long, pointed wings. Small to medium-size (3.5 to 10 inches [9 to 25 cm] long), they have small bills, very short legs, and short tails, or in some, long, forked tails. Their tail feathers are stiffened to support the birds as they cling to vertical surfaces. The sexes look alike: blackish, gray, or brown, often with light markings; many have glossy plumage.

It seems as if swifts fly effortlessly all day, moving in seemingly erratic patterns high overhead, in a wide variety of habitats. They eat flying insects and some spiders found in the air, and occasionally take bugs off the water's surface when they come down for a drink, always in flight. They rarely land except at night, when they come together in large groups gathered on a

WHITE-COLLARED SWIFT
Streptoprocne zonaris
8.5 in (22 cm)
Central America, South America

SHORT-TAILED SWIFT
Chaetura brachyura
4 in (10 cm)
South America

HOUSE SWIFT
Apus nipalensis
6 in (15 cm)
Southern Asia

ASIAN PALM-SWIFT
Cypsiurus balasiensis
5 in (13 cm)
Southern Asia

EDIBLE-NEST SWIFTLET
Aerodramus fuciphagus
4.5 in (12 cm)
Southeast Asia

WHITE-THROATED NEEDLETAIL
Hirundapus caudacutus
8 in (20 cm)
Asia, Australia

CRESTED TREESWIFT
Hemiprocne coronata
9.5 in (24 cm)
Southern Asia

WHISKERED TREESWIFT
Hemiprocne comata
6.5 in (16 cm)
Southeast Asia

vertical cliff face behind a waterfall, inside a hollow tree, in a cave, or among palm-tree fronds. They roost on these vertical surfaces, clinging with their tiny, strong, sharply clawed feet and bracing themselves against the sides of the roost with their stiff tails. A swift spends more time airborne than any other type of bird; some are even suspected of copulating in the air and at least some appear able to "roost" in the air, flying slowly on "automatic pilot" while they snooze. The name swift is apt, as these are the fastest birds in level flight, moving along at up to 70 mph (112 kph); the White-throated Needletail may reach speeds of 100 mph (160 kph). Some of the swifts can echolocate like bats, using this sonarlike ability to navigate in dark roosting caves or at dusk when returning to their roosts.

Swifts are monogamous, likely pair for long periods, and most are colonial breeders. The sexes share breeding chores. Nests, attached to a vertical wall, usually consist of plant pieces, twigs, and feathers glued together with the birds' saliva.

A few species of swifts are known to be rare and are considered threatened; for many others, so little is known that population sizes or vulnerabilities are uncertain. Five are classified as vulnerable and one, the Guam Swiftlet, is endangered. In general, people don't disturb or harm swifts, aside from the species whose nests are collected for soup. Edible-nest swiftlets are small swifts that roost and nest in caves. Their breeding caves are regularly raided and nests gathered as a valuable commercial resource. Not surprisingly, edible-nest swiftlet populations in countries in which they occur, such as Malaysia and Indonesia, have plunged as people settle more areas and new roads open up previously inaccessible regions with breeding caves.

TREESWIFTS (family Hemiprocnidae) are four species of forest and woodland birds that occur over parts of southern Asia and the New Guinea region. They resemble swifts but differ by having head crests, long forked tails, and being more brightly colored; two species have distinctive white head stripes. Treeswifts, like swifts, catch insects on the wing during long periods of sustained flight, but they also will sally out after flying insects from an exposed perch in a tree. Monogamous breeders, treeswifts nest solitarily. The tiny cup nest, attached to the side of a branch, is constructed of bark, feathers, and saliva. None of the treeswifts are threatened.

TREESWIFTS

Distribution:
Parts of southern Asia,
Melanesia

No. of Living
Species: 4

No. of Species
Vulnerable,
Endangered: 0, 0

No. of Species Extinct
Since 1600: 0

GREEN VIOLET-EAR
Colibri thalassinus
4.5 in (11 cm)
Mexico, Central America,
South America

WHITE-TIPPED SICKLEBILL
Eutoxeres aquila
5 in (13 cm)
Central America, South America

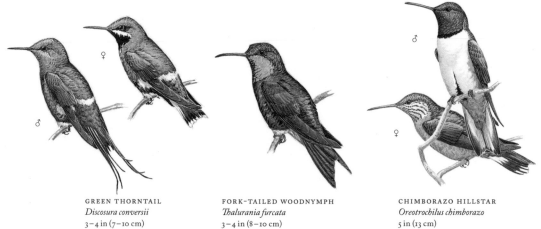

GREEN THORNTAIL
Discosura conversii
3–4 in (7–10 cm)
Central America, South America

FORK-TAILED WOODNYMPH
Thalurania furcata
3–4 in (8–10 cm)
South America

CHIMBORAZO HILLSTAR
Oreotrochilus chimborazo
5 in (13 cm)
South America

GIANT HUMMINGBIRD
Patagona gigas
8.5 in (21 cm)
South America

COLLARED INCA
Coeligena torquata
5.5 in (14 cm)
South America

Hummingbirds

HUMMINGBIRDS are among the most recognized kinds of birds, the smallest birds, and undoubtedly among the most beautiful, albeit on a minute scale. Limited to the New World, the hummingbird family, Trochilidae (in order Apodiformes with the swifts), contains about 330 species. The variety of forms encompassed by the family, not to mention the brilliant iridescence of most of its members, is indicated in the names attached to some of the different subgroups: in addition to the hummingbirds proper, there are emeralds, sapphires, sunangels, sunbeams, comets, metaltails, fairies, woodstars, woodnymphs, pufflegs, sabrewings, thorntails, thornbills, and lancebills. These are mainly very small birds, usually gorgeously clad in iridescent metallic greens, reds, violets, and blues. Most hummingbirds (hummers) are in the range of only 2.5 to 5 inches (6 to 13 cm) long, although a few larger kinds reach 8 inches (20 cm), and they tip the scales at an almost imperceptibly low 0.1 to 0.3 ounces (2 to 9 grams)—some weighing no more than a large paper clip. Bill length and shape varies extensively among species, each bill closely adapted to the precise type of flowers from which a species delicately draws its liquid food. Males are usually more colorful than females, and many of them have gorgets, bright, glittering throat patches. Not all hummers are so vividly outfitted; one group, called hermits (because of their solitary ways) are known for dull, greenish brown and gray plumage.

Hummingbirds occupy a broad array of habitat types, from exposed high mountainsides at 13,000 feet (4,000 m) to midelevation arid areas to sea level tropical forests and mangrove swamps. All get most of their nourishment from consuming flower nectar. They have long, thin bills and

Distribution:
New World

No. of Living
Species: About 330

No. of Species
Vulnerable,
Endangered: 9, 20

No. of Species Extinct
Since 1600: 2

SWORD-BILLED HUMMINGBIRD
Ensifera ensifera
6.5–9 in (17–23 cm)
South America

GORGETED SUNANGEL
Heliangelus strophianus
4.5 in (11 cm)
South America

BOOTED RACKET-TAIL
Ocreatus underwoodii
6–7 in (15–18 cm)
South America

BLACK-THROATED MANGO
Anthracothorax nigricollis
4.5 in (12 cm)
Central America, South America

WHITE-BEARDED HERMIT
Phaethornis hispidus
5 in (13 cm)
South America

BLACK-TAILED TRAINBEARER
Lesbia victoriae
6–10 in (15–26 cm)
South America

LONG-TAILED SYLPH
Aglaiocercus kingi
4–7.5 in (10–19 cm)
South America

PERUVIAN SHEARTAIL
Thaumastura cora
3–5 in (8–13 cm)
South America

LONG-BILLED STARTHROAT
Heliomaster longirostris
4.5 in (12 cm)
Mexico, Central America,
South America

CRIMSON TOPAZ
Topaza pella
5–9 in (13–23 cm)
South America

RACKET-TAILED COQUETTE
Discosura longicauda
3–4 in (8–10 cm)
South America

RED-BILLED STREAMERTAIL
Trochilus polytmus
4.5–12 in (11–30 cm)
West Indies

PLAIN-CAPPED STARTHROAT
Heliomaster constantii
4.5 in (12 cm)
Mexico, Central America

VIOLET-CROWNED HUMMINGBIRD
Agyrtria violiceps
4 in (10 cm)
North America

BUMBLEBEE HUMMINGBIRD
Atthis heloisa
3 in (7 cm)
Mexico

specialized tongues to lick nectar from slender flower tubes, which they do while hovering. Because nectar is mostly a sugar and water solution, hummingbirds need to obtain additional nutrients, such as proteins, from other sources. Toward this end they also eat the odd insect or spider, which they catch in the air or pluck off spiderwebs. Hummers are capable of very rapid, finely controlled, acrobatic flight, more so than any other kind of bird. The bones of their wings are modified to allow for perfect, stationary hovering flight and for the unique ability to fly backward. Their wings vibrate in a figure eight–like wingstroke at up to eighty times per second. To pump enough oxygen and nutrient-delivering blood around their little bodies, hummer hearts beat up to ten times faster than human hearts—six hundred to one thousand times per minute. To obtain sufficient energy to fuel their high metabolism, hummingbirds must eat many times a day. Quick starvation results from an inability to feed regularly. Predators on hummingbirds include small, agile hawks and falcons, as well as frogs and large insects, such as praying mantises, that ambush the small birds as they feed at flowers.

Hummingbirds are polygamous breeders in which females do almost all the work. In some species, a male in his territory advertises for females by singing squeaky songs. A female enters the territory and, following courtship displays, mates. She then leaves the territory to nest on her own. Other species are lek breeders. In these cases, males gather at traditional, communal mating sites called leks. Each male has a small mating territory in the lek, perhaps just a perch on a flower. The males spend hours there each day during the breeding season, advertising for females. Females enter a lek, assess the displaying males, and choose ones to mate with. After mating, females leave the lek or territory and build their nests, which are cuplike and made of plant parts, mosses, lichens, feathers, animal hairs, and spider webbing. Nests are placed in small branches of trees, often attached with spider webbing. Females lay eggs, incubate them, and feed regurgitated nectar and insects to their young.

Habitat loss and degradation are the main threats to these stunning little birds. About twenty-nine hummingbird species in Mexico, Central America, and South America are considered threatened, with about twenty of these currently endangered. Nine of the latter are critically endangered, with tiny ranges and miniscule populations.

Mousebirds

MOUSEBIRDS are common, conspicuous small birds of sub-Saharan Africa that inhabit forest edges, savanna, scrublands, and farms and gardens. Best known for being energetic consumers of fruits and vegetables, flowers and foliage, they frequently anger and frustrate farmers and gardeners, but they are also appreciated for their sociable, playful habits. They are named mousebirds because of their somewhat mouselike appearance (smallish and drab, with long tails) and their typical small-rodentlike behaviors of scuttling through vegetation, living in groups, and huddling together. There are only six mousebirds (called colies in some parts of Africa), comprising family Coliidae (placed in its own order, Coliiformes, the only order wholly endemic to Africa). Mousebirds are noted among ornithologists mainly for their classification: although they resemble perching birds (order Passeriformes) and historically were placed in that group, research over many decades indicates that mousebirds are actually a distinct, probably ancient group, not closely related to other birds, which is why they now have their own order.

All mousebirds (11 to 15 inches [28 to 38 cm] in length, including long tail) are similar in appearance. They have short necks; short, rounded wings; short, robust, somewhat finchlike bills; pronounced crests that flatten when they fly; and long (7 to 11 inches [18 to 28 cm]), stiff tails. Their short legs have curious feet; all four toes can face forward, but one or two are flexible enough to be turned back for gripping. Mousebirds often hang from branches, instead of perching like other birds. Feathers are soft and frilly, and owing to this, their plumage is less waterproof than that of most birds; they easily

Distribution:
Sub-Saharan Africa

No. of Living
Species: 6

No. of Species
Vulnerable,
Endangered: 0, 0

No. of Species Extinct
Since 1600: 0

WHITE-HEADED MOUSEBIRD
Colius leucocephalus
12 in (30 cm)
Africa

SPECKLED MOUSEBIRD
Colius striatus
12–14 in (30–36 cm)
Africa

BLUE-NAPED MOUSEBIRD
Urocolius macrourus
13–14 in (33–36 cm)
Africa

RED-FACED MOUSEBIRD
Urocolius indicus
11.5–14.5 in (29–37 cm)
Southern Africa

WHITE-BACKED MOUSEBIRD
Colius colius
12 in (31 cm)
Southern Africa

become drenched in heavy rains. All are dully colored in gray, brown, or buff. Several have bare skin (red, black, or gray) around the eyes that forms a mask. The sexes look alike.

Mousebirds feed primarily on plants, eating leaves, buds, flowers, and stems, as well as fruit, seeds and nectar, and regularly devour their own body weight in food daily. Scrambling through the vegetation, they are acrobatic feeders and often dangle from one foot while trying to reach a fruit. Some, such as the Spectacled Mousebird, apparently can eat fruits and leaves that are poisonous to other vertebrates (the clay soils they also eat may help detoxify the poisonous compounds). Mousebirds are usually found in groups, often between five and eight birds, but when fruiting trees are bearing prolifically, dozens of mousebirds may gather to feed. Groups move about locally as food availability changes, but they do not undertake extensive migrations. They fly in single-file style from bush to bush and pack together in a tight huddle when resting or sleeping.

The gregarious habits of mousebirds persist throughout the year, although, when breeding, pairs tend to split off from the flock and raise their broods without the group's involvement. However, two or more pairs may build nests near each other, and there are cases of more than one female laying in a nest and of more than two "parents" attending a brood of young— that is, sometimes there are helpers at the nest, which are usually young from previous broods. Monogamy is the usual, but not invariable, breeding arrangement (occasionally males breed with more than one female, and some females have been known to have more than one mate in a single breeding season); many pairs remain together for years. Nests, simple open cups made of grass, twigs, or weed stems, are placed in shrubs or trees, usually well hidden in thick foliage. Both sexes build the nest, incubate eggs, and feed young. They live up to 12 years in captivity.

Mousebirds are common and widespread in Africa; none are threatened. In flocks, they can inflict extensive damage to fruit crops such as figs, grapes, peaches, and plums, and to vegetables such as lettuce and tomatoes. They are persecuted by farmers for this reason, and many are also killed by agricultural pesticides. However, because they are so successful at taking advantage of people's alterations of natural habitats—of their farms, orchards, and gardens—mousebird populations are stable. Many Africans consider them charming birds, watching them with great pleasure.

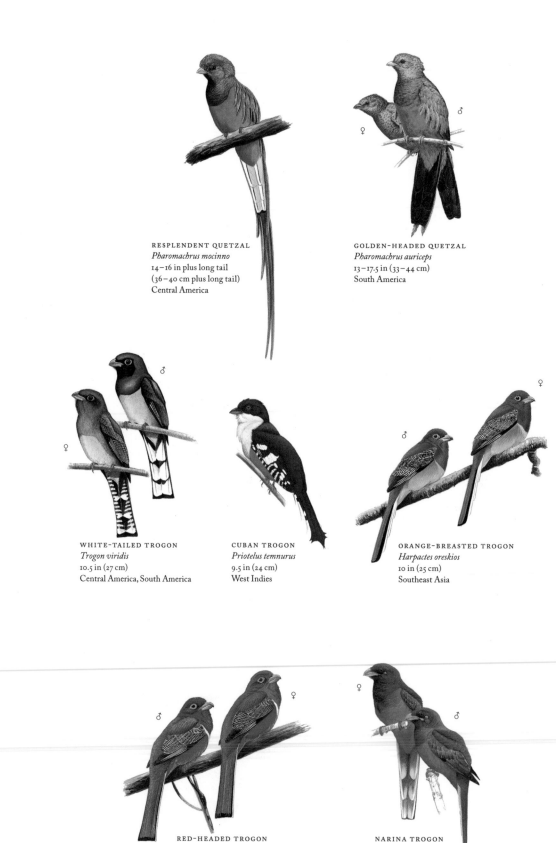

RESPLENDENT QUETZAL
Pharomachrus mocinno
14–16 in plus long tail
(36–40 cm plus long tail)
Central America

GOLDEN-HEADED QUETZAL
Pharomachrus auriceps
13–17.5 in (33–44 cm)
South America

WHITE-TAILED TROGON
Trogon viridis
10.5 in (27 cm)
Central America, South America

CUBAN TROGON
Priotelus temnurus
9.5 in (24 cm)
West Indies

ORANGE-BREASTED TROGON
Harpactes oreskios
10 in (25 cm)
Southeast Asia

RED-HEADED TROGON
Harpactes erythrocephalus
13 in (33 cm)
Southeast Asia

NARINA TROGON
Apaloderma narina
12 in (31 cm)
Africa

Trogons

TROGONS, colorful forest dwellers, are generally regarded by wildlife enthusiasts as among the globe's most beautiful and glamorous of birds. Their family, Trogonidae (in its own order, Trogoniformes), is distributed through tropical and subtropical regions of the Neotropics, Africa, and southern Asia. It consists of thirty-nine species of medium-size birds (9 to 16 inches [23 to 40 cm] long) with compact bodies, short necks, and short chickenlike bills. Considering the broad and widely separated geographic areas over which the species of this family are spread, the uniformity of the family's body plan and plumage pattern is striking. Male trogons have glittering green, blue, or violet heads and chests, with contrasting bright red, yellow, or orange underparts. Females are duller, usually with brown or gray heads, but share the males' brightly colored breasts and bellies. There are some differences among trogons of various regions; for instance, the three African species have small patches of colored skin below the eye, which the twenty-four New World species and twelve Asian species do not. The characteristic trogon tail is long and squared-off, often with horizontal black and white stripes on the underside. Trogons usually sit erect with their distinctive tails pointing straight to the ground.

One trogon stands out from the flock: the regal-looking Resplendent Quetzal. Famously described as "the most spectacular bird in the New World," this quetzal generally resembles other trogons, but the male's emerald-green head is topped by a ridged crest of green feathers, and truly ostentatiously, long green plumes extend up to 2 feet (60 cm) past the

*Distribution:
New World, Africa,
southern Asia, mainly
in the Tropics*

*No. of Living
Species: 39*

*No. of Species
Vulnerable,
Endangered: 0*

*No. of Species Extinct
Since 1600: 0*

end of the male's typical trogon tail. This large trogon was revered or even considered sacred by the ancient inhabitants of Central America, including the Mayans, and is now a national symbol of Guatemala.

Trogons are generally observed either solitarily or in pairs. In spite of their distinctive calls (trogon "cow-cow-cow" calls are one of the characteristic sounds of many New World tropical forests) and brilliant plumage, they can be difficult to locate or to see clearly when spotted perched on a tree branch. Like green parrots, trogons easily meld into dark green overhead foliage. Trogon behavior is not much help because these birds usually perch for long periods with little moving or vocalizing, the better presumably to avoid the notice of predators. Trogons, therefore, are most often seen when flying. This often occurs in sudden bursts as they flip off the branches on which a moment before they sat motionless and sally out in undulatory, short flights to snatch insects, their main food. Bristles at the base of the bill may protect their eyes from flailing bugs. Trogons also swoop to grab small lizards, frogs, and snails, and take small fruits from trees while hovering. Quetzals eat mainly fruit. Trogons rarely fly far, staying mostly in their fairly small territories.

Monogamous breeders, trogons usually nest in tree cavities or in excavations in arboreal ant, termite, or wasp nests (they have been observed taking over a wasp nest, carving a nest hole in it, and then eating their insect neighbors during nesting). Generally the trogon female incubates eggs during the night and the male takes over during the day. Young are fed by both parents.

Trogons mainly inhabit tropical forests, and the continuing loss of these forests is the greatest threat they face. Although most species are still common, they are bound to become less so in the future as more forests are cut and degraded. Ten species are considered near-threatened. Some authorities believe the Resplendent Quetzal is now endangered, primarily owing to destruction of its cloud forest habitat. Large tracts of Central American and southern Mexican cloud forest, its only homes, continue to be cleared for cattle pasture, agriculture, and logging. These quetzals are also threatened because local people illegally hunt them for trade in skins, feathers, and live birds. Because of the quetzal's prominence in Guatemala (the bird is depicted on its state seal and its currency) and as a particularly gorgeous poster animal for conservation efforts, several special reserves have been established for it in Central America.

Kingfishers

KINGFISHERS are handsome, bright birds of forests and woodlands that, in some parts of the world, make their living chiefly by diving into water to catch fish. But many kingfishers, such as most of Australia's, including the famous kookaburras, hunt on land. Classified with the bee-eaters, rollers, and hornbills in the order Coraciiformes, the approximately ninety kingfisher species (family Alcedinidae) range over most temperate and tropical areas of the globe. Most species are rather shy, but many of the ones that chiefly eat fish, including the New World kingfishers, because of their need to perch in open areas around water, are often spotted along rivers and streams or along the seashore.

Kingfishers range in size from small to largish birds, but all are similar in form: large heads with very long, robust, straight bills, short necks, short legs, and, for some, noticeable crests. The smallest species is the 4-inch (10-cm) African Dwarf Kingfisher, and the largest is Australia's Laughing Kookaburra, at up to 18 inches (46 cm). Kingfishers are usually quite colorful, often with brilliant blues or greens predominating (and sometimes both). There are some regional patterns, such as in Australia, where blue and white, along with chestnut-orange is common, and in the New World, where dark green or blue-gray above, and white or chestnut-orange and white below, is the norm. Male and female kingfishers often look alike, but in some the two sexes differ in plumage pattern, especially on the breast.

Many kingfishers, as the name suggests, are fish-eaters. Usually seen hunting alone, they sit quietly, attentively on a low perch, such as a tree

Distribution: All continents except Antarctica

No. of Living Species: 92

No. of Species Vulnerable, Endangered: 11, 1

No. of Species Extinct Since 1600: 0

BELTED KINGFISHER
Ceryle alcyon
11–13 in (28–33 cm)
North America, South America

AMAZON KINGFISHER
Chloroceryle amazona
12 in (30 cm)
Mexico, Central America, South America

WHITE-THROATED KINGFISHER
Halcyon smyrnensis
11 in (28 cm)
Southern Asia

BLUE-BREASTED KINGFISHER
Halcyon malimbica
10 in (25 cm)
Africa

SHINING-BLUE KINGFISHER
Alcedo quadribrachys
6.5 in (16 cm)
Africa

BUFF-BREASTED PARADISE-KINGFISHER
Tanysiptera sylvia
11.5–14.5 in (29–37 cm)
Australia, New Guinea

BLUE-WINGED KOOKABURRA
Dacelo leachii
15.5 in (40 cm)
Australia, New Guinea

SACRED KINGFISHER
Todiramphus sanctus
8.5 in (22 cm)
Australasia, Indonesia

branch over water or a bridge spanning a stream, while scanning the water below. When they locate suitable prey, they swoop and dive, plunging head first into the water (to depths up to 2 feet [60 cm]) to seize it. If successful, they quickly emerge from the water, return to the perch, beat the fish against the perch to stun it, and then swallow it whole. They may also, when they see movement below the water, hover over a particular spot before diving in. Some species, such as North America's common Belted Kingfisher, will fly out a quarter mile (500 m) or more from a lake's shore to hover 10 to 50 feet (3 to 15 m) above the water, searching for fish. Kingfishers that specialize on fish occasionally supplement their diet with tadpoles and insects. Other kingfishers eat mainly insects and their larvae, vertebrates such as small reptiles and mice, and, when near water, freshwater and saltwater crustaceans. Kookaburras specialize on small snakes, lizards, and rodents, but also take larger insects. The laughing call of the Laughing Kookaburra (koo-hoo-hoo-hoo-hoo-ha-ha-ha-HA-HA-hoo-hoo-hoo) is perhaps one of the world's most recognizable bird sounds. It is a territorial vocalization, given usually at dawn and dusk. Kingfishers fly fast and purposefully, usually in straight and level flight, from one perch to another; often they are seen only as flashes of blue or green darting along waterways. They are highly territorial, aggressively defending their territories from other members of their species with noisy, chattering vocalizations, chasing, and fighting. Although mainly forest birds, they occur in a wide variety of habitats worldwide, including deserts and small oceanic islands.

Most kingfishers are monogamous breeders that nest in holes. Both members of the pair help dig nest burrows in earth or river banks or termite mounds, or they help prepare a tree hole. Both parents incubate eggs and feed nestlings and fledglings. Kookaburras are unusual in that young, when old enough to be independent, sometimes do not leave their parent's territory but remain there for up to 4 years, helping their parents defend the territory and raise more young. Many young kingfishers apparently die during their first attempts at diving for food. Some have been seen first practicing predation by capturing floating leaves and sticks.

Most kingfishers are moderately to very abundant; some do quite well around human settlements. Being primarily forest birds, the main threat they face is loss of their forest habitats. Eleven species, most restricted to small islands, are vulnerable; one is endangered (French Polynesia's Marquesan Kingfisher).

BROAD-BILLED TODY
Todus subulatus
4.5 in (11 cm)
Hispaniola

NARROW-BILLED TODY
Todus angustirostris
4.5 in (11 cm)
Hispaniola

PUERTO RICAN TODY
Todus mexicanus
4.5 in (11 cm)
Puerto Rico

JAMAICAN TODY
Todus todus
4.5 in (11 cm)
Jamaica

CUBAN TODY
Todus multicolor
4.5 in (11 cm)
Cuba

Todies

TODIES are a small group of tiny, dazzling forest birds known mainly to globetrotting bird-watchers and residents of Cuba, Jamaica, Hispaniola, and Puerto Rico, the only places they occur. These diminutive birds are appreciated for their bright colors, relative tameness, and insect-catching ways. Now restricted to the West Indies, todies were once more widespread, as evidenced by the discovery of a millions-of-years-old todylike fossil in Wyoming. Thus, the present todies are considered a relict family—a narrowly distributed group of a once widespread group that has died out in the remainder of its former range. The five tody species (Cuba, Jamaica, and Puerto Rico each support a single species; Hispaniola has two) comprise family Todidae, which is included in the order Coraciiformes with the kingfishers, bee-eaters, motmots, and hornbills.

All todies look much alike—in fact, during most of the nineteenth century they were all thought to be single species (then known as the Jamaican Tody). They are emerald green above, with ruby red throats and whitish underparts tinged with yellow or pink. All have relatively large heads, plump bodies, and short, stubby tails. Bills are long, narrow, and straight, dark above, red below, and usually tilted upward when the birds perch. Some individual todies are as small as 3.5 inches (9 cm) in length; the largest, Hispaniola's Broad-billed Tody, ranges up to 4.5 inches (11.5 cm). Male and female todies look alike.

Todies are voracious insect-eaters. Like hummingbirds, their tiny size means they have fast metabolisms and must feed frequently to obtain enough energy; some have been known to eat up to 40 percent of their

Distribution: West Indies

No. of Living Species: 5

No. of Species Vulnerable, Endangered: 0, 0

No. of Species Extinct Since 1600: 0

body weight in a day. Typically a tody sits quietly on a twig with its head cocked upward, alertly scanning the leaves above. Upon spotting an insect, the bird flits upward in an arc, snatches the bug from, often, the bottom of a leaf, then settles on another twig, and begins the process again. Todies in some habitats will also dart out from a perch and catch insects in the air, like flycatchers, and sometimes even hover a bit, like hummingbirds. When a large prey item is caught, a tody will often beat it on a branch to stun it so it can be easily swallowed. Occasionally these birds consume spiders, millipedes, small fruits and berries, and tiny lizards. Todies are territorial and may stay in pairs all year. Sometimes they temporarily join mixed-species foraging flocks with such other birds as tanagers, thrushes, and vireos. Todies frequently give a variety of short buzzy or beeping calls.

Monogamous breeders, todies nest in burrows dug into earthen banks, usually not very high above the ground. The burrows, which range in length from 5 to 24 inches (12 to 60 cm) long, and average about 12 inches (30 cm), are enlarged at the back, where the eggs are deposited; they are also curved so that the eggs cannot be seen from the entrance. Both sexes excavate the burrow using their bills as chisels to loosen soil. Digging a tunnel takes about 2 months. Courtship consists of brief sex chases interspersed with distinctive "wing rattling," loud whirring sounds produced when air moves rapidly through the long wing feathers. Incubation and feeding offspring are shared by the pair. Parents bring insects to their young at one of the highest rates among insect-eating birds; one pair was observed making 420 feeding trips to their nest in a single day, provisioning three older nestlings. Tody breeding pairs sometimes have helpers at the nest. These are other todies, likely related to the parents, which assist in incubating eggs and bringing food to the young.

Four of the five todies are widely distributed and common throughout their ranges. The one exception is the Narrow-billed Tody, which has declined in Haiti due to habitat loss and is considered near-threatened; however, it remains common in the Dominican Republic portion of Hispaniola. Some todies benefit from people's alterations of the environment when they dig their nesting burrows in road banks or the sides of drainage ditches.

Motmots

MOTMOTS, although saddled with a ridiculous name, are among the New World's most visually stunning birds, with their bodies of blended shades of green and soft cinnamon-browns, their crisp black masks, and in some, brilliant blue or turquoise head patches. For lovers of wild birds, seeing some of the motmots, perhaps with sunlight glinting off their bright turquoise-bejeweled crowns, can be a paramount experience. In addition to their beauty, motmots are known for their far-carrying, deep, hooting calls (the "BOO-boop, BOO-boop" calls of the Blue-crowned Motmot are probably the source of the term *motmot*), which are characteristic sounds of many neotropical forests, and for their uniquely shaped tails. The motmot family, Momotidae, with ten species, is included in order Coraciiformes with the kingfishers, bee-eaters, and hornbills, and is confined in its distribution to Mexico and Central and South America. The family's distribution is unusual in that more of its members (eight species) occur in the small area of Central America than in relatively huge South America (five species; some species occur in both regions).

Motmots are colorful, slender, small to medium-size birds (6.5 to 19 inches [16 to 48 cm] long). They have fairly long, broad bills, down-curved at the end. The bills have serrated edges, adapted to grab and hold their animal prey. The most peculiar motmot feature is the tail. In seven of the ten species, two central feathers of the tail grow much longer than others. Soon, feather barbs near the end of these two feathers drop off, usually from

Distribution:
Neotropics

No. of Living
Species: 10

No. of Species
Vulnerable,
Endangered: 1, 0

No. of Species Extinct
Since 1600: 0

RUFOUS MOTMOT
Baryphthengus martii
16.5–18.5 in (42–47 cm)
Central America, South America

BLUE-CROWNED MOTMOT
Momotus momota
15–17 in (38–43 cm)
Mexico, Central America, South America

BROAD-BILLED MOTMOT
Electron platyrhynchum
12–15.5 in (31–39 cm)
Central America, South America

TURQUOISE-BROWED MOTMOT
Eumomota superciliosa
13–15 in (33–38 cm)
Central America

KEEL-BILLED MOTMOT
Electron carinatum
12–15 in (31–38 cm)
Central America

TODY MOTMOT
Hylomanes momotula
6.5 in (17 cm)
Central America

RUSSET-CROWNED MOTMOT
Momotus mexicanus
12–14 in (31–36 cm)
Mexico, Central America

brushing against tree branches and other wearing, resulting in short lengths of barbless vane and, below this area, in racketlike feather tips. Male and female motmots are alike in size and coloring.

Motmots are arboreal residents of, mainly, forests and woodlands, but they also occur in more open habitats, such as orchards, tree-lined plantations, suburban parks, and some drier scrub areas. They are predators on insects (particularly beetles, butterflies, dragonflies, and cicadas), spiders, and small frogs, lizards, and snakes, which they snatch in the air, off leaves, and from the ground. Typically they perch quietly on tree branches, utility wires, or fences, sometimes idly swinging their long tails back and forth, until they spy a suitable meal. They then dart quickly, seize the prey, and ferry it back to the perch. If the item is large or struggling—like a big beetle or lizard—the motmot will hold it tight in its bill and whack it noisily against the perch before swallowing it. Motmots also eat small fruits, up to the size of plums, which they collect from trees while briefly hovering. Motmots occur either solitarily or in pairs and may remain in pairs throughout the year (although the sexes may separate during the day to feed). An unusual feature of their behavior is that motmots typically are active well into twilight, going to sleep later than most other birds.

Following suitable courtship activities (such as male and female calling back and forth high up in the trees or holding bits of leaves in their bills), motmots begin breeding. They are monogamous and nest solitarily in burrows. Both male and female dig the burrow, often placed in the vertical bank of a river or road. Tunnels range up to 13 feet (4 m) long, but most are on the order of 5 feet (1.5 m). Both parents incubate eggs and feed the young. Nestlings are fed insects and, when older, other animal food, such as small lizards.

Several of the motmots are rare over parts of their ranges, but they are not considered threatened owing to their greater abundance in other regions. The Keel-billed Motmot has an extensive range within Central America, but is common only in small, isolated areas within that range, in southern Mexico, Belize, Guatemala, Nicaragua, and Costa Rica; because of this highly fragmented distribution and an apparently small total population, the Keel-billed is considered vulnerable. The forests in which this species occurs are continually reduced and degraded, so habitat loss is the prime threat to its existence.

CHESTNUT-HEADED BEE-EATER
Merops leschenaulti
8 in (20 cm)
Southern Asia

RED-BEARDED BEE-EATER
Nyctyornis amictus
10.5–12 in (27–31 cm)
Southeast Asia

BLUE-TAILED BEE-EATER
Merops philippinus
11.5–13.5 in (29–34 cm)
Southern Asia

LITTLE BEE-EATER
Merops pusillus
6.5 in (17 cm)
Africa

SWALLOW-TAILED BEE-EATER
Merops hirundineus
9 in (23 cm)
Africa

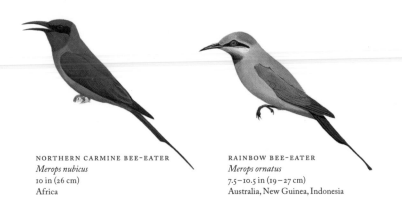

NORTHERN CARMINE BEE-EATER
Merops nubicus
10 in (26 cm)
Africa

RAINBOW BEE-EATER
Merops ornatus
7.5–10.5 in (19–27 cm)
Australia, New Guinea, Indonesia

Bee-eaters

Distribution:
Old World

No. of Living
Species: 25

No. of Species
Vulnerable,
Endangered: 0, 0

No. of Species Extinct
Since 1600: 0

BEE-EATERS are smallish to midsize birds that feed on flying insects, predominantly bees and wasps. Their elegant form and brilliantly colored plumage, combined with an outgoing nature and their conspicuousness in such open areas as parks, gardens, and roadsides, make them favorites of bird-watchers. Aside from the pleasure bee-eaters provide wildlife observers, their main interaction with people is an antagonistic one; they linger around commercial honeybee hives, where, for obvious reasons, they are unwanted. The bee-eater family, Meropidae, contains twenty-five species distributed in southern Europe, southern Asia, Africa, and Australia. Most, eighteen of twenty-five, occur in Africa (either confined there or having that continent included in their range); Australia has a single representative, the Rainbow Bee-eater. The group is included in order Coraciiformes with the kingfishers and hornbills.

Slender, streamlined birds with long, thin, sharply pointed, down-curved bills, bee-eaters are often dazzling in their bright plumage of greens, blues, and reds. They range in length from 6 to 15 inches (16 to 38 cm). Most have thick black eye stripes and long tails. In some, the two central tail feathers grow very long, and these streamers can add up to 4 inches (10 cm) to a bird's length. The sexes look alike or nearly so.

Bee-eaters, birds of warm climates, prefer open habitats such as woodlands, savanna, plantations, and forest edges and clearings; a few inhabit forest interiors. All subsist by hawking for flying insects. They sit alertly on exposed perches, such as bare branches or utility wires, watching their surroundings carefully, then sally out to catch prey that flies by. They snatch insects out of the air with the tip of their bill, then return to a perch, often the same one they just left. They take many different insects, but between 60 percent and 90 percent of their prey consists of bees, wasps, and hornets. To avoid being stung, they grip these insects by their abdomens and beat and rub them against a tree branch or other surface until the venom is discharged; the bird then removes the insect's wings and swallows the body whole. Bee-eaters may also have some immunity to bee venom. Some species in Africa routinely attend brush fires to feast on insects driven up by the flames, and they will follow grazing mammals and even large birds such as Ostriches to catch insects flushed out of hiding places as the large animals walk. Bee-eaters, highly social, are usually in pairs or family groups, or during the winter, in large flocks. They are very vocal; individuals in groups call frequently to each other during foraging sessions. Mostly aerial animals, bee-eaters only occasionally come to the ground. Many species are migratory, probably because insect availability in any one region often changes dramatically through the year.

Bee-eaters nest in horizontal burrows (some up to 8 feet [2.4 m] long) that they excavate in steep hillsides, earthen roadsides, riverbanks, or the ground. Breeding is monogamous, with both sexes contributing to nesting chores. Some species nest solitarily, but many breed in colonies of various sizes, sometimes thousands nesting together. In several species, mostly the ones that nest in colonies, the nesting pair has adult helpers, relatives of the breeding pair, that aid in bringing food to nestlings; in these species, anywhere from two to eight adults tend a single nest. Adults bring insects to the young (after rendering bees and wasps harmless), feeding them at the nest tunnel's entrance; young know to move there when they hear adults arrive with food. In some species, mated pairs stay together for life.

None of the bee-eaters are considered threatened, and many species are very common. These birds are killed in some parts of their range because of their habit of perching near commercial beehives and feeding on the stinging insects. Some forest-dwelling species suffer when their forest habitats are logged or altered for agriculture. For instance, Southeast Asia's Red-bearded Bee-eater has experienced a significant reduction in its range where its tropical forest habitats have been converted to oil palm and rubber plantations.

Rollers;
Ground-Rollers;
Cuckoo-Roller

ROLLERS are among the Old World's glamour birds—handsome, colorful, and often, highly conspicuous in their open habitats as they sit on exposed perches to hunt insects and other small prey. They are called rollers because they roll in flight during their spectacular aerial territorial displays; it is for this behavior that they are perhaps best known. Their vivid plumage of sky blues, purple blues, blue greens, lilacs, and russet browns also make them stand out from the avian crowd. Family Coraciidae consists of the twelve species of "typical" rollers, which are distributed around Eurasia, Africa, and the Pacific. Australia has a single species, as does Europe; eight occur in Africa. Two other kinds of rollers, ground-rollers and cuckoo-rollers, discussed below, were formerly included in the typical roller family, but now are usually placed in separate families. All roller families are included in order Coraciiformes with the kingfishers and bee-eaters.

Stocky, medium-size birds (10.5 to 15 inches [27 to 38 cm] long), rollers have large heads, short necks, and short legs. Some have long tail streamers. Bills, short, broad, and fairly stout, are hooked at the end and vary in color among species from black to yellow to red. Male and female look alike. The widespread Dollarbird is named for the large bluish white dollar-size spot on each wing.

Rollers, warm-climate birds, prefer open country including woodlands, savanna, forest edges, and agricultural sites, but some species inhabit tropical forests. Solitarily or in pairs they perch on high vantage points in open areas, such as on dead tree branches, utility wires, and fence posts, scanning their

ROLLERS

Distribution:
Old World

No. of Living
Species: 12

No. of Species
Vulnerable,
Endangered: 1, 0

No. of Species Extinct
Since 1600: 0

INDIAN ROLLER
Coracias benghalensis
12–13.5 in (30–34 cm)
Southern Asia

DOLLARBIRD
Eurystomus orientalis
10.5–12.5 in (27–32 cm)
Southern Asia, Australia

IMM

LILAC-BREASTED ROLLER
Coracias caudata
11.5 in (29 cm)
Africa

EURASIAN ROLLER
Coracias garrulus
12.5 in (32 cm)
Eurasia, Africa

RUFOUS-CROWNED ROLLER
Coracias naevia
14–15.5 in (35–40 cm)
Africa

PITTA-LIKE GROUND-ROLLER
Atelornis pittoides
10–11.5 in (25–29 cm)
Madagascar

CUCKOO-ROLLER
Leptosomus discolor
15–19.5 in (38–50 cm)
Madagascar

surroundings for likely prey such as insects (mostly beetles and grasshoppers) that fly by, or for scorpions, small lizards, snakes, frogs, and even tiny mammals like mice that move on the ground. They will sally out to catch flying insects but more often fly to the ground to grab prey there. Some species tend to eat while on the ground; others take prey back to an elevated perch. Often the victim is whacked a few times against a branch or other surface to immobilize it and then swallowed. Rollers are often attracted to brush fires, where they snatch insects escaping the flames. Like some other insect-eaters, rollers periodically regurgitate pellets containing hard, indigestible insect parts.

As befitting birds that sometimes chase and catch insects in the air, rollers are agile fliers. Their prowess is especially evident when (in some species) they perform their exuberant "rolling" aerial displays, which are thought to be directed aggressively at other rollers and territorial intruders (including people). They don't actually roll, that is, they do not somersault in the air. Rather, in these displays the bird flies straight up into the air, then dives toward the ground while twisting its trunk this way and that, beating its wings, and giving loud, raucous calls; near the ground it levels out, swoops up again, and repeats the process. The displays may also help maintain pair-bonds between mates.

Rollers, monogamous, nest in tree holes, termite mounds, or occasionally in rock crevices. Eggs are incubated by both parents, and both feed the young. Some rollers may have long lives, estimated at more than 20 years or more. Indonesia's Azure Roller, an island-dwelling forest species, is the only threatened roller, considered vulnerable.

GROUND-ROLLERS are a group of five species (family Brachypteraciidae) of medium size (9.5 to 18 inches [24 to 46 cm]), mostly terrestrial birds that occur only in Madagascar. They have large heads, stout, short bills, short wings, and, in keeping with birds that spend most of their time on the ground, long, strong legs. Some are drab, others fairly colorful. Four are forest species, one inhabits woodland and scrub areas. All are shy and skulking. They eat insects, spiders, worms, centipedes, snails, and small vertebrates. They are monogamous, nesting in burrows they dig in the ground or, in one case, in tree cavities. Three ground-rollers are vulnerable, owing mainly to habitat loss. The CUCKOO-ROLLER (single-species family Leptosomidae) occupies many habitat types in Madagascar and the Comoro Islands. It eats insects and small lizards and nests in tree hollows; it is not threatened.

GROUND-ROLLERS

Distribution: Madagascar

No. of Living Species: 5

No. of Species Vulnerable, Endangered: 3, 0

No. of Species Extinct Since 1600: 0

CUCKOO-ROLLER

Distribution: Madagascar, Comoro Islands

No. of Living Species: 1

No. of Species Vulnerable, Endangered: 0

No. of Species Extinct Since 1600: 0

BROWN HORNBILL
Anorrhinus tickelli
23.5–25.5 in (60–65 cm)
Southeast Asia

WREATHED HORNBILL
Rhyticeros undulatus
29.5–33.5 in (75–85 cm)
Southeast Asia

RHINOCEROS HORNBILL
Buceros rhinoceros
31.5–35.5 in (80–90 cm)
Southeast Asia

HELMETED HORNBILL
Rhinoplax vigil
43.5–47 in (110–120 cm)
Southeast Asia

SILVERY-CHEEKED HORNBILL
Bycanistes brevis
23.5–27.5 in (60–70 cm)
Africa

SOUTHERN YELLOW-BILLED HORNBILL
Tockus leucomelas
15.5 in (40 cm)
Southern Africa

CROWNED HORNBILL
Tockus alboterminatus
19.5 in (50 cm)
Africa

ABYSSINIAN GROUND-HORNBILL
Bucorvus abyssinicus
35.5–39.5 in (90–100 cm)
Africa

Hornbills

HORNBILLS, with their large size, huge bills, often bold colors, and extraordinary breeding, are among the Old World's most distinctive and intriguing birds. For bird-watchers who travel to Africa, southern Asia, or the New Guinea region, the sole places they occur, hornbills are often at or near the top of must-see lists. The sounds hornbills generate are almost as impressive as their striking looks—not only their far-carrying croaking and booming calls but also the loud whoosh-whoosh-whoosh noises they make with their wings as they fly. Their massive, curved, frequently colorful bills suggest a kinship with toucans, but the two groups are not closely related (though they may have developed big bills for the same reason, to cut down and manipulate tree fruit). The hornbill family, Bucerotidae, with fifty-four species, is placed in the order Coraciiformes, with the kingfishers and bee-eaters.

These are medium-size to very large birds (1 to 4 feet [30 to 122 cm] long) with long tails, short legs, and patches of bare skin on their face and throat that are often brightly colored. Their immense bills, often in vibrant reds or yellows, are usually topped with a casque, an extra ridge; in some, the casque is large, in others, reduced. In the Rhinoceros Hornbill, it is shaped like a horn. Bills appear heavy but are actually very light, being constructed of a spongelike substance covered by a thin hornlike material. (One, the Helmeted Hornbill, has a heavier, solid casque.) Male hornbills are usually larger than females, with more pronounced bills and casques, and the sexes often differ in coloring.

Hornbills are forest, woodland, and savanna birds, usually seen in pairs or

Distribution: Africa, southern Asia, Melanesia

No. of Living Species: 54

No. of Species Vulnerable, Endangered: 5, 4

No. of Species Extinct Since 1600: 0

small family groups. Many species spend most of their time in the canopy of forests but will move lower to exploit a good food supply, such as a fruiting tree; some will come to the ground to hunt insects. Two African species, known as ground hornbills, are large, turkey-size black birds with long legs and red and blue throat wattles; they spend most of their time on the ground, flying only reluctantly. Hornbills eat a wide range of plant and animal food, from berries and fruits to insects, lizards, snakes, and small mammals. They can even use their massive bill to delicately gather termites. Small prey is killed with a nip, tossed to the back of the throat, and swallowed; larger prey is passed back and forth along the bill until it is crushed, then eaten.

Hornbill breeding is among the most unusual in the avian world. After a pair selects a cavity in a large tree, the female enters it and, with the male's help, begins to place mud and droppings to make the entrance to the cavity smaller and smaller. Eventually the female is completely walled in save for a tiny vertical slit. She lays eggs and incubates them, dependent on the male to bring her food. When the young hatch, the male brings food for them as well. In at least some species there are extra adults, probably young from previous years, that help feed the female and her young inside their "prison." In some hornbills, the female chips her way out of the cavity before the young are ready to emerge and escapes to help feed them. The young birds instinctively replaster the hole shut with their own droppings; they continue to be fed through the slit until they are ready to fledge. Then they chip away at the plastered hole, usually with no help from the parents, and squeeze out. In other species, the female and chicks emerge from the nest hole together. The function of the plastered-shut hole is probably protection from predators. Breeding adults may remain together for life.

Hornbills generally are thought to be declining throughout their range. They are large and obvious, and in some regions are still hunted for food and the pet trade. However, the chief threat is deforestation: they are increasingly losing their forest homes, especially the large, old trees they use for nest holes, which loggers prefer. Five species are vulnerable and four, endangered (two critically).

Jacamars;
Woodhoopoes;
Hoopoe

JACAMARS are small to medium-size slender birds with very long, fine bills, which they use to catch insects in flight. Their plumage varies among species from fairly drab to brightly colored and iridescent, but most are sufficiently brilliant—with glittering green and blue backs and heads—that the group is considered one of the Neotropics' flashiest. Their overall appearance, including the shimmering plumage, long, delicate bills, and often seemingly excited demeanor, causes many first-time observers to wonder if they might be oversize hummingbirds. But the eighteen jacamar species are not even close relatives of hummingbirds, being most closely related to puffbirds, and the family, Galbulidae, is usually included in the woodpecker order, Piciformes. Jacamars, which occur in Central and South America, range from 5.5 to 13 inches (14 to 34 cm) long. Male and female usually differ slightly in color pattern.

Typically forest dwellers that occur in warmer areas, jacamars are frequently seen along small forest streams and at clearings. They perch on tree limbs, alertly snapping their heads back and forth, scanning for meals. Spotting a flying insect, they dart out to grab it in midair with the tip of the sharp bill. They often then return to the same perch, beat the insect against the perch a few times, then swallow it. After a jacamar grabs an insect such as a large butterfly, its long bill may aid in tightly grasping the insect's body while it attempts escape and also in holding wasps and other stinging insects at safe distances from vulnerable anatomy. Other common diet items are beetles, bees, and dragonflies; one also eats spiders and small lizards.

JACAMARS

Distribution:
Neotropics

No. of Living
Species: 18

No. of Species
Vulnerable,
Endangered: 1, 1

No. of Species Extinct
Since 1600: 0

RUFOUS-TAILED JACAMAR
Galbula ruficauda
7.5–10 in (19–25 cm)
Central America, South America

WHITE-EARED JACAMAR
Galbalcyrhynchus leucotis
8 in (20 cm)
South America

COPPERY-CHESTED JACAMAR
Galbula pastazae
9 in (23 cm)
South America

PARADISE JACAMAR
Galbula dea
10–13.5 in (26–34 cm)
South America

HOOPOE
Upupa epops
10–12.5 in (26–32 cm)
Eurasia, Africa, Madagascar

GREEN WOODHOOPOE
Phoeniculus purpureus
13–14.5 in (33–37 cm)
Africa

ABYSSINIAN SCIMITARBILL
Rhinopomastus minor
9 in (23 cm)
Africa

Monogamous breeders, jacamars nest in short burrows they dig in steep hillsides or in river or stream banks. Both parents incubate eggs and feed insects to the young. In a few species, five or more individuals, presumably family groups, breed together, the extra adults assisting the mated pair in feeding young and perhaps in digging nest burrows. Most jacamars are secure; they are fairly common and have extensive ranges. However, owing to habitat loss and small ranges, two species are threatened; one is considered vulnerable and one endangered.

WOODHOOPOES, medium-size, long-tailed African insect-eaters, are not particularly well known or appreciated. But some of them are spectacular looking, with highly iridescent bluish, greenish, and purplish black plumage and very long, slim, down-curved bills often colored a garish red. There are eight species, all found only in sub-Saharan Africa; two, less flashily attired, are called scimitarbills. The family, Phoeniculidae, is usually placed in the kingfisher order, Coraciiformes. They range from 8 to 15 inches (21 to 38 cm) long; females usually look like males but are often a bit smaller, with shorter bills.

Arboreal birds of forests, woodlands, and savanna, woodhoopoes and scimitarbills are very agile and acrobatic. They climb and forage on tree trunks and branches, often hanging upside down as they use their long, sharp bills to probe and chisel into and around bark for insects and other small arthropods, such as spiders, millipedes, and centipedes. Some also take small lizards and bird eggs, and occasionally berries. Most travel in small, noisy flocks; scimitarbills are quieter and more solitary, found alone or in pairs. Woodhoopoes nest in tree cavities, often old woodpecker holes. All are monogamous, but some, such as the Green Woodhoopoe, breed in cooperative groups; the dominant pair mates and other members of the family group, usually offspring from previous years, help feed the young. Scimitarbills breed in pairs. None of the woodhoopoes are threatened.

The HOOPOE, a striking and unmistakable light brown, black, and white bird with a huge erectile crest, occurs widely in open-country areas of Africa and Eurasia. Its single-species family, Upupidae, is, like the closely related woodhoopoes, included in order Coraciiformes. Medium-size (12 inches [30 cm] long), hoopoes have long, thin, down-curved bills that they use to dig into soft soil and leaf litter for caterpillars, grubs, ants, and beetles. Hoopoes are sometimes in flocks, but nest in tree cavities in lone pairs. The female incubates the eggs alone but is fed regularly by her mate. The Hoopoe is not threatened.

WOODHOOPOES

Distribution:
Sub-Saharan Africa

No. of Living
Species: 8

No. of Species
Vulnerable,
Endangered: 0, 0

No. of Species Extinct
Since 1600: 0

HOOPOE

Distribution:
Eurasia, Africa

No. of Living
Species: 1

No. of Species
Vulnerable,
Endangered: 0, 0

No. of Species Extinct
Since 1600: possibly 1

WHITE-NECKED PUFFBIRD
Notharchus macrorhynchos
10 in (25 cm)
Central America, South America

WHITE-FRONTED NUNBIRD
Monasa morphoeus
8.5–11.5 in (21–29 cm)
Central America, South America

WHITE-WHISKERED PUFFBIRD
Malacoptila panamensis
8 in (20 cm)
Central America, South America

WHITE-CHESTED PUFFBIRD
Malacoptila fusca
7.5 in (19 cm)
South America

SWALLOW-WINGED PUFFBIRD
Chelidoptera tenebrosa
6 in (15 cm)
South America

BLACK-FRONTED NUNBIRD
Monasa nigrifrons
11 in (28 cm)
South America

WHITE-EARED PUFFBIRD
Nystalus chacuru
8.5 in (22 cm)
South America

Puffbirds

PUFFBIRDS, like their close relatives the jacamars, are small to medium-size arboreal birds of neotropical forests that make a living catching insects. There the resemblance ends, because puffbirds differ markedly from jacamars in general form, brightness of coloration, main feeding method, and even in daily demeanor. Jacamars are slender with long, fine, mostly straight bills; puffbirds are stocky with thick, heavy, hooked bills. Jacamars are often radiantly colorful and glossy; puffbirds are covered in subdued browns, grays, black, and white. Jacamars snatch insects from the air, puffbirds take a lot of their prey off leaves, tree trunks and branches, and from the ground. Jacamars have alert, bright, convivial personalities; puffbirds are dull, often sitting quietly and still on tree branches for long periods, even sometimes allowing people quite close before fleeing. The sedate dullness of these birds led many early observers to consider them stupid—and, indeed, this is their local reputation in many regions within their range. They are not dumb; their sometimes slow ways are related to their foraging methods (see below). The thirty-three species in the puffbird family, Bucconidae, some of which are known as nunbirds, nunlets, and monklets, are distributed from southern Mexico to northern Argentina, with most occurring in the Amazon region. The group is usually included in the woodpecker order, Piciformes (but some recent classifications place the puffbirds and jacamars in their own, separate, order, Galbuliformes). Most puffbirds are fairly small, but they range from 5 to 11.5 inches (13 to 29 cm) in length. They have distinctively large heads and loose plumage that, when fluffed out, produces a chubby, puffy appearance.

Distribution:
Neotropics

No. of Living
Species: 33

No. of Species
Vulnerable,
Endangered: 0, 0

No. of Species Extinct
Since 1600: 0

Most puffbirds hunt by perching quietly on tree limbs, waiting patiently for a large insect, spider, or small frog, lizard, or snake to walk, run, hop, or slither by on a nearby trunk or on the ground. The puffbird swoops from its perch and seizes the luckless prey in its bill. The bird then returns to a perch, beats the prey on the perch to stun it, and swallows it. Some puffbirds will also dart out to snatch a flying insect in midair. One, the Swallow-winged Puffbird, the most bizarre-looking of the group because it seems to have no tail and, with its long, tapered wings, resembles a swallow, obtains all its food this way. Occasionally a family group of puffbirds—especially nunbirds—will follow a troop of monkeys or a flock of large birds and catch insects scared into flight by movements of the larger animals. Because puffbirds sit in the open for long periods looking for prey, it is thought that their stillness, and perhaps their "fluff-ball" appearance, help camouflage them from predators. Some puffbirds are quite social, often found in groups of up to ten, perched in a row on a branch or utility wire. And not all are quiet. Black-fronted Nunbirds, for example, often sit together in family groups, on a branch high in a tree, and sing their loud, raucous songs for minutes at a time.

Relatively little is known about puffbird breeding. All species may be monogamous, and most appear to breed, and perhaps live together year-round, as territorial pairs. But several species breed in cooperative groups, numerous adults appearing to feed the young in a single nest. Helpers are probably relatives, usually young of the mated pair from previous nests. Some nest in burrows on the forest floor, some dig nesting tunnels in sandy soil that are up to 6 feet (1.8 m) deep, some excavate burrows in arboreal termite nests, and some may nest in tree cavities. The White-fronted Nunbird surrounds its ground-tunnel entrance with dead leaves and twigs to hide the hole. Male and female share in tunnel excavation, incubation, and feeding the young.

None of the puffbirds are considered threatened, but two or three species are very poorly known and have small ranges in South America; they may be near-threatened. One problem with studying or censusing these birds is the unobtrusiveness of many of the species; they are often quiet and still, and can be quite stealthy in their behavior and movements.

Barbets

BARBETS are tropical, largely arboreal birds noted for their colors, vocalizations, and fruit-centered diets. They are favorites of bird-watchers because of their beauty, sometimes melodic songs, and because they can be fun to try to identify, as, in some regions, several species look much alike, with only subtle differences in color patterns. There are about eighty-two barbet species distributed broadly in tropical South and Central America, Africa, and southern Asia; some range into subtropical regions. The family, Capitonidae, is usually placed in the order Piciformes, with the woodpeckers and toucans; some authorities divide the New World and Old World barbets into separate families.

Rather husky, small to medium-size birds (ranging in length from 4 to 14 inches [10 to 35 cm]), barbets are big-headed and short-necked, and possess sturdy, large (but not long) bills, sometimes notched to help them grip food. The barbet name derives from the bristles that surround the base of the bill. Their colorful, often spotted and streaked plumage is usually combinations of black, white, yellow and red or orange, but some species, especially in Africa, come in more cryptic browns. Many Asian barbets are mainly bright green with multicolored heads of red, yellow, blue, and black. African tinkerbirds, named for their ringing calls, are diminutive barbets clad chiefly in streaked browns, yellows, black, and white. Many barbets show differences between the sexes in plumage patterns.

Barbets are birds of forests and woodlands, although some, chiefly several of the African species, prefer forest edges and more open areas such as scrublands and gardens. All usually require dead trees (or live trees with

Distribution:
South and Central
America, Africa,
southern Asia

No. of Living
Species: 82

No. of Species
Vulnerable,
Endangered: 0, 1

No. of Species Extinct
Since 1600: 0

GREAT BARBET
Megalaima virens
13.5 in (34 cm)
Southern Asia

BLUE-EARED BARBET
Megalaima australis
6.5 in (17 cm)
Southeast Asia

COPPERSMITH BARBET
Megalaima haemacephala
6.5 in (16 cm)
Southern Asia

FIRE-TUFTED BARBET
Psilopogon pyrolophus
11.5 in (29 cm)
Southeast Asia

YELLOW-RUMPED TINKERBIRD
Pogoniulus bilineatus
4.5 in (11 cm)
Africa

ACACIA PIED BARBET
Tricholaema leucomelas
6.5 in (17 cm)
Southern Africa

BLACK-SPOTTED BARBET
Capito niger
7 in (18 cm)
South America

TOUCAN BARBET
Semnornis ramphastinus
8 in (20 cm)
South America

SCARLET-BANDED BARBET
Capito wallacei
8 in (20 cm)
South America

dead branches) where they live because barbets excavate cavities in wood in which to roost and nest. Many species feed in the forest canopy, but others tend to forage lower down in trees and bushes and some will even hop about on the ground. They have a reputation as fruit-eaters, and they commonly eat fruits, especially figs, but many are actually fairly omnivorous, also taking (to varying degrees according to species) plant buds and seeds, insects, spiders, scorpions, centipedes, snails, and worms. Barbets are agile feeders, clinging to bark and hanging at odd angles to reach particularly fine fruits or berries; often they use their tails to support themselves on tree trunks and branches. Some regularly join mixed-species feeding flocks, typically high in the canopy. They often stay in mated pairs, and these pairs can be very aggressive, defending their territories and chasing other barbets from fruiting trees. Others are quite gregarious and live in family groups, usually of three to five birds. Larger groups occasionally gather at fruit-laden trees. Many African barbets are known for their loud, raucous duets. A pair perches close together and, bobbing up and down, wagging their tails, and flicking their wings, bursts into song—sometimes alternating notes, sometimes singing simultaneously, but in such a coordinated manner that it sounds like one song and it is impossible to tell which bird is producing which sounds.

Barbets are apparently all monogamous, although the breeding habits of several species are poorly known. They nest in tree cavities, usually ones they excavate themselves using their powerful bills; sometimes they take over holes dug by woodpeckers. A few nest in tunnels dug into earthen banks, termite mounds, or even into the ground. Both sexes incubate eggs and feed young. Nestlings are fed insects or insects and fruit. Some species breed in cooperative groups, with several helpers, probably young from the previous year, aiding the mated pair; the helpers assist in defending the territory and feeding young. Some African species nest in colonies, occasionally with more than a hundred nest holes in a single tree.

Only a single barbet is endangered, Colombia's White-mantled Barbet. It has a small, fragmented population and is losing its remaining forest habitat to farming, ranching, and other development. Nine other barbets, in South America, Africa, and Asia, are considered near-threatened. Deforestation is clearly the greatest threat barbets face. These pretty birds are also occasionally persecuted because farmers and orchardists believe they damage or steal tree fruit, and some are still captured for the cage-bird trade—especially those considered enchanting singers.

PALE-MANDIBLED ARACARI
Pteroglossus erythropygius
17–19 in (43–48 cm)
South America

GOLDEN-COLLARED TOUCANET
Selenidera reinwardtii
13.5 in (34 cm)
South America

PLATE-BILLED MOUNTAIN-TOUCAN
Andigena laminirostris
18–20 in (46–51 cm)
South America

WHITE-THROATED TOUCAN
Ramphastos tucanus
21–23 in (53–58 cm)
South America

EMERALD TOUCANET
Aulacorhynchus prasinus
12–14.5 in (30–37 cm)
Mexico, Central America, South America

CURL-CRESTED ARACARI
Pteroglossus beauharnaesii
16.5–18 in (42–46 cm)
South America

TOCO TOUCAN
Ramphastos toco
21.5–24 in (55–61 cm)
South America

SPOT-BILLED TOUCANET
Selenidera maculirostris
13–14.5 in (33–37 cm)
South America

Toucans

TOUCANS, with their splendid colors and enormous, almost cartoonish bills, are, along with penguins, ducks, and a few others, among the world's most widely recognized birds. They have become popular symbols of the American tropics and are usually the birds visitors to this region—be they bird-watchers or more traditional tourists—most want to see. The group's most distinguishing feature, the spectacular, usually colorful, disproportionately large bill, is actually very light, mostly hollow, and used for cutting down and manipulating the diet staple, tree fruit.

The toucan family, Ramphastidae, has about forty species (toucans plus the usually smaller toucanets and aracaris), all restricted to the Neotropics; most are confined to tropical areas, but the ranges of a few extend into subtropical South America. The group is usually included in order Piciformes with the woodpeckers and barbets. Toucans range in length from 12 to 24 inches (30 to 61 cm). They have long tails and areas of bare skin, typically brightly colored, around their eyes. Most have multicolored plumage, typically with patterns of green, yellow, and red/orange. But some, especially the largest ones such as the Toco Toucan, are mainly black and white. Their serrated bills (helpful in manipulating fruit) can be vividly colorful or fairly drab, as in some of the toucanets. The Curl-crested Aracari is unusual, with curled, plasticlike feathers on its head. Within a species, male bills are usually longer than females', and the two sexes often differ somewhat in plumage color pattern.

Arboreal, gregarious forest birds, toucans are usually observed in flocks of three to twelve, occasionally up to twenty. They are primarily fruit-eaters,

Distribution: Neotropics

No. of Living Species: About 40

No. of Species Vulnerable, Endangered: 0, 1

No. of Species Extinct Since 1600: 0

their long bill allowing them to perch on heavier, stable branches and reach a distance for hanging fruits. They snip a fruit off, hold it at the tip of the bill, and then, with a forward flip of the head, toss the fruit into the air and swallow it. They increase their protein intake by consuming the odd insect, spider, or small reptile, or even bird eggs or nestlings. Sometimes individual fruit trees are defended by a mated toucan pair from other toucans or from other fruit-eating birds—defended by threat displays and even, against other toucans, by bill clashes; they may eat smaller fruit-seeking birds they catch at fruiting trees. When a flock of toucans land in a tree, their aggressive natures often cause other birds there, even large parrots, to leave. The individuals in a toucan group follow each other one by one, in strings, from one tree to another, usually staying in the high canopy; they will only occasionally fly down to feed at shrubs, or to pluck a lizard from the forest floor. In their slow, undulatory flight, toucans often look awkward or unbalanced, probably because the large bill seems to be leading and pulling the bird behind it. Several toucan species are known to play, grasping each other's bills in apparent contests and tossing fruit to one another. Toucans communicate acoustically with a variety of croaks, honks, grunts, squawks, and rattles, and also by snapping their bill or tapping it against trees or other surfaces.

Toucans nest in tree cavities, either natural ones or those hollowed out by woodpeckers; sometimes they take over already-occupied cavities. Nests can be any height above the ground, up to 100 feet (30 m) or more. Both sexes incubate eggs and feed young. Chicks are usually fed insects at first, then insects and fruit. Toucans are monogamous, but some of the aracaris seem to breed cooperatively in small family groups of three or four; that is, other family members, in addition to the breeding pair, help raise the young in a single nest.

Toucans are common residents in the various regions in which they occur, except where there is extensive deforestation, for instance, in certain parts of Central America. Also, some toucan species may be scarce locally due to hunting because they are taken for food, as pets, and for the cage-bird trade. Where they raid fruit orchards, they are sometimes killed as pests. Peru's Yellow-browed Toucanet, which has a small population and a tiny range, is considered endangered; no other toucans are threatened.

Honeyguides

HONEYGUIDES, smallish, rather drab, mostly African birds, are named for the extraordinary guiding behavior of the Greater Honeyguide, which leads people to honeybee hives. The guiding is mutually advantageous: in some rural African communities, people obtain honey from the hives, and the guides get some of the beeswax, a diet staple for them. Aside from this intriguing relationship with people, honeyguides are known among ornithologists for their unusual breeding behavior: they are brood parasites, the females laying their eggs in the nests of other species, the "host" species then raising honeyguide young.

The seventeen honeyguide species (family Indicatoridae) are placed in order Piciformes with the woodpeckers and barbets. Fifteen occur in sub-Saharan Africa, two in southern Asia. All are mostly dull gray, olive, or greenish brown; the Yellow-rumped Honeyguide, an Asian species, has some bright yellow patches. Ranging from 4 to 8 inches (10 to 20 cm) long, honeyguides have small heads and short bills. Their skin is particularly thick, presumably to help protect them from angry bees whose hives they raid. Males and females look alike or nearly so.

Honeyguides are arboreal birds of forests and woodlands. The diet of most is centered on beeswax, the yellow substance secreted by honeybees and used for building honeycombs; honeyguides possess enzymes that help digest this unusual food source. They gather wax by visiting abandoned hives, ones that have been damaged and opened by mammals, and intact hives that have openings large enough to enter or peck through. Honeyguides are not

Distribution:
Sub-Saharan Africa,
southern Asia

No. of Living
Species: 17

No. of Species
Vulnerable,
Endangered: 0, 0

No. of Species Extinct
Since 1600: 0

GREATER HONEYGUIDE
Indicator indicator
8 in (20 cm)
Africa

LYRE-TAILED HONEYGUIDE
Melichneutes robustus
7 in (18 cm)
Africa

LESSER HONEYGUIDE
Indicator minor
6.5 in (16 cm)
Africa

WAHLBERG'S HONEYBIRD
Prodotiscus regulus
5 in (13 cm)
Africa

YELLOW-RUMPED HONEYGUIDE
Indicator xanthonotus
6.5 in (16 cm)
Himalayas, Myanmar

immune to honeybee venom and some die from multiple bee stings. So even with their thick skins, they usually avoid well-defended hives, or they visit them in the early morning, when lower temperatures render resident bees inactive. They also eat larval and adult insects, spiders, and occasional fruit; some species replace beeswax in their diet by eating scale insects, which have waxy coverings. Honeyguides are considered tough and aggressive, and they need to be. They must defend discovered beehives from other honeyguides wanting to feed and also must be able to place eggs in host-species nests when hosts try to prevent them from doing so.

The guiding behavior of a Greater Honeyguide works like this: When it sees an opportunity, the bird attempts to catch the attention of a person by giving a distinctive guiding call (a chattering sound) from a nearby tree. If the person responds by approaching the bird, it flies to the next tree in the direction of the beehive and continues calling there, and so on. If it isn't followed, it returns to the person it has targeted and tries again. The hive may be half a mile (1 km) or more away, and the route followed is seldom direct. When the bird finally remains calling in one tree, the hive is close by. People who regularly work with honeyguides often whistle to a bird as they follow it, and they always leave part of the honeycomb behind for the bird. It is mainly honeyguides in rural areas that guide people, and as development spreads in Africa, guiding behavior diminishes. Supposedly, Greater Honeyguides will also lead to beehives mammals known as Honey Badgers, which break into beehives to consume honey and bee larvae, but such behavior is not well documented.

Male and female honeyguides do not have long-term pair-bonds; probably both males and females mate with several individuals in a breeding season. Males sing from tree perches to attract females. After mating with a male, a female goes off to lay her eggs in the nests of host species. Many species serve as hosts, but some honeyguides specialize on parasitizing nests of barbets and woodpeckers. The female honeyguide often punctures or removes one of the host's eggs before laying her own. The honeyguide egg usually hatches first, and the honeyguide chick then damages the host's own eggs or, if they hatch, injures the young, and so receives all food from the adults tending the nest. None of the honeyguides are threatened, but three species, including the two Asian ones, are considered near-threatened.

HISPANIOLAN WOODPECKER
Melanerpes striatus
8.5 in (22 cm)
Hispaniola

CUBAN GREEN WOODPECKER
Xiphidiopicus percussus
9 in (23 cm)
Cuba

ECUADOREAN PICULET
Picumnus sclateri
3.5 in (9 cm)
South America

SPOT-BREASTED WOODPECKER
Colaptes punctigula
8 in (20 cm)
South America

CRIMSON-MANTLED WOODPECKER
Piculus rivolii
8.5–11 in (22–28 cm)
South America

CRIMSON-CRESTED WOODPECKER
Campephilus melanoleucos
13–15 in (33–38 cm)
Central America, South America

ANDEAN FLICKER
Colaptes rupicola
12.5 in (32 cm)
South America

RED-NECKED WOODPECKER
Campephilus rubricollis
12 in (31 cm)
South America

Woodpeckers

WOODPECKERS are familiar to most people by reputation and sometimes by sound, if not by sight. They are highly specialized birds of forests and woodlands; where there are trees in the world, there are woodpeckers (excepting only Australia, New Zealand, and some other island nations). Perhaps most famous for their Woody Woodpecker and other cartoon incarnations and for their use as advertising symbols, they are also known for their habits of tapping on wood and nesting in tree holes. The group is big and diverse, encompassing about 216 species, from large woodpeckers up to 22 inches (55 cm) long to tiny piculets at about 3.5 inches (9 cm). The woodpecker family, Picidae, is placed, along with the toucans and barbets, in order Piciformes. Picus, in ancient Roman myths, was a god of the forests, and was transformed by a sorceress into a woodpecker.

Woodpeckers have strong, straight, chisel-like bills, very long barbed tongues, and sharp toes that spread widely for clinging to tree trunks and branches. All but the small piculets have stiffly reinforced tail feathers that support them as they climb on vertical surfaces. To accommodate their constant banging and drumming with their bills on wood, woodpeckers have an array of skeleton and muscle modifications of the head and neck to absorb shock waves. They are mostly olive green, brown, gray, and black and white, frequently streaked or barred; many have small but conspicuous head or neck patches of red or yellow. A few are actually quite showy, with some of Asia's, such as the Common Flameback, taking high honors; many have striking black and white stripes on chest or back. Some have conspicuous

Distribution:
All continents except
Antarctica and
Australia

No. of Living
Species: 216

No. of Species
Vulnerable,
Endangered: 7, 2

No. of Species Extinct
Since 1600: 1

WHITE WOODPECKER
Melanerpes candidus
9.5–11.5 in (24–29 cm)
South America

CAMPO FLICKER
Colaptes campestris
12 in (30 cm)
South America

PALE-CRESTED WOODPECKER
Celeus lugubris
9.5 in (24 cm)
South America

BLACK-BACKED WOODPECKER
Picoides arcticus
9.5 in (24 cm)
North America

RED-COCKADED WOODPECKER
Picoides borealis
8.5 in (22 cm)
North America

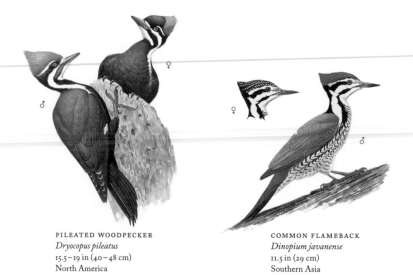

PILEATED WOODPECKER
Dryocopus pileatus
15.5–19 in (40–48 cm)
North America

COMMON FLAMEBACK
Dinopium javanense
11.5 in (29 cm)
Southern Asia

HEART-SPOTTED WOODPECKER
Hemicircus canente
6.5 in (16 cm)
Southern Asia

GREAT SLATY WOODPECKER
Mulleripicus pulverulentus
19.5 in (50 cm)
Southeast Asia, Himalayas

GREATER YELLOWNAPE
Picus flavinucha
13 in (33 cm)
Southern Asia

ORANGE-BACKED WOODPECKER
Reinwardtipicus validus
12 in (30 cm)
Southeast Asia

RUFOUS-NECKED WRYNECK
Jynx ruficollis
7.5 in (19 cm)
Africa

CARDINAL WOODPECKER
Dendropicos fuscescens
6 in (15 cm)
Africa

GROUND WOODPECKER
Geocolaptes olivaceus
8.5–12 in (22–30 cm)
Southern Africa

crests. The sexes usually look alike, but males may be brighter, for instance, with larger red patches.

Woodpeckers occupy diverse habitats and employ various feeding methods. They are associated with trees and are adapted to cling to a tree's bark and to move lightly over its surface, searching for insects; they also drill holes in bark and wood into which they insert their long tongues, probing for hidden insects (tongues are often sticky-coated to extract insects from holes). They usually move up tree trunks in short steps, using their stiff tail as a prop. Woodpeckers eat many kinds of insects that they locate on or in trees, including larval ones. They will also take insects on the wing and many supplement their diets with fruits, nuts, and nectar from flowers. Some eat a lot of fruit, some will eat nestlings of other bird species, and two (called wrynecks) have ants as their main food. The sapsuckers use their bills to drill small holes in trees that fill with sap, which is then eaten. Some woodpeckers live in grassland areas, foraging on the ground (but they require trees for nesting). Woodpeckers hit trees with their bills for three reasons: drilling bark or wood to get at insect food; excavating holes for roosting and nesting; and drumming (rapid beats often on hollow surfaces to amplify the sounds) to send signals to other woodpeckers. They typically weave up and down as they fly, a behavior suited to make it more difficult for predators to track the birds' movements.

Monogamous, some woodpeckers remain paired throughout the year; others live in family groups of up to twenty. Before nesting, a mated male and female carve a hole in a tree, sometimes lining it with wood chips. Both sexes incubate eggs, males typically taking the night shift. Young are fed insects by both parents. After fledging, juveniles remain with parents for several months, longer in those species in which families associate throughout the year.

Birds of forests, woodpeckers are continually pressured as loggers and developers cut and degrade forests. Nine species are currently considered threatened (seven, including the United States' Red-cockaded Woodpecker, are vulnerable; two are endangered). Two of the largest species are now extinct or almost so: northern Mexico's Imperial Woodpecker and the United States' and Cuba's Ivory-billed Woodpecker. A report in 2005 of Ivory-bill sightings in Arkansas excited bird-watchers and scientists alike. Woodpeckers damage trees and buildings and also eat fruit from gardens and orchards, so in some parts of the world they are considered significant pests and treated as such.

Pittas

The little-known PITTAS are some of nature's most dazzling avian creations. They are medium-size ground birds of Old World tropical forests, and it is their coloring that most attracts and fascinates; some of them are luminous in bright reds, sky blues, mustard yellows, and forest greens. At one time, owing to their radiance, they were called jewel-thrushes. Their beauty, elusiveness, and typical rarity combine to create a mystique that renders the pittas perennial favorites of bird-watchers, *favorite* in this case meaning "highly sought after but rarely spotted." The pitta family, Pittidae, includes about thirty species distributed from West Africa through India, Southeast Asia, Melanesia, and parts of Australia. Most occur in Southeast Asia; only a few occur in Africa and Australia. The name *pitta* arises not from a representation of one of their calls but from an Indian word for bird.

Ranging in length from 6 to 11 inches (15 to 28 cm), pittas appear stocky and almost tailless, and have longish, stout legs. Their bright colors are often difficult to appreciate because they are mainly on the birds' underparts and because of the dark shadows of the deep forest floor that these birds inhabit. Easily overlooked because of their forest settings and shyness, pittas would be almost invisible except for their haunting call notes and whistles; they are far more often heard than seen. Occasionally, however, they hop out into the brilliant sunshine of an open forest path and make themselves very noticeable. Males and females look alike in some pitta species; in others, there are color differences.

Pittas, usually found alone or in pairs, are terrestrial foragers in rainforests,

Distribution:
Old World tropics

No. of Living
Species: 32

No. of Species
Vulnerable,
Endangered: 8, 1

No. of Species Extinct
Since 1600: 0

BLUE PITTA
Pitta cyanea
9 in (23 cm)
Southeast Asia

GURNEY'S PITTA
Pitta gurneyi
8.5 in (21 cm)
Southeast Asia

♂

♀

BANDED PITTA
Pitta guajana
8.5 in (22 cm)
Southeast Asia

HOODED PITTA
Pitta sordida
7 in (18 cm)
Southern Asia, New Guinea

NOISY PITTA
Pitta versicolor
8 in (20 cm)
Australia, New Guinea

RAINBOW PITTA
Pitta iris
6.5 in (17 cm)
Australia

mangroves, and scrub areas. They hop about on the ground, tossing and turning leaves with their bills, searching for insects, spiders, snails, worms, and small frogs, lizards and crabs. Hard prey, such as snails and large insects, are often bashed on rocks or logs, the broken shells discarded before the insides are eaten. Pittas probably hunt by sight, sound, and especially by smell (they appear to have one of the bird world's most developed olfactory organs). Secretive most of the time, occasionally they can be quite noisy as they scratch at the leaf litter and send vegetation flying with their bills and powerful feet and legs. When they are alarmed, pittas bound away by foot or fly up to a low branch or a fallen tree trunk to get a better look at whatever has invaded their territory. At night they roost in trees, and they will often perch high up in the forest canopy to vocalize. Most pittas are sedentary, but some species are migratory; for example, the Indian Pitta breeds in the Himalayan region but winters in southern India.

Relatively little is known about social organization and breeding biology of pittas because they are so difficult to observe. All are thought to be monogamous, and they nest solitarily, pairs remaining together at least through the breeding season. The nest is a large, rough dome of sticks, rootlets, and other vegetation with a side entrance. It is lined with fibers and leaves and built on the ground next to a tree base or occasionally low off the ground (to a maximum height of about 10 feet [3 m]) in a vine tangle or on a fallen trunk. Both parents incubate eggs and feed nestlings. Young are probably independent fairly quickly after they fledge.

Owing to their bright colors, pittas, wherever they occur, are often kept as local pets. But they are most threatened by deforestation, which is still occurring at high rates throughout the family's range. Eight species, mainly in the Philippines, Indonesia, and New Guinea area, are considered vulnerable. Gurney's Pitta, critically endangered, is now confined to a single forest reserve in peninsular Thailand, and there may be fewer than fifty individuals left. Once thought likely extinct, it was rediscovered in 1986. The cause of its decline is undoubtedly destruction of the lowland forest habitat to which it is restricted, compounded by trapping for the cage-bird trade. Similarly, Sumatra's Schneider's Pitta was once thought extinct, but multiple recent sightings show it still survives in its high-elevation forest habitats.

DUSKY BROADBILL
Corydon sumatranus
10 in (26 cm)
Southeast Asia

BLACK-AND-YELLOW BROADBILL
Eurylaimus ochromalus
5.5 in (14 cm)
Southeast Asia

LONG-TAILED BROADBILL
Psarisomus dalhousiae
10 in (25 cm)
Southeast Asia

BLACK-AND-RED BROADBILL
Cymbirhynchus macrorhynchos
8.5 in (22 cm)
Southeast Asia

WHITEHEAD'S BROADBILL
Calyptomena whitehaedi
10 in (26 cm)
Southeast Asia

SCHLEGEL'S ASITY
Philepitta schlegeli
5 in (13 cm)
Madagascar

RIFLEMAN
Acanthisitta chloris
3 in (8 cm)
New Zealand

Broadbills; Asities; New Zealand Wrens

BROADBILLS are small to medium-size, thickset, often exquisitely marked forest birds. They are renowned and named for their large, flat, hooked bills and very wide gapes, which aid them in capturing and consuming animal prey. Also noted for their coloring, which ranges from striking red and black to bright green, these beautiful birds are favorites of globetrotting bird-watchers. The family, Eurylaimidae, has fifteen species: four in Africa, nine in Southeast Asia, and two in the Philippines. Ranging from 4.5 to 11 inches (12 to 28 cm) long, broadbills have largish heads, big eyes, rounded wings, and strong feet; tails vary from short to longish. Not all are brightly colored—some are solid brown or streaked brown and white. Broadbill sexes look alike or differ slightly in coloring.

Birds chiefly of forests, some broadbills venture into more open habitats such as woodlands, forest edges, bamboo strands, and even scrub areas. Most take whatever small animals they can find, mainly insects, but also spiders, snails, tree frogs, and lizards; some also eat small fruits and berries. Foraging is either by perching quietly in foliage and then flying to snatch detected prey (on vegetation or from the ground), or taking prey from leaves and branches as the birds search through foliage. Three mostly green species of Southeast Asia primarily eat fruit, and have, through evolution, lost their broad bills, no longer required for catching small animals. Many broadbills are gregarious, occurring in small flocks of up to ten or more individuals. Monogamous breeders, broadbills build pear-shaped nests of vegetation (grasses, vines, twigs, bark, rootlets) that hang over open areas from tips of tree branches. The female alone or both sexes build nests and incubate eggs;

BROADBILLS

Distribution:
Africa, Southeast Asia

No. of Living
Species: 15

No. of Species
Vulnerable,
Endangered: 3, 0

No. of Species Extinct
Since 1600: 0

ASITIES

Distribution:
Madagascar

No. of Living
Species: 4

No. of Species
Vulnerable,
Endangered: 0, 1

No. of Species Extinct
Since 1600: 0

NEW ZEALAND
WRENS

Distribution:
New Zealand

No. of Living
Species: 2

No. of Species
Vulnerable,
Endangered: 0

No. of Species Extinct
Since 1600: 2

both sexes feed the young. Most broadbills still have healthy populations, but one in Central Africa and the two Philippine species are considered vulnerable; loss of forest habitat is the primary threat.

ASITIES are a tiny group of small, stout forest birds with very short tails, recognized for their bright body colors, remarkable bill shapes, and highly restricted distribution; they occur only on the island of Madagascar. The family, Philepittidae, includes four species. Two, known as sunbird-asities, have long, slim, down-curved bills, used to gather nectar from flowers. The other two, the Velvet Asity and Schlegel's Asity, are fruit-eaters, and have relatively short, slightly down-curved bills. The breeding male Velvet Asity is a plump, mainly black bird with bright green and blue skin wattles over each eye; other male asities, all with yellow underparts, also have blue or blue and green eye wattles. Female asities are less brightly colored than males.

Asities are birds mainly of Madagascar's rainforests. The Velvet Asity and Schlegel's Asity eat fruit (which they take from plants in the forest canopy down to the forest floor), but also nectar and occasional insects. The sunbird-asities, which sometimes join mixed-species foraging flocks, eat nectar and some insects. Asities are seen alone, in pairs, or in small groups. Breeding is probably polygamous, with males on territories displaying and calling to attract females; after mating, females probably build nests, incubate eggs, and perhaps feed young by themselves. Nests are usually messy woven structures of vegetation, ball- or pear-shaped, suspended from a tree branch. Only the Yellow-bellied Sunbird-asity is threatened, classified as endangered.

NEW ZEALAND WRENS are very small greenish or brownish birds restricted to the islands of New Zealand. The family, Acanthisittidae, has two species, the Rifleman and Rock Wren. At three to four inches (7.5 to 10 cm) long, they have slender bills (slightly upturned in the Rifleman), short, rounded wings, very short tails, and long legs. The sexes differ, females being drabber and a bit larger than males. The Rifleman is chiefly an arboreal forest dweller; the Rock Wren is a terrestrial/arboreal resident of alpine and subalpine scrub and rock habitats. Both eat insects, spiders and occasional fruit. In mated pairs they maintain year-round territories. Breeding is monogamous, the Rifleman being a cooperative nester, with other adults assisting breeding pairs. The Rock-wren is near-threatened. One New Zealand wren, which was flightless, became extinct around 1894, and another did so around 1972.

Woodcreepers

WOODCREEPERS are small to medium-size brown birds of the Neotropics that pursue a mostly arboreal existence. Although like woodpeckers in some ways (and once thought to be woodpeckers), they are typically inconspicuous, and go about their lives little noticed and generally unappreciated by people. Like woodpeckers, they seek insects by climbing quickly over tree trunks and branches, clinging to vertical and angled surfaces with powerful feet with sharp, curved claws, and with their stiff tail feathers acting as props. But unlike woodpeckers, woodcreepers are mainly small birds with drab plumage and retiring personalities, and most inhabit dark interiors of forests—all traits fostering inconspicuousness. Further, whereas woodpeckers are noisy, pecking at and hammering trees, woodcreepers are relatively quiet, singing simple, often fairly soft songs and lacking the tree-drilling and -drumming behaviors of woodpeckers. Bird-watchers, probably the largest group who regularly notice them, regard woodcreepers as problematic, because in the low light levels of their typical forest habitats, and against the brown tree surfaces that are their usual backdrops, they can be difficult to identify to species (when discernible, bill size, shape and color, body size, and type of plumage streaking are used as distinguishing traits). The family, Dendrocolaptidae, consists of fifty-two species and occurs from central Mexico southward to central Argentina. Most (about nineteen) occur in the Amazon region. Five species, with particularly long, down-curved bills, are known as scythebills.

Fairly uniform in size, shape, and coloring, most woodcreepers are slender birds, 8 to 14 inches (20 to 36 cm) long (but a few are as small as 5

Distribution: Neotropics

No. of Living Species: 52

No. of Species Vulnerable, Endangered: 1, 0

No. of Species Extinct Since 1600: 0

WHITE-STRIPED WOODCREEPER
Lepidocolaptes leucogaster
9 in (23 cm)
Mexico

IVORY-BILLED WOODCREEPER
Xiphorhynchus flavigaster
8–10 in (20–26 cm)
Mexico, Central America

TAWNY-WINGED WOODCREEPER
Dendrocincla anabatina
7 in (18 cm)
Central America

LONG-BILLED WOODCREEPER
Nasica longirostris
14 in (35 cm)
South America

SPOT-CROWNED WOODCREEPER
Lepidocolaptes affinis
8.5 in (21 cm)
Mexico, Central America

RED-BILLED SCYTHEBILL
Campylorhamphus trochilirostris
8.5–11 in (22–28 cm)
South America

NARROW-BILLED WOODCREEPER
Lepidocolaptes angustirostris
8 in (20 cm)
South America

inches [13 cm]). The sexes look alike, with plumage mostly of various shades of brown, chestnut, or tan; wings and tails are predominantly reddish brown. Most have some spotting, streaking, or banding, particularly on the head, back, and underparts. A few are mainly solid grayish and brown. Bills vary extensively, from shortish and slender to long and robust; some are strongly down-curved. Their wings are relatively short, broad, and rounded.

Woodcreepers are common birds of forests but also of forest edges, woodlands, and some other semiopen areas. There is usually at least one species present in any forest within the family's range. Woodcreepers forage by creeping upward on tree trunks and also horizontally along branches, peering under bark and into moss clumps and epiphytes, using their bills to probe and snatch prey in tight nooks and crannies. The foraging procedure is more or less standardized: a woodcreeper flies to the base of a tree and then spirals up the trunk, using its stiff, spiny tail as a third foot to brace itself in a vertical posture against the tree; it checks for prey as it climbs. At the top, the bird flies down to the base of the next tree and repeats the process. Some species dispense with large trunks and forage instead on smaller trees, large shrubs, or on bamboo; some spend much time feeding high in the forest canopy; some descend to the ground to feed; and some will also fly out from trees to catch insects on the wing.

The group eats primarily insects but also takes spiders, as well as small lizards and amphibians. Many woodcreepers are frequent participants, with antbirds, tanagers, and motmots, among others, in mixed-species foraging flocks, some of which specialize on following swarms of army ants, taking prey that rush out from hiding places to avoid the voracious ants. Woodcreepers are most often observed singly or in pairs, but occasionally in small family groups. Some have reputations for being extremely aggressive toward other species, for instance, for harassing and evicting roosting or nesting woodpeckers from tree cavities.

Most woodcreepers practice standard monogamy, with the sexes paired during the entire year and sharing equally in nesting chores. In some, however, no real pair-bonds are formed, and after mating, females nest alone. Nests are usually in tree crevices or hollow trunks and occasionally in arboreal termite nests. Parents line nests with rootlets, dried leaves, or wood chips.

Most woodcreepers are common birds. Only eastern Brazil's Moustached Woodcreeper is considered vulnerable. Woodcreepers are not hunted; the main threat to their populations is habitat loss.

BAR-WINGED CINCLODES
Cinclodes fuscus
6–7.5 in (15–19 cm)
South America

PALE-LEGGED HORNERO
Furnarius leucopus
7 in (18 cm)
South America

ANDEAN TIT-SPINETAIL
Leptasthenura andicola
6.5 in (17 cm)
South America

PEARLED TREERUNNER
Margarornis squamiger
6.5 in (16 cm)
South America

STREAKED XENOPS
Xenops rutilans
4.5 in (12 cm)
Central America, South America

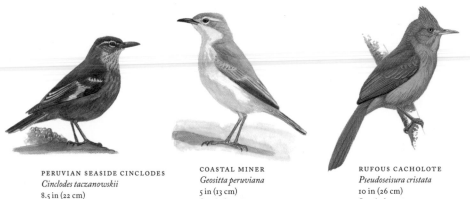

PERUVIAN SEASIDE CINCLODES
Cinclodes taczanowskii
8.5 in (22 cm)
South America

COASTAL MINER
Geositta peruviana
5 in (13 cm)
South America

RUFOUS CACHOLOTE
Pseudoseisura cristata
10 in (26 cm)
South America

Ovenbirds

OVENBIRDS are an interesting, if paradoxical, neotropical group; its member species are quite numerous, and diverse in many ways, but also, with a few significant exceptions, largely unnoted and unvalued in the regions in which they occur. They are mainly insect-eating brown birds that live and forage on or near the ground. Leading to their inconspicuousness are their drab, often cryptic plumage; their typical occupation of habitats in which the camouflaging effect is maximized; their unobtrusive, usually stealthy behavior; and their usually unmelodic, undistinguished vocalizations. Although visually and vocally nondescript, ovenbirds are widely recognized by ornithologists and naturalists for their ecological diversity (their varied foraging behavior, for instance, and their occupation of essentially all terrestrial habitats of South America) and, especially, for their nests. These nests are variable in architecture, construction materials, and placement (see below), but some are quite impressive, particularly the mud/clay and straw Dutch oven–like nests that provide the group's common name. (Because of its large oven nests that it often places on human-crafted structures such as fence-posts, utility poles, and roofs, the plain-looking Rufous Hornero is very familiar to local people over a broad swath of South America; indeed, it is Argentina's national bird.)

The ovenbird family, Furnariidae, has 240 species, distributed from southern Mexico to southern South America. Ovenbirds certainly vary somewhat in physical form, but their fairly uniform coloring, with various shades of brown, reddish brown, and gray, some with streaks or spots, and typically skulking behavior, render the group a frustrating one for bird-

Distribution:
Neotropics

No. of Living
Species: 240

No. of Species
Vulnerable,
Endangered: 15, 12

No. of Species Extinct
Since 1600: 0

watchers. These birds are hard to observe and sometimes difficult to tell apart. The family, which includes birds called miners, earthcreepers, cinclodes, spintails, thornbirds, canasteros, xenops, treerunners, treehunters, foliage-gleaners, and leaftossers, among others, is most diverse in the Andes and in south-central South America (Brazil supporting about 100 species).

Ranging from 4 to 10 inches long (10 to 26 cm), ovenbirds have short, rounded wings, often long tails (that may be "spiny" at the end), and strong legs and feet. Many have wing bands and tail patches that are seen mainly when the birds fly. The sexes within a species look alike.

Ovenbirds occur in all land habitats within their range, from sea level to the highest mountain elevations that support vertebrate life, and from wet forests to dry deserts. But the majority are tropical forest residents. Most ovenbirds live low in their habitats or even on the ground, but a few are regular in the canopy of tall forests. Some will perch acrobatically or move quickly along branches to reach food items; many will use a foot to clamp food against a branch or the ground before eating it. They all consume insects, mainly taking their prey from the undersides of leaves and branches. Some (especially cinclodes) also regularly eat mollusks; some (treehunters) take small lizards and frogs; and some (miners) eat many seeds. Most ovenbirds apparently mate for life and stay year-round in pairs.

All ovenbirds are probably monogamous, with members of a pair sharing nest construction, egg incubation, and provisioning young. Virtually all species build closed nests with side entrances. Some, such as xenops, make nest tunnels in vertical banks; others use tree cavities or construct elaborate structures of twigs or mud. For instance, Pale-legged Horneros build domed nests of clay, about 8 inches (20 cm) high, usually placed on tree branches. The sun bakes the clay hard and few predators can break into these strong, durable nests, which can last several years. Thornbirds build enormous, bulky, hanging nests of interlaced dry twigs placed in large bushes or trees. Shaped like a boot, the single entrance is in the "toe"; these nests usually have at least two chambers inside. One chamber serves as the incubating area for eggs and nestlings; and the others as dormitories or perhaps baffles for intruding predators.

Most ovenbirds are common and secure. Habitat loss, including deforestation and conversion of wild areas to agricultural use, is the prime threat they face. Several also have very small ranges, especially some in Brazil's Atlantic coastal forests and in Peru. Fifteen species are vulnerable and twelve are endangered (three critically).

Antbirds

ANTBIRDS are small to medium-size, rather drab inhabitants of the lower parts of forest vegetation in the Neotropics. They are recognized mainly for an intriguing feeding specialization: many regularly follow army ant swarms, snatching small creatures that leave their hiding places to avoid the predatory ants. The large antbird family, Formicariidae, distributed from southern Mexico to central Argentina, encompasses about 272 species. Some classifications consider only the 63 predominantly ground-dwelling antbirds, known as antpittas and antthrushes, to comprise family Formicariidae; and the 209 "typical" antbirds, which tend to perch and forage in low vegetation (including the antshrikes, antvireos, antwrens, antbirds, bare-eyes, and bushbirds), to constitute a separate family, Thamnophilidae. The strange compound names of these birds, such as antwrens and antthrushes, apparently arose because the naturalists who named them could not ascertain local names and believed they were similar in some ways to North American and Eurasian wrens and thrushes.

Like several other neotropical families, such as woodcreepers and ovenbirds, antbirds are clearly an ecologically significant group: they are species-diverse, widespread, ubiquitous in some major habitats, and abundant. Yet they are usually entirely unknown among the majority of people who live in the regions in which they occur. One reason this is true is that antbirds are clad predominantly in inconspicuous browns and grays, which allows them to blend into their chosen landscapes. Furthermore, they are often secretive and live amid dense forest vegetation (although a few exhibit bolder behavior and live at forest edges, so are sometimes seen

Distribution: Neotropics

No. of Living Species: 272

No. of Species Vulnerable, Endangered: 16, 20

No. of Species Extinct Since 1600: 0

BARRED ANTSHRIKE
Thamnophilus doliatus
6.5 in (16 cm)
Central America, South America

SCALED ANTPITTA
Grallaria guatimalensis
6.5 in (17 cm)
Central America, South America

GREAT ANTSHRIKE
Taraba major
8 in (20 cm)
Central America, South America

WHITE-PLUMED ANTBIRD
Pithys albifrons
4.5 in (12 cm)
South America

BLACK-SPOTTED BARE-EYE
Phlegopsis nigromaculata
6.5 in (17 cm)
South America

MATO GROSSO ANTBIRD
Cercomacra melanaria
6.5 in (16 cm)
South America

CHESTNUT-CROWNED ANTPITTA
Grallaria ruficapilla
7.5 in (19 cm)
South America

YELLOW-BROWED ANTBIRD
Hypocnemis hypoxantha
4.5 in (12 cm)
South America

around settlements). Consequently, they are often difficult to observe for any length of time; most, with their simple if sometimes loud songs, are typically heard but almost never seen. This, and the fact that many species look much alike, renders the group an especially difficult one for bird-watchers; much effort can be expended to catch even brief glimpses of these birds.

Antbirds range in length from 3 to 13.5 inches (7.5 to 34 cm). Generally, smaller ones are called antwrens and antvireos; midsize ones are called antbirds; and larger species are called antshrikes, antthrushes, and antpittas. Many of these birds vary in appearance and some are boldly patterned. Generally, however, males are dressed in understated browns and grays, and black and white; females are usually duller, with olive brown or chestnut predominating. Some species have bushy crests, bright red eyes, or patches of bright bare skin around the eye. Antbird bills vary in length, but many are fairly long and robust, and some are hooked. Their wings are short and their legs are strong (and long in ground dwellers).

Confined generally to thick vegetation in forest areas, antbirds are active animals that mainly eat insects, although some of the larger ones also consume fruit, small lizards, snakes, and frogs. Some are ant-followers, foraging in mixed-species flocks that pursue army ants. Others are ground or foliage feeders, rummaging through leaf litter, tossing dead leaves aside with their bills as they search for insects, or investigating the undersides of leaves and branches. Some stay on the shady forest floor, but many inhabit the lower to middle parts of forests, while a few are canopy birds. When those that follow ants for a living breed, they temporarily cease ant-following to establish and defend breeding territories; others maintain year-round territories.

Most antbirds, monogamous breeders, appear to pair for life. Many build cup nests from pieces of vegetation (grasses, twigs, dead leaves) woven together and supported in or suspended from a fork of branches low in a tree or shrub. Some species nest in tree cavities. Male and female antbirds share in nest-building, incubation of eggs and feeding young. Some ant-following species have abbreviated breeding, apparently to facilitate their following the very mobile army ant colonies. In several species, family groups remain together, male offspring staying with the parents, even after acquiring mates themselves.

About thirty-six antbird species, mostly Brazilian, are threatened (sixteen are considered vulnerable, sixteen are endangered, and four are critically endangered). Antbirds are not hunted or commercially exploited; the main threat to them is loss of their forest habitats.

BLACK-THROATED HUET-HUET
Pteroptochos tarnii
9 in (23 cm)
South America

CRESTED GALLITO
Rhinocrypta lanceolata
8.5 in (21 cm)
South America

WHITE-BREASTED TAPACULO
Scytalopus indigoticus
4.5 in (11 cm)
South America

OLIVE-CROWNED CRESCENT-CHEST
Melanopareia maximiliani
6 in (15 cm)
South America

CHESTNUT-CROWNED GNATEATER
Conopophaga castaneiceps
5.5 in (14 cm)
South America

BLACK-BELLIED GNATEATER
Conopophaga melanogaster
6 in (15 cm)
South America

Tapaculos;
Gnateaters

TAPACULOS are small to midsize, mostly dark-colored ground birds of the Neotropics. Many inhabit South America's Andes region; only one lives as far north as Costa Rica. Many in the family, Rhinocryptidae (with fifty-five species, including birds called huet-huets, gallitos, bristlefronts, and crescent-chests, among others), are obscure, having drab plumage, shy habits, and unmelodic vocalizations. They are noted for this obscurity, their diverse representation in southern South America, the small geographic ranges of many of them, and the way they often hold their tails in an upright, or cocked, position. The name *tapaculo* may have arisen from a Spanish word that refers to their cocked tails (or it may sound like the call of one of the Chilean tapaculos).

Ranging from 4 to 9 inches (10 to 23 cm) long, tapaculos have short, relatively heavy bills, short, round wings, and mostly short tails (but some have long tails). They have long legs with large, strong feet used for scratching at soil and moving leaf litter. They are clad in shades of brown and gray, plus, sometimes, black and white; some are mainly black. In most species, the sexes look alike, but in some, they differ conspicuously.

Tapaculos predominantly inhabit forests and woodlands, but a few species occur in more open habitats such as scrub areas and tussock grasslands. They tend to be secretive and are usually observable for only short periods, thus their habits are poorly known. They run and sometimes walk or hop along the ground, utilizing their weak flight abilities infrequently and then only briefly. Meals consist of insects and spiders, but also occasional centipedes, snails,

TAPACULOS

Distribution:
Neotropics

No. of Living
Species: 55

No. of Species
Vulnerable,
Endangered: 1, 3

No. of Species Extinct
Since 1600: 0

berries, and seeds. Most find food either by scratching the ground with their feet, exposing hidden prey, or taking it from the ground or low vegetation as they move. Most tapaculos stay in year-round territorial pairs.

The breeding behavior of only a few tapaculos is known. They are monogamous, with the males and females apparently sharing many, if not all, routine nesting chores. Some nests are deep cups of vegetation (rootlets, moss, twigs, grass), while others are globular structures with side or top entrances. Some are hidden in a bush, in tall grass, or in a rotten stump, but most are at the ends of burrows in the ground, which the tapaculos excavate themselves. A few nest in hollow tree trunks. Tapaculos are not often hunted; the main threat to their populations is habitat loss. Most species are numerous in the places they occur, but many have very limited distributions. Currently one species is vulnerable and three, all Brazilian, are endangered (two critically).

GNATEATERS

Distribution:
South America

No. of Living
Species: 8

No. of Species
Vulnerable,
Endangered: 0, 0

No. of Species Extinct
Since 1600: 0

GNATEATERS are small, stocky, short-tailed birds of forests and woodlands of the northern half of South America. They are perhaps best known, mainly among ornithologists, for all that is unknown about them: they are a small, obscure group, with a propensity for staying low in dim forest vegetation. They are little studied and are probably among the least-noticed of all bird groups. Bird-watchers are the only sizable collection of people concerned with gnateaters. Eight species comprise the family, Conopophagidae. They have slightly hooked, flattened bills, short, rounded wings, and relatively long legs. All are outfitted in various shades of brown and gray, and black and white; most have a conspicuous white stripe behind the eye. The sexes differ somewhat in appearance, particularly in that females lack the males' black patches.

Little is known of gnateater ecology and behavior. They are chiefly birds of the forest understory; some frequent woodlands, thickets, or forest edges. They are almost always observed within 5 feet (1.5 m) of the ground or on the ground. Food consists of small insects and spiders, and perhaps the occasional berry or tiny frog. They forage by pouncing from a perch to take prey on the ground or by reaching from a perch to take prey from nearby foliage. Gnateaters are apparently monogamous; nests are shallow cups of vegetation placed in a fork of thin branches or plant stalks. All gnateater species, within appropriate habitats, are common or fairly common birds; none are threatened.

Cotingas

COTINGAS are undeniably among the New World tropics' glamour birds, not only for their flashy colors but also for their diverse array of shapes, sizes, ecologies, and breeding systems. The group encompasses tiny warbler-size birds and large crow-size birds (in fact, the cotingas include some of the smallest and largest passerines, or perching birds, ranging from 3 to 19 inches [8 to 48 cm]). The group includes species that eat fruit and insects but also some that eat only fruit, which is rare among birds. In some species of cotingas the sexes look alike; however, in many the males are spectacularly attired in bright spectral colors and females are plain. The group also includes territorial species that breed monogamously and lekking species that breed promiscuously (see below), and without doubt, some of the strangest-looking birds of neotropical forests. The family, Cotingidae, which is closely allied with the manakins and American flycatchers, contains about seventy species, and occurs from southern Mexico south to northern Argentina. (The Sharpbill, a small olive and yellow bird with a sharply pointed bill, is sometimes placed in its own family, Oxyruncidae, but here I follow classifications that consider it a cotinga. Likewise, the three species of plantcutters, smallish finchlike birds, sometimes placed in a separate family, Phytotomidae, are included here.)

The diversity of the cotingas is reflected in their names. In addition to species called cotingas, there are bellbirds (named for the loud bell-like vocalizations of one species), cocks-of-the-rock, purpletufts, umbrellabirds, fruiteaters, fruitcrows, mourners, pihas, and the Capuchinbird. Perhaps the only generalization that applies to all is that they have short legs and

Distribution: Neotropics

No. of Living Species: about 70

No. of Species Vulnerable, Endangered: 11, 6

No. of Species Extinct Since 1600: 0

BLACK-NECKED RED-COTINGA
Phoenicircus nigricollis
9 in (23 cm)
South America

PLUM-THROATED COTINGA
Cotinga maynana
7.5 in (19 cm)
South America

RED-CRESTED COTINGA
Ampelion rubrocristatus
8.5 in (21 cm)
South America

GREEN-AND-BLACK FRUITEATER
Pipreola riefferii
7 in (18 cm)
South America

BARE-NECKED FRUITCROW
Gymnoderus foetidus
13–15 in (33–38 cm)
South America

GUIANAN COCK-OF-THE-ROCK
Rupicola rupicola
10.5 in (27 cm)
South America

POMPADOUR COTINGA
Xipholena punicea
8 in (20 cm)
South America

SHARPBILL
Oxyruncus cristatus
6.5 in (17 cm)
South America

relatively short, rather wide bills, the better to swallow fruits. Males of many species are quite ornate, with patches of gaudy plumage in unusual colors. For instance, some of the typical cotingas are lustrous blue and deep purple, and some are all white; others are wholly black, or green and yellow, or largely red or orange. Females are usually duller and plainer. Among the most unusual-looking cotingas are the bellbirds, some of which have wormlike wattles hanging from their heads, and the umbrellabirds, which have umbrella-shaped black crests.

Cotingas live primarily in tropical and subtropical forests, usually staying high in the canopy. They are fruit specialists, taking small to medium-size fruits from trees, often while hovering. Some, such as the pihas and fruitcrows, supplement their fruit diet with insects taken from treetop foliage, and a few catch flying insects. Others, particularly the fruiteaters and bellbirds, feed exclusively on fruit. Pure fruit eating has benefits: fruit is conspicuous, abundant, easy to eat, and in the tropics, usually always available. Fruit eating can also be problematic: fruit is relatively low in protein so nestlings fed only fruit grow very slowly. Cotingas feed heavily from palms and laurels and from plants of the blackberry/raspberry family. Some cotingas may live solitarily; others apparently live in pairs or small groups; fruitcrows, for example, live in highly social flocks of five to ten individuals.

Some cotingas pair up, defend territories, and breed conventionally in apparent monogamy. But others, such as umbrellabirds, bellbirds, cocks-of-the-rock, and pihas, are lekking species, in which males individually stake out display trees and repeatedly perform vocal and visual displays to attract females. Females enter display areas, assess the jumping and calling males, and choose ones with which to mate. Females leave after mating and then nest and rear young alone. Nests, usually placed in trees or bushes, are generally small, open, and inconspicuous. Some nest cups are made of loosely arranged twigs, some of mud, and some of pieces of plants.

Many cotingas have very small geographical ranges. This, in concert with the rapid forest loss that has occurred in many parts of their range during the last half century, has severely depressed the populations of many species, including some spectacular ones such as the Turquoise Cotinga, Three-wattled Bellbird, and Bare-necked and Long-wattled Umbrellabirds. In total, eleven cotingas are vulnerable and six endangered; one of the latter, the smallest cotinga, Brazil's Kinglet Calyptura, is critically endangered, with a wild population of fifty or fewer.

ROUND-TAILED MANAKIN
Pipra chloromeros
4.5 in (11 cm)
South America

LONG-TAILED MANAKIN
Chiroxiphia linearis
4.5 in (12 cm)
Central America

WHITE-CROWNED MANAKIN
Pipra pipra
4 in (10 cm)
Central America, South America

THRUSH-LIKE MANAKIN
Schiffornis turdinus
6.5 in (16 cm)
Central America, South America

HELMETED MANAKIN
Antilophia galeata
5.5 in (14 cm)
South America

BAND-TAILED MANAKIN
Pipra fasciicauda
4.5 in (11 cm)
South America

Manakins

MANAKINS are a neotropical group of small, compact, stocky birds with short bills and tails, big eyes, and two attention-grabbing features: brightly colored plumage and some of the most elaborate courtship displays practiced by birds. Some male manakins are outstandingly beautiful, predominantly glossy black with brilliant patches of bright orange-red, yellow, or blue on their heads, throats, or both. Some have deep blue on their undersides and/or backs. The exotic appearance of male manakins is sometimes enhanced by long, streamerlike tails, up to twice the length of the body, produced by the elongation of two of the central tail feathers. Females, in contrast, are duller and less ornate, often olive green, yellowish, and/or gray. To accompany courtship displays, the wing feathers of some species, when moved in certain ways, make whirring or snapping sounds. Manakins, family Pipridae (closely related to the cotingas and American flycatchers), occur only in southern Mexico and Central and South America; there are about fifty-three species, including some called tyrant-manakins and piprites. They range in length mainly from 3.5 to 6 inches (9 to 16 cm); the male Long-tailed Manakin of Central America, when its long tail streamers are included, is a few inches longer.

Manakins are very active birds, occurring chiefly in warmer, lowland forests, although some range up into cloud forests; most manakins are tropical but some inhabit subtropical regions. Residents of the forest understory, they eat mostly small fruits, which they pluck from bushes and trees while in flight, and they also take insects from foliage. They are fairly social during feeding and other daily activities, but males and females

Distribution:
Neotropics

No. of Living
Species: 53

No. of Species
Vulnerable,
Endangered: 2, 2

No. of Species Extinct
Since 1600: 0

do not pair. Rather, males mate with more than one female, and females probably mate with multiple males. After mating, females build nests and rear young by themselves. During the breeding season, males, singly, in pairs, or in small groups, stake out display sites on tree branches, in bushes, or on cleared patches of the forest floor, and spend considerable amounts of time giving lively vocal and visual displays, trying to attract females. An area that contains several of these performance sites is called a lek.

At the lek, male manakins "dance," performing rapid, repetitive, acrobatic movements, sometimes making short up and down flights, sometimes rapid slides, twists, and turns, sometimes hanging upside down on a tree branch while turning rapidly from side to side and making snapping sounds with their wings. The details of a male's dance differ from species to species. Females, attracted to leks by the sounds of male displays and by their memories of lek locations—the same traditional forest sites are used from one year to the next—examine the energetically performing males with a critical eye and then choose the ones with which they want to mate. Sometimes the females make the rounds several times before deciding. In a few species, two and sometimes three males join together in a coordinated dance on the same perch. In these curious cases, one male is dominant, and only the dominant of the duo or trio eventually gets to mate with interested females. After mating the female builds a shallow cup nest that she weaves into a fork of tree branches, 3 to 50 feet (1 to 15 m) off the ground. She incubates her eggs and rears the nestlings herself, bringing them fruit and insects.

Some manakins have very small ranges in South America and this, combined with rapid forest loss and degradation, is placing some species in jeopardy. Only four, however, are currently threatened (two vulnerable and two endangered). Two of them are interesting cases. The Golden-crowned Manakin is known only from a tiny area of forest south of the Amazon River in central Brazil. It is rarely if ever seen; one was collected in 1957, and what biologists know about the species comes from this single specimen. The Araripe Manakin of eastern Brazil was first discovered and named in 1998; because its population is tiny and the area in which it occurs is under pressure from recreational development, it is considered critically endangered.

New World Flycatchers

NEW WORLD FLYCATCHERS comprise a huge group of birds broadly distributed over most habitats from Alaska and northern Canada to the southern tip of South America. The family, Tyrannidae, is considered among the most diverse of avian groups. With between 390 and 425 species (more than any other bird family), depending on which classification scheme is followed, flycatchers typically contribute a hefty percentage of the avian biodiversity in every locale where they occur. For instance, flycatchers make up fully a tenth of the land bird species in South America, and a quarter of Argentinean species. The group is also called tyrant flycatchers or American flycatchers, which differentiates it from Old World flycatcher families.

New World flycatchers include birds called elaenias, pewees, tyrannulets, spadebills, attilas, monjitas, kingbirds, pygmy-tyrants, tit-tyrants, tody-tyrants, chat-tyrants, ground-tyrants, becards, and tityras, among others. They employ various feeding methods but are united in being predominantly insect-eaters. Certainly, though, they are best known for making aerial sallies to catch insects on the wing, that is, for flycatching. Their bills are usually broad and flat, the better to snatch flying bugs from the air. Tail length is variable, but some species have long, forked tails, which probably facilitate rapid insect-catching maneuvers. Flycatchers vary in length from 2.5 to 12 inches (6 to 30 cm); a few, such as the Scissor-tailed Flycatcher, have extra-long tails that can add another six or more inches (15 cm). At their smallest, flycatchers are some of the world's tiniest birds, smaller than some hummingbirds; the Short-tailed Pygmy-Tyrant is the smallest passerine, or perching, bird.

Distribution:
New World

No. of Living
Species: 425

No. of Species
Vulnerable,
Endangered: 14, 12

No. of Species Extinct
Since 1600: 0

BLACK PHOEBE
Sayornis nigricans
7 in (18 cm)
North America, South America

FORK-TAILED FLYCATCHER
Tyrannus savana
11–15.5 in (28–40 cm)
Central America, South America

WHITE-RUMPED MONJITA
Xolmis velata
7.5 in (19 cm)
South America

SIRYSTES
Sirystes sibilator
7 in (18 cm)
South America

SPOTTED TODY-FLYCATCHER
Todirostrum maculatum
4 in (10 cm)
South America

BLACK-BACKED WATER-TYRANT
Fluvicola albiventer
5.5 in (14 cm)
South America

EASTERN KINGBIRD
Tyrannus tyrannus
8 in (20 cm)
North America, South America

PINE FLYCATCHER
Empidonax affinis
5.5 in (14 cm)
Mexico, Central America

GREAT KISKADEE
Pitangus sulphuratus
9 in (23 cm)
North America, South America

PLAIN-CAPPED GROUND-TYRANT
Muscisaxicola alpina
7.5 in (19 cm)
South America

SCISSOR-TAILED FLYCATCHER
Tyrannus forficatus
7.5–14 in (19–35 cm)
North America, Central America

LONG-TAILED TYRANT
Colonia colonus
7–10 in (18–25 cm)
Central America, South America

ROYAL FLYCATCHER
Onychorhynchus coronatus
6.5 in (17 cm)
Central America, South America

VERMILION FLYCATCHER
Pyrocephalus rubinus
6 in (15 cm)
North America, South America

DUSKY FLYCATCHER
Empidonax oberholseri
6 in (15 cm)
North America

SULPHUR-BELLIED FLYCATCHER
Myiodynastes luteiventris
8 in (20 cm)
North America, South America

TUFTED FLYCATCHER
Mitrephanes phaeocercus
5 in (13 cm)
Mexico, Central America, South America

GREATER PEWEE
Contopus pertinax
8 in (20 cm)
North America, Central America

CARIBBEAN ELAENIA
Elaenia martinica
6.5 in (17 cm)
West Indies

LOGGERHEAD KINGBIRD
Tyrannus caudifasciatus
10 in (25 cm)
West Indies

MASKED TITYRA
Tityra semifasciata
8.5 in (21 cm)
Mexico, Central America, South America

ROSE-THROATED BECARD
Pachyramphus aglaiae
7 in (18 cm)
North America, Central America

Most flycatchers are dull shades of gray, brown, or olive green; many species have some yellow in their plumage; relatively few are quite flashily attired in, for example, bright red. One set of common flycatchers, of which the best known is the broadly distributed Great Kiskadee, share a common, bright color scheme, with yellow breasts and black and white or dark heads. A great many of the smaller, drabber flycatchers, clad in olives and browns, are extremely difficult to tell apart in the field, even for experienced bird-watchers. Flycatcher sexes are usually similar in size and coloring.

Flycatchers occur in a large array of habitats, from high-elevation mountainsides, to lowland moist forests and rainforests, to grasslands, marshes, and mangrove swamps. Most obtain the majority of their food by employing the classic flycatching technique of perching motionless on tree or shrub branches or on fences or utility wires, then darting out in short, swift flights to capture insects in the air. They then return time and again to the same perch to repeat the process. Many also take insects from foliage as they fly through vegetation, and some supplement their diets with berries and seeds. Larger flycatchers often will also take small frogs and lizards, and even small fish and tadpoles from the edges of lakes and rivers. Some species change diet with season; the Eastern Kingbird, for instance, eats insects in summer in North America, but after migrating to the Amazon region, switches to mostly fruit in winter. A few species eat only fruit. Most flycatchers inhabit exclusive territories that mated pairs defend for all or part of the year. Many species are migratory, breeding in temperate regions, including southern South America, but spending nonbreeding months in the tropics.

Most flycatchers are monogamous. Many are known for spectacular courtship displays, males impressing females by engaging in aerial acrobatics. Some build cup nests, roofed nests, or globular hanging nests placed in trees or shrubs. Others construct mud nests that they attach to vertical surfaces such as rock walls, and some nest in holes in trees or rocks. Only females incubate eggs, but usually both sexes construct nests and feed young.

New flycatcher species are still being discovered and described scientifically, some as recently as the late 1990s. Although many flycatchers are abundant, with wide ranges, several species are undergoing precipitous declines. This is particularly true in South America, as flycatchers' habitats are increasingly altered for development and agriculture, and trees are cut for lumber and firewood. Currently fourteen are vulnerable and twelve endangered (two critically, both Brazilian tyrannulets).

BROWN CREEPER
Certhia americana
5 in (13 cm)
North America, Central America

WHITE-THROATED TREECREEPER
Cormobates leucophaeus
6.5 in (17 cm)
Australia

BROWN TREECREEPER
Climacteris picumnus
6.5 in (17 cm)
Australia

BLACK-TAILED TREECREEPER
Climacteris melanura
7 in (18 cm)
Australia

STRIPE-SIDED RHABDORNIS
Rhabdornis mysticalis
6.5 in (16 cm)
Philippines

Creepers; Australasian Treecreepers; Rhabdornises

CREEPERS, family Certhiidae, are small forest birds of north temperate regions that also occur in some forested parts of Africa. They are most often recognized for their characteristic foraging method. A bird flies to the bottom of a tree trunk, spirals its way up it nearly to the top, then drops down to the bottom of another, nearby, tree, and spirals up again. Creepers' long toes and deeply curved claws allow them to cling tightly to vertical trunks. There are seven species, five of which are called treecreepers. Two are very abundant, widespread, and well known: the Brown Creeper, which occurs over most of North America and is the New World's only family member, and the Eurasian Treecreeper, which is broadly distributed from Britain eastward to Japan.

About 5 to 6 inches (13 to 15 cm) in length, creepers are brown and black striped above and light below, with very long, slender, down-curved bills and long tails with pointed feathers. Males and females look alike. The Spotted Creeper of India and Africa is slightly different, also dark above and light below, but spotted all over (it was formerly placed in its own family). With their slender, curved bills, creepers probe into crevices in tree bark for insects and spiders. They place their cup nests of twigs and other vegetation behind loose pieces of bark or in crevices. Only the female incubates; both sexes feed young. None of the creepers are threatened.

AUSTRALASIAN TREECREEPERS, family Climacteridae, are small (5 to 7.5 inches. [13 to 19 cm]), stocky, brownish forest birds with tan wing bars, slender, ever-so-slightly down-curved bills, and long toes, which make their livings searching for insects on tree trunks and branches. There are only

CREEPERS

Distribution:
North America,
Eurasia, Africa

No. of Living
Species: 7

No. of Species
Vulnerable,
Endangered: 0, 0

No. of Species Extinct
Since 1600: 0

AUSTRALASIAN
TREECREEPERS

Distribution:
Australia, New Guinea

No. of Living
Species: 7

No. of Species
Vulnerable,
Endangered: 0, 0

No. of Species Extinct
Since 1600: 0

RHABDORNISES

Distribution:
Philippines

No. of Living
Species: 3

No. of Species
Vulnerable,
Endangered: 0, 0

No. of Species Extinct
Since 1600: 0

seven species, six of which are endemic to Australia and one to New Guinea. Male and female look slightly different.

Treecreepers hop and run over tree trunks and branches, using their slender bills to poke and probe for insects, especially ants. They forage in predictable patterns, spiraling up tree trunks, then flying down to near the bottom of the trunk of the next tree and repeating the process, or moving along thick tree branches from the base outward. Their strong legs and large feet and claws permit them to grip bark so tightly that they can even move along the undersides of horizontal branches. Some also forage on the ground. Treecreepers are sedentary birds, holding permanent territories on which they forage and breed.

Most treecreepers breed in small communal groups (although at least one species breeds in solitary pairs). "Auxiliary" males, presumably related to the main breeding pair, help feed the young (and sometimes the incubating female). Both sexes build the nest, a saucer-shaped mat of leaves, bark, fur, and feathers placed in a tree hollow. Usually only the female incubates eggs. None of the Australasian treecreepers are threatened, although three of the six Australian species are considered vulnerable in various Australian states owing to their dependence on forest habitats that are increasingly cleared or degraded.

RHABDORNISES, once called Philippine Creepers, are small forest and forest-edge birds that are endemic to the Philippines. There are three species, comprising family Rhabdornithidae. They are about 6 inches (16 cm) long, brown above, light below, and streaked, with straight, narrow, dark bills that have slightly hooked tips. Males and females look alike or almost so.

The appellation "creeper" for these birds is now frowned upon because rhabdornises do not usually creep. Rather they feed by hopping and jumping along tree branches, eating insects and some seeds and small fruit, and perhaps the odd small frog. Sometimes they catch insects in the air, and groups have been seen feeding on flying termites in this manner. Usually occurring in small groups or sometimes combined with mixed-species foraging flocks, rhabdornises are very active and usually remain high in the leafy parts of trees. Before nightfall, some gather into large roosting groups, at times numbering in the hundreds. Rhabdornis breeding is not well known, but they are thought to be monogamous and to nest in tree cavities. None of the three rhabdornises are considered threatened, although two of them are fairly uncommon.

Lyrebirds;
Scrub-birds

LYREBIRDS are large, rather plain-looking brown and gray forest birds that nonetheless manage to be on the viewing wish lists of most bird-watchers who visit Australia, the only place they occur. The interest stems from their spectacular long tails, which are used in world-renowned courtship displays, their limited distribution, and their secretiveness, which makes spotting them quite challenging. The lyrebird family, Menuridae, has only two species. Both are pheasant-size, ranging up to 35 inches (90 cm) long, with big, powerful feet. The males and females look alike, but females are a bit smaller with less elaborate tails. The Superb Lyrebird, which occurs along the southern half of Australia's eastern seaboard, is fairly common; Albert's Lyrebird is uncommon and occurs only in one small portion of the eastern seaboard.

Foraging on the forest floor either alone or in small parties, lyrebirds use their feet to dig into soil (to 5 inches [12 cm] deep) to find and eat worms, spiders, insects and insect larvae, and millipedes, among other invertebrates. They also tear apart rotting wood on the forest floor, looking for food. Brown lyrebirds are camouflaged on the ground and further protect themselves with their shy ways. When alarmed, they tend to run away; they are weak flyers, usually managing only short, clumsy flights. They do, however, roost overnight high in trees, jumping up, branch by branch. Lyrebirds are most famous for their courtship displays and vocal mimicry. They sing long, loud songs that often include mimicked parts of the vocalizations of such birds as whipbirds, kookaburras, cockatoos, and currawongs. Lyrebirds are largely sedentary, males maintaining territories during breeding periods.

LYREBIRDS

Distribution:
Australia

No. of Living
Species: 2

No. of Species
Vulnerable,
Endangered: 1, 0

No. of Species Extinct
Since 1600: 0

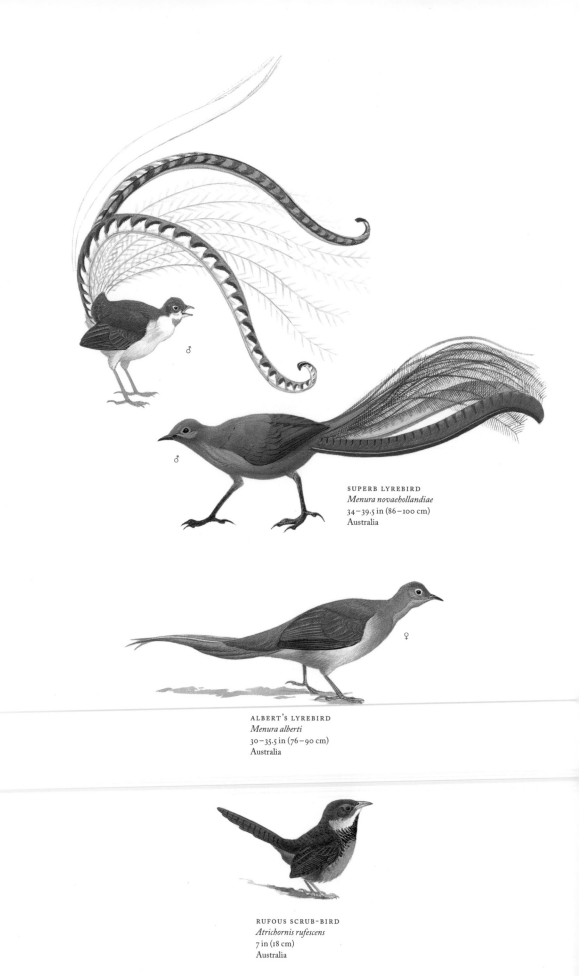

SUPERB LYREBIRD
Menura novaehollandiae
34–39.5 in (86–100 cm)
Australia

ALBERT'S LYREBIRD
Menura alberti
30–35.5 in (76–90 cm)
Australia

RUFOUS SCRUB-BIRD
Atrichornis rufescens
7 in (18 cm)
Australia

Lyrebirds are promiscuous breeders. Males in their territories give vocal and visual displays to attract females. They strut around on the ground, on low tree branches, and, in the Superb Lyrebird, on raised earthen mounds, singing and displaying their gaudy tail (which resembles a Greek lyre). The central part of the long tail is held, spread and fanlike, over the head, and the two large, boldly patterned side feathers point out to either side. When a female approaches, the male quivers the tail feathers, and there is much jumping and circling. After mating, the female departs to nest and raise the single young on her own. She builds a nest of sticks, moss, bark, and rootlets on or near the ground in a vegetation tangle, on an earthen bank, or in a treefern, among other places.

The lyrebirds are restricted to relatively small areas of moist forest that are increasingly cleared or degraded. Albert's Lyrebird is considered vulnerable because it occurs over a very small range and its numbers (of only a few thousand) are believed to be declining. Lyrebirds were killed in great numbers during the nineteenth century for their tail feathers.

SCRUB-BIRDS, like the lyrebirds, are endemic to Australia, and bird-watchers seek them out because of their rarity. The family, Atrichornithidae, contains only the Rufous Scrub-bird and Noisy Scrub-bird, midsize brown birds with black markings. They both occur over extremely small areas and have small populations, thus rendering the Atrichornithidae one of the globe's rarest bird families. Scrub-birds are 6.5 to 9 inches (16 to 23 cm) long, with stout, pointed bills and sturdy legs; females are duller and smaller than males.

Secretive, fast, and alert, scrub-birds run and creep through dense undergrowth on the forest floor. They poke about in the leaf litter, turning pieces over with their bill, taking mainly insects, but also occasional small lizards and frogs. They are weak flyers, rarely moving more than a few yards at a time in the air. Scrub-birds breed monogamously or polygynously (some males mating with more than one female). Females appear to undertake all the nesting chores: building the domed nest of vegetation hidden in low undergrowth, incubating eggs, and feeding young. Although both scrub-birds have small populations, only the Noisy Scrub-bird, of southwestern Australia, is considered vulnerable; conservation efforts during the past decades have increased its chances of survival.

SCRUB-BIRDS

Distribution:
Australia

No. of Living
Species: 2

No. of Species
Vulnerable,
Endangered: 1, 0

No. of Species Extinct
Since 1600: 0

TOOTH-BILLED BOWERBIRD
Scenopoeetes dentirostris
10.5 in (27 cm)
Australia

GREEN CATBIRD
Ailuroedus crassirostris
9.5–12.5 in (24–32 cm)
Australia

REGENT BOWERBIRD
Sericulus chrysocephalus
10–12 in (25–30 cm)
Australia

SATIN BOWERBIRD
Ptilonorhynchus violaceus
11–12.5 in (28–32 cm)
Australia

GREAT BOWERBIRD
Chlamydera nuchalis
13.5–15 in (34–38 cm)
Australia

WESTERN BOWERBIRD
Chlamydera guttata
10–11 in (25–28 cm)
Australia

FIRE-MANED BOWERBIRD
Sericulus bakeri
10.5 in (27 cm)
New Guinea

Bowerbirds

If you were to travel to Australia, listen in on the excited dinner-table conversation at a nature lodge, and hear that a "bower" was located that day, you might wonder just what these people were talking about. Bowers, elaborate courtship structures, and the birds that make them, are one of the preeminent Australian attractions for bird-watchers and other nature enthusiasts. Male BOWERBIRDS, sometimes spectacularly colored, actually build bowers on their territories to attract females and convince them to mate. Some, for instance, erect walls of twigs that are stuck into the ground, walls that form a structure that resembles an actor's stage or a marriage bower; the walls may be painted by the male with his saliva that has been colored with compounds such as charcoal or leaf juices. The male may place around the bower small objects, both natural and artificial, that he has collected to impress females and enhance his courtship displays. With his bower constructed (or spruced up, if he is using an old one), a male vocalizes and cavorts to attract passing females. A female detects a male, approaches, evaluates his bower and courtship displays and, if convinced that he is a high-quality individual, mates. In some ways, this interaction can be viewed as the male bowerbird using sophisticated "tools" to get what he wants, and the comparison to human behavior (males attempting to secure mates by tempting them with real estate and offering them collected objects, including, perhaps, attractively colored rocks) has been made more than a few times. Owing to these birds' incredible bower construction and associated mating behavior, some biologists consider bowerbirds to be among the most advanced of birds.

Distribution:
Australia, New Guinea

No. of Living
Species: 20

No. of Species
Vulnerable,
Endangered: 1, 0

No. of Species Extinct
Since 1600: 0

There are twenty bowerbird species (family Ptilonorhynchidae). Ten occur in Australia; eight are endemic, and two are shared with New Guinea, where the remaining species occur. Bowerbirds are medium-size to fairly large birds (8 to 15 inches [21 to 38 cm]), chunky, with shortish tails and legs; their bills are short, heavy, and, in some, slightly down-curved or hooked. Females in certain species are plainly colored, mainly brown and streaked, but in others, the sexes look alike.

Bowerbirds are mostly denizens of wet forests but a few live in drier habitats, including scrubland, grassland, and open woodlands. They forage in trees and on the ground. Their diets tend mostly toward fruit and berries, but other plant materials, such as leaves and shoots, are also eaten. They also take insects, spiders, earthworms, and perhaps occasional small frogs. Most bowerbird males maintain exclusive territories, in the center of which they build their bowers. Many species form small communal flocks after breeding, sometimes raiding fruit orchards; others appear to be solitary during nonbreeding periods.

Bowerbirds that build bowers are divided into two types. *Maypole builders* construct single or twin towers of sticks; *avenue builders* make walled avenues of sticks, with cleared areas, or "platforms," at both ends. There is a relationship between bower complexity and male plumage: the brighter the plumage, the simpler the bower. All bower-builders decorate their bowers with collected objects placed on the ground. The types of objects collected, and their colors, depend on species. Males appear to collect things that more or less match their plumage colors. The striking blue black male Satin Bowerbird builds a simple avenue bower of sticks, usually painted black with saliva mixed with charcoal. He adorns the bower platforms with blue objects: berries, feathers, flowers and, these days, bottle tops and pen caps. Research shows that the more blue objects a male has, the more matings he obtains. Most bowerbirds are promiscuous breeders; after mating at a bower, a female goes off and nests by herself; males continue to advertise at their bowers, attempting to attract and mate with multiple females each breeding season.

None of Australia's bowerbirds are currently threatened, although many of them are confined to the country's eastern wet forests, which are increasingly cleared or degraded. One New Guinea species, the Fire-maned Bowerbird, is considered vulnerable because it occurs only over a small mountainous region and its population is shrinking.

Fairy-wrens and Grasswrens

FAIRY-WRENS, brightly colored and sometimes easily observed, are among the most charming avian inhabitants of the Australian region. They are small to medium-size insect-eaters that chiefly occupy shrubs, thickets, and undergrowth. A few are common denizens of forest edge areas, parklands, and picnic grounds, so are commonly seen and appreciated. The family, Maluridae, contains about twenty-eight species, including the less colorful emu-wrens and GRASSWRENS; twenty-two occur in Australia and the remainder in the New Guinea area. The group is considered closely related to the honeyeaters and thornbills.

These birds are 5.5 to 8.5 inches (14 to 22 cm) long, with small, short bills, short wings, and long tails (comprising half or more of each bird's total length) that are usually held stiffly upward. They often stand out because the males have patches of bright, iridescent blue. Male Splendid Fairy-wrens, mostly blue, and male Variegated Fairy-wrens, blue, black, and red, are considered by many bird fanciers to be among the world's most beautiful small birds. Females in the family look different from males; female fairy-wrens, for instance, are usually brownish. Grasswrens are mainly black and brown above with white streaking; some have white underparts. Emu-wrens, brownish with (in males) blue faces and upper breasts, are so-named because their long tail feathers are thought to be Emu-like, that is, they look coarse, loose, and messy. This look is due to the lack of tiny hooks that, in most feathers, hold feather barbs together, forming and stiffening the feathers.

Fairy-wrens occupy all kinds of terrestrial habitats, from rainforests to

Distribution: Australia, New Guinea

No. of Living Species: 28

No. of Species Vulnerable, Endangered: 2, 0

No. of Species Extinct Since 1600: 0

SUPERB FAIRY-WREN
Malurus cyaneus
5 in (13 cm)
Australia

VARIEGATED FAIRY-WREN
Malurus lamberti
4.5–5.5 in (11–14 cm)
Australia

RED-BACKED FAIRY-WREN
Malurus melanocephalus
4–5 in (10–13 cm)
Australia

WHITE-WINGED FAIRY-WREN
Malurus leucopterus
4.5 in (12 cm)
Australia

SPLENDID FAIRY-WREN
Malurus splendens
5 in (13 cm)
Australia

RED-WINGED FAIRY-WREN
Malurus elegans
6 in (15 cm)
Australia

STRIATED GRASSWREN
Amytornis striatus
6–7.5 in (15–19 cm)
Australia

desert grasslands and rocky hillsides, but they are most typical of grassland, thickets, and shrub areas. Grasswrens and emu-wrens primarily inhabit grasslands and shrublands. They forage on the ground and in shrubs, hopping as they search for insects and spiders to eat; grasswrens also eat seeds. Some also search for food on tree trunks and branches. Flying is kept to a minimum, most flights limited to within vegetation or from shrub to ground. Emu-wrens and grasswrens are more cryptic in their coloring and behavior than are fairy-wrens and so are less frequently seen.

Members of this family are mainly sedentary. Most live in highly social, communal groups of three to twelve individuals (but some species may live in monogamous pairs), usually consisting of a dominant male and female that breed and subordinate nonbreeders that help maintain the year-round territory and assist during nesting. These helpers are often male offspring from previous nests that stay with their parents for a year or more to help out before striking out on their own. Groups often sit on branches or utility wires side by side, touching each other, and they roost overnight like this; they also will groom each other. Groups move through their territory during the day, stopping to forage in various spots; in hot regions, most foraging is accomplished in early morning and late afternoon. When alarmed, these birds flee quickly into dense vegetation, then apparently drop to the undergrowth and sneak quietly away.

A breeding female fairy-wren builds a domed nest in a shrub or low thicket on her own, and incubates her eggs. Nests are constructed of leaves, twigs, grass, and bark. The female's mate and other members of her social group help feed and guard young. One danger to these birds is that cuckoos (which are brood parasites, laying their eggs in nests of other species, and having the host species raise their young) often lay in fairy-wren nests, effectively destroying many of their breeding efforts.

Most fairy-wrens, grasswrens, and emu-wrens are common birds. But for several species, not much is known about the health of their populations. Two Australian species with very limited ranges are considered vulnerable: the Mallee Emu-wren of southern Australia, which has had more than half its historical scrub and grassland habitat cleared for agriculture and livestock grazing, and the White-throated Grasswren, which is restricted to rocky escarpment habitats in northern Australia. The latter species has a small and fragmented population and is thought to be especially susceptible to brush fires.

WHITE-THROATED GERYGONE
Gerygone olivacea
4.5 in (11 cm)
Australia, New Guinea

BROWN GERYGONE
Gerygone mouki
4 in (10 cm)
Australia

YELLOW-THROATED SCRUBWREN
Sericornis citreogularis
5 in (13 cm)
Australia

WHITE-BROWED SCRUBWREN
Sericornis frontalis
4.5 in (12 cm)
Australia

YELLOW-RUMPED THORNBILL
Acanthiza chrysorrhoa
4.5 in (12 cm)
Australia

YELLOW THORNBILL
Acanthiza nana
4 in (10 cm)
Australia

SHY HEATHWREN
Hylacola cauta
5 in (13 cm)
Australia

SOUTHERN WHITEFACE
Aphelocephala leucopsis
4.5 in (12 cm)
Australia

SPOTTED PARDALOTE
Pardalotus punctatus
3.5 in (9 cm)
Australia

Thornbills,
Scrubwrens,
and Pardalotes

Nature-oriented visitors to the Australian region quickly notice that there are numerous kinds of tiny birds, including THORNBILLS, SCRUBWRENS, PARDALOTES, and others, that flit about the local trees and shrubs, and on the ground. Owing to their small sizes, agile natures, and mostly dull plumage, they are not often observed for long and are difficult to identify, even for journeymen bird-watchers, who often give up trying to differentiate the various species. Some in the group are warblerlike in their looks and behavior, and thus the entire assemblage is sometimes called the *Australian warblers*.

The family in question, Acanthizidae, contains about sixty-nine species and is distributed in Australia, New Guinea, Southeast Asia, New Zealand, and some Pacific islands; most of these, forty-eight, occur in Australia. These birds range in length from the tiny Weebill (Australia's smallest bird) at 3 inches (8 cm) long to bristlebirds at up to 10.5 inches (27 cm); most are 4 to 5 inches (10 to 13 cm). They tend to be dully, cryptically colored. Browns, olives, and grays predominate, so much so that American birders in the region are usually tempted to call them LBJs (little brown jobs) or, in the local vernacular, WLBBs, (wretched little brown birds). Some species usually hold their tails erect, like wrens, and so are called scrubwrens, heathwrens, or fieldwrens. Pardalotes (of which there are four species, all endemic to Australia; sometimes separated into their own family, Pardalotidae), also called diamond birds, are perhaps the most distinctive of the group, having very short bills and tails and being brightly patterned with patches of white,

Distribution: Australia, New Guinea, New Zealand

No. of Living Species: 69

No. of Species Vulnerable, Endangered: 3, 3

No. of Species Extinct Since 1600: 1

black, brown, red, and yellow. Male and female in all these species generally look alike or almost alike.

These birds inhabit mainly woodlands, rainforests, and mangroves, but some occur in shrubland, thickets, or grassland. Most species are arboreal insect-eaters; many also eat some plant materials, especially seeds. Scrubwrens and heathwrens feed on the ground or in lower parts of vegetation, often taking snails and crustaceans; they are seen alone, in pairs, or in small groups. Gerygones (also called fairy warblers) feed singly or in pairs on insects they find in the outer foliage of woodland trees. Thornbills, always very active, feed, usually in small groups, on insects and occasional seeds; some specialize on the tree canopy, some at lower levels of the forest, and some are more terrestrial feeders, foraging on fallen tree limbs and even on the ground. Pardalotes scurry about the high leaves of trees, especially those along watercourses, searching for insects; they often hang upside down to search a leaf or grab a bug. Whitefaces have short, strong bills for seed eating, but also take some insects. Other species in the family are called scrubtit, pilotbird, rock-warbler, and mouse-warbler, among other names. Many of these birds are sedentary, but some, such as the pardalotes, gather into flocks after breeding and make migratory or nomadic movements.

Most of these Australian warblers construct small domed nests of vegetation (grass, bark, stems, roots) that are placed on or in the ground, in tree cavities, or in shrubs or tree foliage. Some species have been well studied, but precise breeding information is unknown for several. Most are monogamous, but a good number are cooperative breeders, in which grown young from previous nests stay with their parents and help them maintain their breeding territories and feed young in subsequent nests. In many (gerygones and thornbills, for instance), the female builds the nest and incubates alone, but the male and, often, other members of the family group, help feed the young. In other groups, male and female share nesting duties more equally. Pardalotes breed in monogamous pairs or in small groups that include nest helpers. A cup or domed nest is placed in a burrow in the ground or in a tree hollow; male and female dig the burrow, build the nest, and usually share incubation.

These birds generally are secure. A few species are in trouble, mostly because of habitat destruction, but also sometimes owing to harmful effects of introduced plants and animals. Three species are considered vulnerable, and three, one bristlebird, one gerygone, and one pardalote, are endangered.

Honeyeaters and Australian Chats

Tourists in the Australian countryside who are at all aware of wildlife will certainly have their attention drawn to an abundant group of very noisy, active, aggressive, usually plain-looking birds known as HONEYEATERS. Comprising Australia's largest bird family, they are everywhere on the continent, occupying essentially all terrestrial habitats. These arboreal birds are so abundant and successful that in many woodland areas there are ten or more species present. The family, Meliphagidae (*meli* means "honey"; *phag* means "eater"), contains about 170 species that are distributed mostly in the Australia/New Guinea region, but also in parts of Indonesia, in New Zealand, and on many Pacific islands; about seventy, including five within a subgroup known as AUSTRALIAN CHATS (sometimes placed in their own family, Epthianuridae) occur in Australia, about sixty-five in New Guinea. The reason for this group's great success in the Australian region is related to its chief food source, plant nectar. These birds are specialists on feeding on nectar (also taking some fruit and insects), and nectar, which is mostly a sugar-water solution, is abundant in most habitats across the region.

Honeyeaters, which range in length from 5 to greater than 14 inches (13 to 35+ cm), are slender, streamlined birds with long, slim, down-curved bills. Most are attired in dull gray, greenish olive, or brown, often with streaks, so they are not the most visually glamorous of birds. But many have small, contrasting patches of yellow, and one group, genus *Myzomela*, is largely red. Many, such as the friarbirds, have areas of bare colored skin on the face, and

Distribution:
Australasia, Indonesia

No. of Living
Species: About 170

No. of Species
Vulnerable,
Endangered: 6, 4

No. of Species Extinct
Since 1600: 6

RED WATTLEBIRD
Anthochaera carnunculata
13–14 in (33–36 cm)
Australia

NOISY FRIARBIRD
Philemon corniculatus
12–14 in (31–35 cm)
Australia, New Guinea

SILVER-CROWNED FRIARBIRD
Philemon argenticeps
10.5–12.5 in (27–32 cm)
Australia

BLUE-FACED HONEYEATER
Entomyzon cyanotis
12 in (31 cm)
Australia, New Guinea

BELL MINER
Manorina melanophrys
7.5 in (19 cm)
Australia

NOISY MINER
Manorina melanocephala
10 in (26 cm)
Australia

YELLOW-THROATED MINER
Manorina flavigula
10 in (26 cm)
Australia

WHITE-NAPED HONEYEATER
Melithreptus lunatus
5.5 in (14 cm)
Australia

BRIDLED HONEYEATER
Lichenostomus frenatus
8.5 in (21 cm)
Australia

VARIED HONEYEATER
Lichenostomus versicolor
8 in (20 cm)
Australia, New Guinea

FUSCOUS HONEYEATER
Lichenostomus fuscus
6.5 in (16 cm)
Australia

BROWN HONEYEATER
Lichmera indistincta
5 in (13 cm)
Australia, New Guinea

BAR-BREASTED HONEYEATER
Ramsayornis fasciatus
5.5 in (14 cm)
Australia

SCARLET HONEYEATER
Myzomela sanguinolenta
4.5 in (11 cm)
Australia

RED-HEADED HONEYEATER
Myzomela erythrocephala
4.5 in (12 cm)
Australia, New Guinea

RUFOUS-BANDED HONEYEATER
Conopophila albogularis
4.5 in (12 cm)
Australia, New Guinea

SPINY-CHEEKED HONEYEATER
Acanthagenys rufogularis
9.5 in (24 cm)
Australia

WHITE-FRONTED HONEYEATER
Phylidonyris albifrons
6.5 in (17 cm)
Australia

EASTERN SPINEBILL
Acanthorhynchus tenuirostris
6.5 in (16 cm)
Australia

GIBBERBIRD
Ashbyia lovensis
5 in (13 cm)
Australia

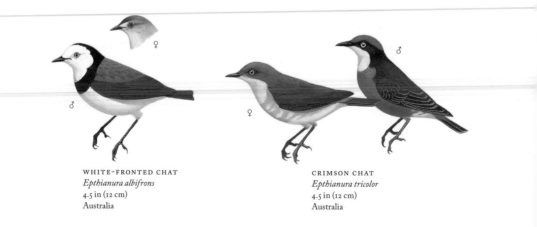

WHITE-FRONTED CHAT
Epthianura albifrons
4.5 in (12 cm)
Australia

CRIMSON CHAT
Epthianura tricolor
4.5 in (12 cm)
Australia

some, such as the wattlebirds, have hanging protuberances from the ear or eye area. In most, male and female look alike, but females are usually smaller. Australian chats are small (4 to 5 inches [10 to 13 cm]), brightly colored birds of open country that forage by walking along the ground, picking up seeds and insects in their slim, straight bills.

Mainly birds of forests and shrublands, honeyeaters are gregarious, pugnacious birds that forage in flowering trees and shrubs usually in small groups. They often squabble over feeding areas, sometimes chasing each other away from good food sources. Most honeyeaters have very long tongues that can be thrust deep into long, tubular flowers to collect nectar and into cracks between pieces of tree bark to gather other fluid foods. The tongue is tipped with a brushlike structure that soaks up nectar and other liquids like a mop; it is then "wrung out" against the roof of the mouth. Many of these birds can flick their tongues into flowers at rates of up to ten licks per second, thus quickly emptying flowers of nectar. Probably all honeyeaters include in their diet, to varying degrees, some insects (for protein). Some also take a good amount of fruit and berries. Other sugary substances that are parts of the diets of various species are honeydew (a sugary solution produced by some small insects after they feed on plant juices), manna (sugary granules from damaged eucalypt leaves), and lerp (sugary coatings of some sap-sucking insects).

Most honeyeater species nest alone in monogamous pairs but some nest in colonial groups of up to twenty or more pairs. Only the female builds the nest, a rough cup made of twigs and bark that is placed in a fork in a tree or shrub or hung from small branches. She also is the sole incubator; males help feed the young.

Most honeyeaters are abundant; a few are rare in parts of their ranges. Honeyeaters that eat fruit sometimes damage orchards and are occasionally persecuted for these actions. Currently six species are vulnerable and four endangered; several of these are restricted to islands, where they are subject to harmful effects of introduced species as well as habitat loss and degradation. About five honeyeaters in historic times occurred in Hawaii. They were mainly striking black and yellow birds with long tails and long, down-curved bills. As of the early 1980s, two species, known as 'o'os, still existed, one on Kauai, one possibly on Maui; all are now extinct, victims of habitat loss and introduced diseases.

JACKY-WINTER
Microeca fascinans
5 in (13 cm)
Australia, New Guinea

GRAY-HEADED ROBIN
Heteromyias albispecularis
6.5 in (17 cm)
Australia

RED-CAPPED ROBIN
Petroica goodenovii
4.5 in (12 cm)
Australia

HOODED ROBIN
Melanodryas cucullata
6.5 in (16 cm)
Australia

SOUTHERN SCRUB-ROBIN
Drymodes brunneopygia
8.5 in (22 cm)
Australia

SCARLET ROBIN
Petroica multicolor
5 in (13 cm)
Australia, Melanesia, Fiji

WESTERN YELLOW ROBIN
Eopsaltria griseogularis
6 in (15 cm)
Australia

WHITE-BREASTED ROBIN
Eopsaltria georgiana
6 in (15 cm)
Australia

Australasian Robins

AUSTRALASIAN ROBINS are a group of mostly small, commonly seen arboreal songbirds spread over Australia, New Guinea, New Zealand and some Pacific islands. In Australia, these birds are known mainly as that continent's perch-and-pounce insect-eaters. The great majority of them sit high or low on tree branches or other perches, flicking their wings and often moving their tails up and down or side to side, watching for potential prey. When a likely meal is spotted, the bird flies quickly and easily down, grabs the prey (chiefly insects and other invertebrates) from the ground or other surface, then returns to its perch to feast. These robins are not closely related to the New World robins, which are in the thrush family. Rather, they comprise their own midsize family, Petroicidae, with forty-five species; twenty occur in Australia, about twenty-five occur in New Guinea, and three occur in New Zealand.

Generally plumpish, mostly 4.5 to 6.5 inches (11 to 17 cm) long, Australasian robins have large, rounded heads, large eyes, square tails, and, in most, white wing bars. Their short bills are surrounded by small "whiskers" that presumably help funnel flying or running insects into the mouth. In Australia especially, these robins are often divided into several groups based on plumage color. There is a red group, including the Scarlet and Red-capped Robins, in which males have bright red or pinkish breasts (presumably furnishing the group name robin, after the European red-breasted robins); a yellow group, including the Western Yellow Robin; a brown group, including the Gray-headed Robin; and a miscellaneous group,

Distribution:
Australia, New Guinea,
New Zealand, some
Pacific islands

No. of Living
Species: 45

No. of Species
Vulnerable,
Endangered: 0, 1

No. of Species Extinct
Since 1600: 0

including the black and white Hooded Robin and blue gray White-breasted Robin. In most species, the sexes look alike; the main exception is the red robins, in which females are brownish.

Australasian robins occur mainly in forests and woodlands. In Australia, they occupy essentially all wooded habitats, but some inhabit savanna and even very open habitats such as shrublands, grasslands, and rocky sites. A few larger species (8 inches [21 cm] long) known as scrub-robins forage on the ground. They have relatively long legs and walk and run about, flipping over leaf litter to look for insects. Another group (genus *Microeca*), the most flycatcher-like of the family, spends a lot of time flying, chasing and grabbing insects on the wing; one of the six *Microeca* flycatchers is the gray brown Jacky-winter, which is common and familiar throughout most of rural Australia. Some robins, in pairs, maintain year-round territories; others spend nonbreeding periods in moving mixed-species flocks of insect-eating birds. Robins are graceful in flight, often flying low and in fast undulatory spurts.

Generally monogamous, Australasian robins breed in pairs, each couple defending a territory in which to nest. In most species, the female builds the nest and incubates the eggs, but in some, the male also incubates. Both sexes feed young. Nests are small, shallow, cup-shaped affairs, placed among a tree's small branches or in a fork (or on the ground, in the case of scrub-robins). Some nests are decorated with pieces of bark and lichens, perhaps to better camouflage them. The nests of the *Microeca* flycatchers are very small (the smallest among Australia's birds, for instance); *microeca* means "small house."

Most Australasian robins are common. Some, such as Australia's Gray-headed Robin, have very small distributions (in this case, in the limited rainforests of the country's far northeast), so may be vulnerable in the future. The only member of the family currently considered endangered is New Zealand's Black Robin, a small insect-eater that feeds on the ground and on low branches. Although it has a tiny population, the species is considered a conservation success story. Historically restricted to the Chatham Islands east of New Zealand's main islands, the Black Robin population crashed when cats and rats (introduced by human travelers) reached these remote outposts and killed adult birds and destroyed nests. By 1980 fewer than ten individuals existed. With intensive conservation efforts, including transferring the few survivors to nearby, mammal-free islands, the population is now above a hundred and increasing.

Whistlers

Distribution:
Australia, New Guinea,
Southeast Asia, New
Zealand, some Pacific
islands

No. of Living
Species: 57

No. of Species
Vulnerable,
Endangered: 1, 1

No. of Species Extinct
Since 1600: 1

WHISTLERS and their relatives are a midsize group of stout, mainly tree-living songbirds of Australasia, known for their insect-eating ways and their vocalizations. Many of them are considered among the region's outstanding singers. The family Pachycephalidae contains about fifty-seven species of whistlers, shrike-thrushes, shrike-tits, pitohuis, and bellbirds, among others, and is distributed throughout Australia, New Guinea, New Zealand, many Pacific islands, and parts of Southeast Asia. Twenty-four species occur in New Guinea, sixteen in Australia.

Known collectively as whistlers or thickheads, these birds are small to medium-size (5.5 to 10 inches [14 to 26 cm] long), with robust bodies, thick, rounded heads, and thick, strong bills. Some have bills with down-curved tips, like shrikes; hence the names shrike-thrushes and shrike-tits. Most of these birds (some whistlers and the shrike-thrushes and bellbirds) are outfitted in dull grays, browns, or reddish brown, whereas others (shrike-tits, some whistlers) are boldly marked with bright yellows and black and white. The Golden Whistler, for example, which is broadly distributed in Australia and on many Pacific islands, is quite striking and attractive. Some whistlers have crests or bare colored skin (wattles) hanging from the face. Male and female generally look slightly different, females being more dully colored. New Zealand's representatives of the family are three species known as mohouas (with the common names Whitehead, Yellowhead, and Pipipi); they are forest and scrub birds slightly smaller (5 to 6 inches [13 to 15 cm]) than most others in the group.

BORNEAN WHISTLER
Pachycephala hypoxantha
6.5 in (16 cm)
Borneo

GOLDEN WHISTLER
Pachycephala pectoralis
7 in (18 cm)
Australia, Melanesia, Fiji

CRESTED SHRIKE-TIT
Falcunculus frontatus
7 in (18 cm)
Australia

RUFOUS WHISTLER
Pachycephala rufiventris
6.5 in (17 cm)
Australia, New Caledonia

RUFOUS SHRIKE-THRUSH
Colluricincla megarhyncha
7 in (18 cm)
Australia, New Guinea

SANDSTONE SHRIKE-THRUSH
Colluricincla woodwardi
10 in (25 cm)
Australia

CRESTED BELLBIRD
Oreoica gutturalis
8.5 in (22 cm)
Australia

WHITEHEAD
Mohoua albicilla
6 in (15 cm)
New Zealand

Within their range, whistlers inhabit most areas with trees. They forage mainly for insects and insect larvae on tree trunks and branches, in tree foliage, and in some species, such as the Crested Bellbird, in leaf litter on the ground. The Crested Shrike-tit and shrike-thrushes use their powerful bills to strip pieces of bark away from trees, then look underneath for bugs. Larger species, such as the shrike-thrushes, also eat small frogs, lizards, bird eggs, and even tiny birds and mammals. One Australian whistler occupies mangrove areas and eats small crabs; at least one New Guinea species eats a lot of fruit. Many of the whistlers are solitary during nonbreeding periods; others apparently occupy year-round territories in pairs; and some, such as the shrike-tits and bellbirds, often travel around in small family groups.

Many whistlers are noted for their bursts of rich, strong, whistled songs. Some of these songs, because they are so loud and common, are quite familiar to local people; the Rufous Whistler's song, for instance, is a characteristic sound of spring over many parts of Australia. Crested Bellbirds (not to be confused with the New Zealand Bellbird, which is in the honeyeater family; nor with the New World bellbirds, which are cotingas) are known especially for the male's loud, far-carrying, liquid-noted song.

Whistlers are generally monogamous, breeding in pairs. An exception is the Crested Shrike-tit, which breeds often in small communal groups, assisted by helpers that are presumably grown young from previous nests of the breeding pair. Nests are usually coarse cups of twigs, bark, grass, moss, and rootlets placed in a tree fork or crevice; ground species nest in rock cavities. Both sexes typically carry out all or most nesting duties. The Crested Bellbird has an unusual nest-decorating habit: it squeezes hairy caterpillars to incapacitate them, and then, for reasons known only to the bellbird, hangs them from the nest rim.

Most species of whistlers are common and secure. A few are at risk in some parts of their ranges owing to destruction of their forest or woodland habitats. Only two are currently threatened. New Zealand's Yellowhead is considered vulnerable, with a declining population. Since European settlement, it has disappeared from about 75 percent of its range, and introduced mammal predators—especially stoats (weasels)—continue to destroy many of its nests. The Sangihe Shrike-thrush, a forest species, is critically endangered; its tiny population of less than two hundred is confined to a single small island in Indonesia that has been almost completed cleared of its forests for agricultural purposes. The Piopio, or New Zealand Thrush, thought now to be a whistler, became extinct in about 1894.

GRAY-CROWNED BABBLER
Pomatostomus temporalis
11.5 in (29 cm)
Australia, New Guinea

WHITE-BROWED BABBLER
Pomatostomus superciliosus
8 in (20 cm)
Australia

♂

♀

LOGRUNNER
Orthonyx temminckii
7.5 in (19 cm)
Australia, New Guinea

WHITE-WINGED CHOUGH
Corcorax melanorhamphos
17.5 in (44 cm)
Australia

APOSTLEBIRD
Struthidea cinerea
12.5 in (32 cm)
Australia

KOKAKO
Callaeas cinerea
15 in (38 cm)
New Zealand

SADDLEBACK
Creadion carunculatus
10 in (25 cm)
New Zealand

Logrunners;
Australasian Babblers;
Australian Mudnesters;
New Zealand
Wattlebirds

Profiled here are four small families restricted to the Australasian region.

There are just two species of LOGRUNNERS, family Orthonychidae. Both occur in Australia in that continent's eastern wet forests, and one of them, the Logrunner, also occurs in New Guinea. Both are chunky, ground-dwelling, midsize songbirds (7 to 11 inches [17 to 28 cm] long), mostly brown and white, with short bills and large feet. Another name for the Logrunner is Spine-tailed Logrunner, after the "spines," or short pieces of bare feather shafts, that protrude from the end of the tail in both species. The other logrunner is known as the Chowchilla, because some of its calls sound like this word.

Logrunners are noisy rainforest birds that live on territories in permanent communal groups of, typically, five or six. They use their large feet to scratch and scrape the ground for insects, snails, and other forest-floor invertebrates, including leeches. They fly little, foraging and escaping danger mostly by rapid running and hopping. They breed in pairs or communal groups. The female builds the domed nest of sticks and moss on the ground or in low vegetation, and she incubates the eggs. Males, with or without other members of the communal group, help feed the young. Neither logrunner is threatened.

AUSTRALASIAN BABBLERS are highly social, active, noisy birds that forage chiefly on the ground for insects. The family, Pomatostomidae, has five species, four of which occur in Australia (three endemic, one shared with New Guinea) and two in New Guinea. They resemble the Eurasian/African babblers somewhat in behavior and appearance, but the two groups

LOGRUNNERS

Distribution:
Australia

No. of Living
Species: 2

No. of Species
Vulnerable,
Endangered: 0, 0

No. of Species Extinct
Since 1600: 0

AUSTRALASIAN
BABBLERS

*Distribution:
Australia and New
Guinea*

*No. of Living
Species: 5*

*No. of Species
Vulnerable,
Endangered: 0, 0*

*No. of Species Extinct
Since 1600: 0*

AUSTRALIAN
MUDNESTERS

*Distribution:
Australia*

*No. of Living
Species: 2*

*No. of Species
Vulnerable,
Endangered: 0, 0*

*No. of Species Extinct
Since 1600: 0*

NEW ZEALAND
WATTLEBIRDS

*Distribution:
New Zealand*

*No. of Living
Species: 2*

*No. of Species
Vulnerable,
Endangered: 0, 1*

*No. of Species Extinct
Since 1600: 1*

are not closely related. They are medium-size birds (7 to 11.5 inches [18 to 29 cm] long) with long, down-curved bills, brown with boldly patterned white and brown faces and white-tipped tails.

Australasian babblers spend their days in small groups of up to a dozen related individuals and sleep together in domed stick "dormitories" that they build in trees. These groups are fairly sedentary and defend communal territories. Babblers forage on the ground, picking up bugs but also pushing their pointed bills into the soil to dig for insect larvae. They also search lower parts of shrubs and tree trunks and branches for food. Although mostly taking insects, they are considered omnivorous, also snacking on spiders, tiny frogs and reptiles, crustaceans, and even some seeds and fruit. Babblers fly quickly, low to the ground, from place to place, or run from danger. They breed communally, the dominant male and female doing the actual reproducing, the other members of the group (mostly young from previous nests) helping to feed the incubating female and then the young. The breeding pair builds the domed stick nest in a tree. None of these babblers are threatened, but the Gray-crowned Babbler has experienced population declines in several Australian regions.

The AUSTRALIAN MUDNESTERS are the Apostlebird and White-winged Chough, the only members of endemic Australian family Corcoracidae. The two species, related to crows, are vaguely crowlike in appearance. They are large (12 to 18 inches [30 to 46 cm] long), dull-plumaged birds that live in communal groups (usually five to fifteen individuals) on year-round territories; they roost and nest in trees but forage on the ground for insects and other small invertebrates, some small vertebrates, and seeds. Nests are mud cups (hence the family's common name) placed on tree branches, and all members of a group help with nesting and raising young. Neither mudnester is threatened.

The two species of NEW ZEALAND WATTLEBIRDS, family Callaeidae, restricted to New Zealand, are medium-size (10 to 15 inches [25 to 38 cm] long) with short wings, strong legs, long tails, and limited flight capabilities. Both have fleshy wattles hanging from the base of the bill, reddish in the Saddleback, blue or orange in the Kokako. Although they spend much time moving about in trees, they also forage extensively on the ground for insects, other small invertebrates, and fruit. Wattlebirds remain year-round in territorial pairs and breed monogamously. The Kokako is endangered, the Saddleback considered near-threatened. A third wattlebird, the Huia, became extinct in the early 1900s.

Whipbirds, Quail-thrushes, and Jewel-babblers

WHIPBIRDS, QUAIL-THRUSHES, JEWEL-BABBLERS, and a few others comprise a small group of midsize insect-eating songbirds restricted to Australia, New Guinea, and parts of Southeast Asia. They are probably best known for the year-round territorial song of the male Eastern Whipbird, which ends with a brief, very loud burst of sound that powerfully resembles the crack of a long whip and is one of the most striking, characteristic sounds of eastern Australia's wet forests. ("Whip-crack" vocalizations are lacking in the other whipbirds, Australia's Western Whipbird and New Guinea's little-known Papuan Whipbird.) Although the various birds in this group, for instance, the whipbirds versus the quail-thrushes, differ significantly in form, color patterns, and many behaviors, they do have similarities. Many species of the group inhabit woodlands or forests, spend most of their time on or near the ground, and include insects in their diet.

The family, Cinclosomatidae, includes fifteen species; in addition to the kinds listed above there are wedgebills (in Australia), the Blue-capped Ifrita (of New Guinea), and the Malaysian Rail-babbler; most, eight species, occur in Australia. Australian whipbirds (including the wedgebills), 7.5 to 12 inches (19 to 31 cm) long, are crested, brown or olive brown, and have long white-tipped tails and wedge-shaped bills. The Papuan Whipbird is crestless and dark olive green. Quail-thrushes, 7 to 11 inches (18 to 28 cm) long, are boldly patterned in brown, gray, black, and white. Jewel-babblers (8 to 9 inches [20 to 23 cm] long) are restricted to New Guinea and are the most colorful members of the family, ranging from brown above and olive and blue below to being overall mostly blue; all have bold black and white facial markings.

Distribution:
Australia, New Guinea,
Southeast Asia

No. of Living
Species: 15

No. of Species
Vulnerable,
Endangered: 0, 0

No. of Species Extinct
Since 1600: 0

EASTERN WHIPBIRD
Psophodes olivaceus
10–12 in (25–31 cm)
Australia

WESTERN WHIPBIRD
Psophodes nigrogularis
8–10 in (20–26 cm)
Australia

SPOTTED QUAIL-THRUSH
Cinclosoma punctatum
10–11 in (25–28 cm)
Australia

CHESTNUT QUAIL-THRUSH
Cinclosoma castanotus
8–10 in (20–26 cm)
Australia

CHIMING WEDGEBILL
Psophodes occidentalis
8.5 in (21 cm)
Australia

Whipbirds forage on or close to the ground for insects, insect larvae, seeds, and the occasional vertebrate, such as a small lizard. Remaining year-round in mated pairs on territories, the male and female call back and forth to each other in brief duets as they forage in dense forest, thickets, or scrub habitats. Wedgebills occur in communal family groups of up to twenty or more individuals that defend small territories in their shrub and grassland habitats. They forage on the ground for insects and seeds. Quail-thrushes, fairly secretive, wary, and highly camouflaged, live on the ground in forest, woodland, or scrub areas, in pairs or small groups. They pick up food, mainly insects and seeds, from the ground, and also sometimes use their strong feet to dig for food. Like quail, they often freeze in place when alarmed; when approached closely, they will burst from cover suddenly and noisily, and fly away low and fast. Jewel-babblers forage on the forest floor or in low vegetation in pairs or small parties; like the quail-thrushes, when disturbed they burst out and fly away, with loud calls and wing sounds. The Malaysian Rail-babbler is a shy bird that stalks about the forest floor, moving its head back and forth like a chicken; it eats insects and seeds, but also some fruit.

Whipbirds breed in monogamous pairs. The nest is a bulky cup of woven twigs, bark, and grass placed in a shrub or thicket. The female incubates while the male feeds her; both sexes feed young. At 10 to 11 days old, nestlings hop out of the nest although not yet flight capable. They hide in the undergrowth for about a week, while the parents bring them food, until they can fly. Quail-thrushes breed in monogamous pairs. Nests are loose cups of bark, leaves, and grass placed on the ground by a tree, stump, shrub, or rock. Some nest in depressions in the ground that are lined with vegetation. Females incubate unassisted by their mates; both sexes feed young.

None of the birds in this family are threatened, but two are near-threatened: the Western Whipbird occurs only in several small, fragmented populations in southern Australia, in habitats that were much reduced during Australia's agricultural development, and the Malaysian Rail-babbler occurs in some forested parts of Thailand, Malaysia, and Indonesia, areas that have undergone extensive deforestation in recent decades.

AFRICAN BLUE-FLYCATCHER
Elminia longicauda
5.5 in (14 cm)
Africa

LEADEN FLYCATCHER
Myiagra rubecula
6 in (15 cm)
Australia, New Guinea

BLACK-FACED MONARCH
Monarcha melanopsis
7 in (18 cm)
Australia, New Guinea

WILLIE-WAGTAIL
Rhipidura leucophrys
8.5 in (21 cm)
Australia, New Guinea

NORTHERN FANTAIL
Rhipidura rufiventris
7 in (18 cm)
Australia, New Guinea

BROAD-BILLED FLYCATCHER
Myiagra ruficollis
6 in (15 cm)
Australia, New Guinea

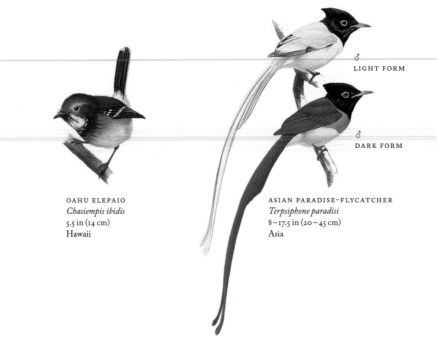

LIGHT FORM

DARK FORM

OAHU ELEPAIO
Chasiempis ibidis
5.5 in (14 cm)
Hawaii

ASIAN PARADISE-FLYCATCHER
Terpsiphone paradisi
8–17.5 in (20–45 cm)
Asia

Monarch Flycatchers and Fantails

MONARCH FLYCATCHERS are small to midsize forest and woodland birds that catch insects for a living. Many utilize the classic, time-tested flycatching method: they perch motionless (or with tail vibrating) on tree branches, utility wires, or other exposed spots, then dart out in short, swift flights to snatch insects from the air, returning quickly to the same perch to consume their catch. Others pursue "gleaning": moving quickly through the foliage of trees or shrubs, checking leaves and twigs, grabbing insects that they detect or that flush from cover at the bird's movement. Aside from their insect-eating ways, monarch flycatchers are perhaps best known for a small subgroup known as paradise-flycatchers, handsome crested birds with very long, streamerlike tails; and for the FANTAILS, a large subgroup whose members, usually highly active, characteristically flick their tails open repeatedly into a fan shape and wave them as they forage in trees and foliage, probably to disturb insects and flush them from hiding spots. The family, Monarchidae, has about 140 member species, distributed around Africa, southern and eastern Asia, and the Australia/New Guinea region. (The 43 fantails are sometimes considered separately as family Rhipiduridae.)

Monarch flycatchers range in length from 5 to 12 inches (12 to 30 cm); most are only 5 to 8 inches (12 to 20 cm), but some—the paradise-flycatchers—have very long tail streamers (elaborate central tail feathers) that add several inches to overall length. Most in the family have small, flat, broad bills that are surrounded by small bristles (that presumably help funnel flying bugs into the mouth), and many have slight crests. Tails are generally

Distribution:
Africa, southern
and eastern Asia,
Australasia

No. of Living
Species: About 140

No. of Species
Vulnerable,
Endangered: 10, 11

No. of Species Extinct
Since 1600: 2

relatively long, to aid in highly maneuverable, insect-chasing flight; feet are small and weak. Colors in the group are mainly gray, blue, brown, black, or white, but some are more brightly clad (in yellow and black, for instance); several are reddish brown above and white below. In many species, male and female differ extensively in appearance.

Flycatchers and fantails usually forage solitarily or in pairs, but some occur in small family groups. They eat mainly insects but also spiders; a few consume small crabs, worms, centipedes, and/or snails. Although flycatching and gleaning are the feeding methods used most often by the group, other techniques are also employed. Some, especially those called monarchs, hop over tree trunks and branches in search of bugs, and the Willie-wagtail, a common black and white fantail plentiful in urban and suburban areas of Australia, pursues insects, especially butterflies, on or near the ground. (Willie-wagtail's name, as you might expect, arises with the bird's movements; whether active or at rest, these birds almost constantly sway their bodies and wag their tails.) All these birds breed in monogamous pairs, establishing breeding-season or year-round territories that are often aggressively defended from other members of the same species. Willie-wagtails, for instance, are celebrated for their pugnacity, often chasing much larger birds, such as kookaburras, from their territories.

Both male and female monarch flycatchers usually participate in all aspects of nesting, helping to build nests, incubate eggs, and feed young. They construct cup-shaped nests of made of grasses or other vegetation (rootlets, bark, leaf pieces) that often held together with spider-webbing or mud, and placed on a tree branch.

Most of the monarch flycatchers are very common, but a number are threatened. One of ten species classified as vulnerable is a small reddish brown bird endemic to the Hawaiian Islands, the Elepaio (also called Hawaiian Flycatcher). Predominantly a bird of higher-elevation native forests, it is one of the few native forest songbirds that a visitor to Hawaii, with a little time and effort, can actually locate and see. The Elepaio on the islands of Kauai and the Big Island of Hawaii are secure, but the population on the main island of Oahu has declined. (Some authorities consider the Elepaio on these islands to be three separate species.) Among eleven monarch flycatchers that are endangered, five are critically endangered, and among these are two of the exquisite paradise-flycatchers. Most threatened monarch flycatchers occur only on islands.

Drongos

DRONGOS are bold, noisy, mostly all-black birds of the Old World tropics. Observed frequently because they favor open habitats and, often, exposed perches, such as the tops of trees, bushes and utility poles, drongos are perhaps most recognized for their tails. These are long and forked, often with the outer few feathers curving outward (and sometimes upward, resulting in a curling effect), and in some species quite ornate, with wirelike extensions and racket-shaped tips. The elaborate tails, in combination with long, pointed wings, aid the birds with aerial maneuverability when they chase flying insects in acrobatic flight. The family, Dicruridae, has twenty-three species and occurs throughout sub-Saharan Africa, Madagascar, most of southern Asia, New Guinea, and parts of Australia. The term *drongo* apparently arose in Madagascar, as the name given to the local species (the Crested Drongo) by some of the native peoples.

Drongos are medium-size, varying from 7 to 15 inches (18 to 38 cm) long, not including the extra-long tail shafts of species such as the Greater Racket-tailed Drongo, which can add another 10 inches (25 cm) to total length. They have stout, broad bills slightly hooked at the tip, which are encircled by small bristles that may guide the entry of flying bugs into the mouth, and short legs with weak feet. They are almost all attired totally in black, usually with a glossy sheen; a few have white or reddish brown patches. Some have crests, most have red eyes. Drongo sexes look alike.

Arboreal birds of wooded areas, drongos occur in habitats ranging from rainforest edges to more open sites with scattered trees, including

*Distribution:
Sub-Saharan Africa,
southern Asia,
Australasia*

*No. of Living
Species: 23*

*No. of Species
Vulnerable,
Endangered: 0, 2*

*No. of Species Extinct
Since 1600: 0*

HAIR-CRESTED DRONGO
Dicrurus hottentottus
12 in (31 cm)
Southern Asia

BLACK DRONGO
Dicrurus macrocercus
11 in (28 cm)
Southern Asia

GREATER RACKET-TAILED DRONGO
Dicrurus paradiseus
12.5–25.5 in (32–65 cm)
Southern Asia

FORK-TAILED DRONGO
Dicrurus adsimilis
10 in (25 cm)
Africa

SPANGLED DRONGO
Dicrurus bracteatus
12 in (31 cm)
Australia, Philippines

agricultural lands. They feed primarily on insects, from tiny flies to large butterflies and beetles, which they capture by flying out from their exposed perches and grabbing the bugs in the air. Usually they fly back to a perch and dismantle a captured insect with their bill, often while holding it down with their feet. They have powerful jaws, which enable them to eat larger, hard-cased insects, such as beetles. Drongos occasionally fly to the ground to catch insects there, and also, opportunistically, take lizards and even small birds. They often take advantage of insects flushed by large mammals such as cattle; some drongos regularly perch on the backs of these animals, watching for insect movement. Some supplement their insect diet with flower nectar. A few species have been observed to dabble in food piracy, robbing other birds of their prey. Forest-dwelling species often join mixed-species flocks of foraging birds. Drongos are sometimes known for their beneficial effects on reducing insect populations; for instance, India's Black Drongo, common and conspicuous throughout most of the subcontinent's agricultural districts, is widely appreciated for its contributions to insect control. Drongo flight is generally slow and undulating.

Most drongos are solitary or found in pairs (although some, such as the Hair-crested Drongo, will gather in a flock to feed on nectar at a flowering tree). They are often aggressive and apparently fearless, quick to chase and attack other birds, even large crows and raptors (not to mention cattle) that enter their territories or approach their nests. In some regions, small species of birds often nest close to drongo nests, apparently as a security measure; the drongos, by chasing potential enemies from their nest, simultaneously protect the nests of other birds nesting nearby.

Drongos breed in monogamous pairs. Their small nest is hung in the fork of narrow branches high in the crown of a tree or at the base of tall palm tree fronds. It is made of small twigs, rootlets, leaves, and grasses, and in many species is held together with spider webbing and lichens. Both parents help construct the nest, incubate the eggs, and feed the young.

Two drongos, endemic to small Indian Ocean islands (the Comoros) near Madagascar, are endangered; both have total populations below 250. Loss and degradation of their forest habitats (which are cleared for agriculture and logging) are probably the major causes of their declines. Many of the drongos are common, even abundant, birds.

BROWN JAY
Cyanocorax morio
15.5–17.5 in (39–44 cm)
North America, Central America

AZURE-HOODED JAY
Cyanolyca cucullata
10–12 in (26–31 cm)
Mexico, Central America

GREEN JAY
Cyanocorax yncas
10 in (26 cm)
North America, South America

YUCATAN JAY
Cyanocorax yucatanicus
12.5 in (32 cm)
Central America

TURQUOISE JAY
Cyanolyca turcosa
12.5 in (32 cm)
South America

VIOLACEOUS JAY
Cyanocorax violaceous
14.5 in (37 cm)
South America

WHITE-TAILED JAY
Cyanocorax mystacalis
13 in (33 cm)
South America

PURPLISH JAY
Cyanocorax cyanomelas
14.5 in (37 cm)
South America

Jays, Crows, Ravens, and Magpies

JAYS, CROWS, RAVENS, and MAGPIES are members of family Corvidae, a group of strikingly handsome, often conspicuous songbirds, of about 118 species, that occurs just about everywhere in the world. These birds, as a group called corvids, are known for their intelligence and adaptability; in several ways, the group is considered to be the most highly developed, or evolved, of birds. In spite of this, some in the family—mainly crows and ravens—have often been considered by people as ill omens. This undoubtedly traces to the birds' frequently all-black plumage and habit of eating carrion, both putatively sinister traits. Ravens, in particular, have long been associated in many northern cultures with evil or death. Crows and some of the other corvids are often quite abundant, and where these smart birds occur in force in agricultural regions, they are usually considered pests. Many in the family are also very noisy.

Corvids range in length from 8 to 28 inches (20 to 71 cm), many near the higher end, which is large for songbirds (indeed, the all-black Common Raven, which ranges over most of the Northern Hemisphere, is the largest species in the huge order Passeriformes). They have stout, fairly long bills, robust legs, and strong, large feet. Many, especially crows, ravens, rooks, and jackdaws, are all or mostly black. Jays are mainly attired in bright blues, purples, greens, yellows, and white. New World jays tend to be blue, and many jays, such as Steller's Jay, have conspicuous crests. Some in the family have very long tails, and a few of these, such as Mexico's Black-throated Magpie-Jay and Asia's Blue Magpie, are among the group's most comely birds.

Distribution:
All continents except
Antarctica

No. of Living
Species: 118

No. of Species
Vulnerable,
Endangered: 8, 5

No. of Species Extinct
Since 1600: 0

GRAY JAY
Perisoreus canadensis
11.5 in (29 cm)
North America

STELLER'S JAY
Cyanocitta stelleri
11–12.5 in (28–32 cm)
North America, Central America

BLACK-BILLED MAGPIE
Pica pica
17–19.5 in (43–50 cm)
Eurasia, Africa, North America

CLARK'S NUTCRACKER
Nucifraga columbiana
12 in (31 cm)
North America

FLORIDA SCRUB-JAY
Aphelocoma coerulescens
11 in (28 cm)
Florida

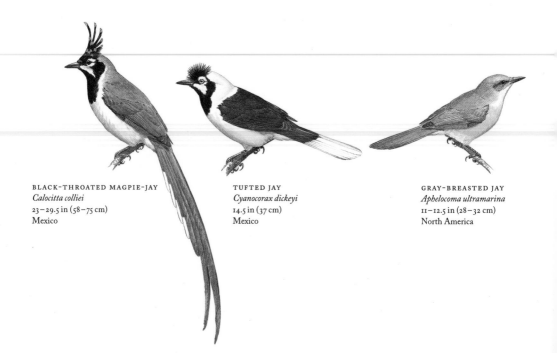

BLACK-THROATED MAGPIE-JAY
Calocitta colliei
23–29.5 in (58–75 cm)
Mexico

TUFTED JAY
Cyanocorax dickeyi
14.5 in (37 cm)
Mexico

GRAY-BREASTED JAY
Aphelocoma ultramarina
11–12.5 in (28–32 cm)
North America

BLUE MAGPIE
Urocissa erythrorhyncha
25.5–27 in (65–68 cm)
Southern Asia

GREEN MAGPIE
Cissa chinensis
15 in (38 cm)
Southeast Asia

RUFOUS TREEPIE
Dendrocitta vagabunda
18–19.5 in (46–50 cm)
Southern Asia

BLACK MAGPIE
Platysmurus leucopterus
15.5 in (40 cm)
Southeast Asia

LARGE-BILLED CROW
Corvus macrorhynchos
19–23 in (48–59 cm)
Asia

PIED CROW
Corvus albus
18–19.5 in (46–50 cm)
Africa, Madagascar

WHITE-NECKED RAVEN
Corvus albicollis
19.5–21.5 in (50–54 cm)
Africa

Corvids are primarily birds of woodlands, but many, especially among the crows, prefer more open habitats, including human-altered areas such as parks, gardens, and agricultural sites. They eat a large variety of foods (and try to eat many others). They feed on the ground and in trees, taking fruits and nuts, insects (including some in flight), bird eggs and nestlings, and carrion; visits to garbage dumps also yield meals. Bright and versatile, they are quick to take advantage of new food sources and to find food in agricultural and even residential environments. Important scavengers, jays, crows, and other corvids help break down dead animals so that the nutrients bound up in them are recycled into food webs. Corvids are major predators on bird nests and are considered responsible for a significant amount of the nest predation on many songbird species, particularly those with open-cup nests. Corvids typically use their feet to hold food down while tearing it with their bills. Hiding food for later consumption, caching, is widely practiced by the group.

Usually quite social, many corvids live in family parties, which maintain and defend year-round territories on which a family forages, roosts, and breeds. Young and nonbreeding adults may form nomadic flocks. Corvids are usually raucous and noisy, giving varieties of harsh, grating squawks and croaks as their foraging flocks straggle from tree to tree; among jays, some calls sound like "jay."

Generally monogamous, some corvids nest alone, while others nests in loose colonies. Nests, constructed by both sexes, are large, bulky, and bowl-shaped, built of sticks and twigs, and placed usually in trees but sometimes in rock crevices. Only the female incubates the eggs, while the male hunts and brings her food. Either both sexes feed the nestlings or the female alone does it, with food brought by the male. In larger species, pairs may mate for life. Several species of jays raise young cooperatively: generally the oldest pair in the group breeds and the other members, who are relatives, serve only as helpers, assisting in nest construction and feeding the young.

Most corvids are common. Some adjust well to people's activities, indeed often expanding their ranges when they can feed on farm crops. Eight species, including the Florida Scrub-jay, are vulnerable, five are endangered. The two most threatened are Pacific island species, the Hawaiian and Mariana Crows; the Hawaiian species is now probably extinct in the wild.

Birds-of-Paradise

BIRDS-OF-PARADISE are tropical rainforest birds that are celebrated for two aspects of their biology: the fantastic and bizarre plumage of some of them, with tremendously long tail feathers and, sometimes, head plumes, and the breeding displays males use to attract females and convince them to mate. In New Guinea, where they mainly occur, birds-of-paradise were, and still are, an important part of the cultures of indigenous groups. Several species are venerated, their long plumes considered prizes, indications of wealth, and sometimes as flamboyant fashion accessories. The family, Paradisaeidae, closely related to bowerbirds, is comprised of about forty-four species. In addition to New Guinea (with about thirty-six species), birds-of-paradise occur on nearby islands and in parts of eastern Australia. Included in the family are birds called paradise-crow, manucode, paradigalla, astrapia, parotia, riflebird, sicklebill, and melampitta.

Midsize birds, birds-of-paradise have a great range of lengths, owing to their long tails; among the smallest is Wilson's Bird-of-paradise at about 7 inches (18 cm) long. Some have central tail feathers up to 3.3 feet (1 m) long, yielding total lengths of up to 4.4 feet (1.35 m). Several in the family are all black or clad in dull browns, but a good number are richly endowed with glossy bright reds, oranges, or yellows, sometimes with patches of iridescent green. Bills are generally stout and fairly long; the sicklebills and riflebirds have very long down-curved bills. Within a species, males generally have more colorful and ornate plumage than females, but in a few (such as manucodes and paradigallas), the sexes look alike.

Distribution:
New Guinea, Australia

No. of Living
Species: 44

No. of Species
Vulnerable,
Endangered: 4, 0

No. of Species Extinct
Since 1600: 0

PARADISE RIFLEBIRD
Ptiloris paradiseus
11.5 in (29 cm)
Australia

CRESTED BIRD-OF-PARADISE
Cnemophilus macgregorii
9.5 in (24 cm)
New Guinea

STEPHANIE'S ASTRAPIA
Astrapia shephaniae
21–33 in (53–84 cm)
New Guinea

WAHNES' PAROTIA
Parotia wahnesi
14–17 in (36–43 cm)
New Guinea

KING OF SAXONY BIRD-OF-PARADISE
Pteridophora alberti
8–19.5 in (20–50 cm)
New Guinea

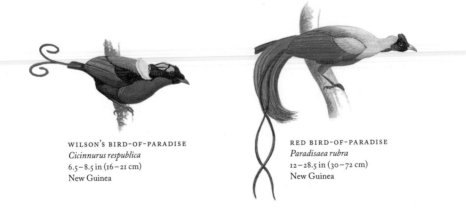

WILSON'S BIRD-OF-PARADISE
Cicinnurus respublica
6.5–8.5 in (16–21 cm)
New Guinea

RED BIRD-OF-PARADISE
Paradisaea rubra
12–28.5 in (30–72 cm)
New Guinea

The primary habitat of birds-of-paradise is rainforest, but many at least occasionally utilize other habitats, including forest edges, woodlands, savanna, and gardens; most occur in New Guinea's forested mountains. These birds are chiefly consumers of fruits, berries, and insects. They usually fly from fruiting tree to fruiting tree, staying in the middle and higher canopy, and often hanging at odd angles to reach tasty morsels. But they will also fly down to feed on tree trunks, stumps, and fallen logs. Some move through foliage, seeking insects and spiders; some use their long, strong bills to turn over bark and dig into dead wood in search of bugs or to push deeply into fruits. Frogs and lizards are occasionally taken, and at least a few species are known to add some leaves, flowers, or nectar to their diets. Birds-of-paradise often forage solitarily, but aggregations may occur at fruit-laden trees.

In many species, males seem to hold territories year-round, feeding, and eventually, displaying for females within the territories. Most are promiscuous breeders. During breeding seasons, males spend most of each day at display sites, usually horizontal branches high in large trees. They vocalize and display to attract the attention of passing females. Male displays vary among species, but they usually involve moving the head up and down, stretching the neck, rhythmically swaying the body, hopping from side to side, and extending the wings. When a female approaches, a male's antics increase in intensity and he may encircle the female with his out-stretched wings and dance around her. When she is satisfied, presumably of the male's good genes, the two mate. The female then leaves to nest on her own. The male returns to his mate-attraction displays and will mate with as many females as he can attract. The female builds a bulky bowl nest of dried leaves, twigs, ferns, and vine pieces high in a tree fork or in a vegetation tangle, lays and incubates eggs, and feeds her young. Species in which the sexes look alike tend to be monogamous, with the sexes sharing nesting chores.

Only four of the birds-of-paradise, all New Guinea species, are currently threatened (considered vulnerable), but eight others are near-threatened. Loss of their rainforest habitats is the prime threat to these splendid birds. Additionally, some riflebirds are killed routinely in agricultural areas where they raid fruit crops. Birds-of-paradise are still sometimes targeted in parts of New Guinea for food and for trade in feathers, skins, or live birds.

WHITE-BREASTED WOODSWALLOW
Artamus leucorhynchus
6.5 in (17 cm)
Australia, New Guinea, Indonesia, Philippines

DUSKY WOODSWALLOW
Artamus cyanopterus
6.5 in (17 cm)
Australia

BLACK-FACED WOODSWALLOW
Artamus cinereus
7 in (18 cm)
Australia

LITTLE WOODSWALLOW
Artamus minor
5 in (13 cm)
Australia

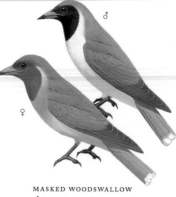

MASKED WOODSWALLOW
Artamus personatus
7 in (18 cm)
Australia

BORNEAN BRISTLEHEAD
Pityriasis gymnocephala
10 in (26 cm)
Borneo

Woodswallows

WOODSWALLOWS are fairly small, gregarious, dully colored birds that somewhat resemble swallows in form and feeding behavior, although the two groups are not closely related. Despite their size and coloring, woodswallows are often conspicuous in the regions in which they occur because they typically perch huddled together in small groups on bare tree branches or utility poles or wires; and when not huddling, they engage for long periods in circling, back-and-forth flight, like swallows, as they forage in the air for insects. There are twelve species in the family, Artamidae, which is related to the butcherbirds and crows. All are called woodswallows except one, the Bornean Bristlehead, a bizarre bird only recently recognized as a woodswallow by analysis of its DNA. Woodswallows are restricted to southern Asia and Australasia.

Husky birds, mostly 5 to 8 inches (12 to 20 cm) long, woodswallows are mainly black, brown, gray, and white, and often bicolored, for example gray and white or gray and brown. They all have long, broad, pointed wings, blunt, short tails, and short legs and toes. Their stout, slightly down-curved bill and wide gape are helpful for catching flying insects. Bills are usually gray blue, with dark tips. The Bornean Bristlehead, endemic to Borneo, is stocky, about 10 inches (26 cm) long, black with a brightly colored head and massive, hooked black bill; the top and sides of its head are covered with stubby, bristlelike feathers.

Woodswallows occur in many habitats: forest edges and clearings, scrub areas, agricultural lands, even some deserts. Their major habitat requirement is that insects be plentiful. They forage in pairs or small groups,

Distribution: Australasia, southern Asia

No. of Living Species: 12

No. of Species Vulnerable, Endangered: 0, 0

No. of Species Extinct Since 1600: 0

typically catching, and usually consuming, insects in flight. Some species, in addition to aerial insect-catching, will take bugs from the ground or from foliage. Although they subsist almost exclusively on insects, woodswallows occasionally also take nectar and pollen from flowers. They have brushy-tipped tongues that aid in this endeavor. They are graceful flyers, moving along with much gliding. They are one of the few kinds of songbirds that can soar, like hawks or seabirds, using only spread wings and rising air currents to stay aloft.

Woodswallows are often aggressive, sometimes attacking and chasing much larger birds, such as crows and hawks, which approach their roosting or nesting sites. They are very social, roosting and perching in huddled parties (often four to six strong), especially in cold weather, and preening each other. Larger aggregations of a hundred or more individuals will huddle together on trees on particularly cold days. Roosting overnight, most woodswallows cluster together in close contact, in "knots," often clinging to the side of a vertical tree trunk or in a tree hollow. Some species are fairly sedentary throughout the year; others are migratory; a few are nomadic. Two Australian species, the Masked and White-browed Woodswallows, are highly nomadic, moving frequently over long distances in large flocks and breeding when temperature and rainfall conditions, and so insect availability, are favorable. They move generally northward in fall and southward in spring to breed. The two species often mix in flocks during movements and even in nesting colonies.

The Bornean Bristlehead not only looks different from other woodswallows, it forages differently, hunting for large insects by searching for them amid the leaves of tall trees in old-growth rainforests of Borneo; it also takes small vertebrates. The Bristlehead occurs in small family groups and is known especially for its loud honking, mewing, and chortling calls.

Woodswallows, monogamous breeders, nest in loose colonies in pairs or small groups. Nests are flimsy, made of twigs, grasses, roots, and bark placed usually in trees or bushes, or on tree stumps; a few nest in rock crevices or holes. Both sexes build the nest, incubate eggs, and feed chicks. In several species, other adults in a group, presumably relatives, help the breeding pair feed their young.

Most of the woodswallows are abundant, widespread birds, and none are threatened. The Bornean Bristlehead, however, is considered near-threatened. It occurs sparsely only in lowland old-growth forests of Borneo, and these forests are rapidly being cut, burned, and otherwise degraded.

Butcherbirds and Currawongs; Magpie-larks

BUTCHERBIRDS and CURRAWONGS are mid- to large-size, highly successful forest and woodland birds of Australia and New Guinea that frequently share bicolored plumage patterns and, to varying degrees, predatory natures. They prey on larger insects but also on small vertebrates. Currawongs are also called crow-shrikes and bell-magpies; the word *currawong* sounds like one of the calls of the Pied Currawong. The family, Cracticidae, with twelve species, includes the Australasian Magpie and two New Guinea species known as peltops.

Butcherbirds, 9.5 to 17 inches (24 to 44 cm) long, are outfitted in black, white, and gray. They have relatively large, robust, straight bills, used to capture, kill, and dismember their prey. The three currawongs, all endemic to Australia, are large (to 20 inches [50 cm]), mostly black or gray, with white wing and tail patches. They have long tails and long, straight bills with a sharp, hooked point. They resemble crows, but are more slender and have yellow eyes. Australasian Magpies, to 17 inches (44 cm) long, are black and white, with white, black, or gray backs. They have robust bills that they use to dig into soil for food. Peltops, about 8 inches (20 cm) long, are black with white cheek and back patches and red at the base of the tail. Male and female in the family generally look alike, although males' bills are usually a bit larger than females.

Butcherbirds, usually seen alone, are predators of forests, woodlands, and agricultural districts. They perch, generally a few yards above the ground, and scan for prey that includes large insects, crustaceans, small reptiles and

BUTCHERBIRDS
AND CURRAWONGS

Distribution:
Australia, New Guinea

No. of Living
Species: 12

No. of Species
Vulnerable,
Endangered: 0, 0

No. of Species Extinct
Since 1600: 0

PIED BUTCHERBIRD
Cracticus nigrogularis
12.5–14 in (32–36 cm)
Australia

GRAY BUTCHERBIRD
Cracticus torquatus
9.5–12 in (24–30 cm)
Australia

AUSTRALASIAN MAGPIE
Gymnorhina tibicen
15–17.5 in (38–44 cm)
Australia, New Guinea

PIED CURRAWONG
Strepera graculina
16.5–19.5 in (42–50 cm)
Australia

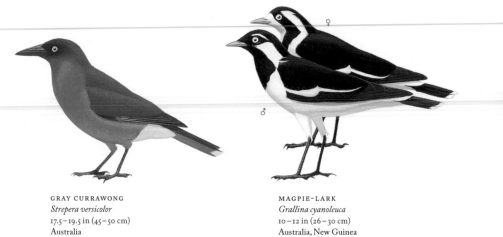

GRAY CURRAWONG
Strepera versicolor
17.5–19.5 in (45–50 cm)
Australia

MAGPIE-LARK
Grallina cyanoleuca
10–12 in (26–30 cm)
Australia, New Guinea

mammals, and birds, especially young ones. They pounce on prey, kill it with their powerful bill, then take it and either wedge it into a tree fork or cranny or impale it on a thorn or other sharp object. Thus immobilized, the prey is easily torn, dismembered, and eaten, or alternatively, cached for later consumption. In this "butchering" behavior these birds are similar to many shrikes. At least some butcherbirds live on permanent territories, in small family groups of three to five. Currawongs forage in trees, including on trunks and fallen logs, and on the ground, looking chiefly for insects but also for small vertebrates. They also eat fruit and berries and will scavenge in garbage. Currawongs are mostly sedentary, although one, the Pied Currawong, is known for forming into large flocks that roam, during nonbreeding periods, over wide areas in search of abundant insects.

Australasian Magpies, essentially butcherbirds that forage on the ground, occur mainly in small territorial groups of from three to twenty or so, in all types of open habitats, including open forests but also agricultural areas and towns. The birds roost together at night, then fly to fields in the morning to start the day's foraging for insects. They walk over the ground searching, dig into the soil with their bills, and even peer under cattle droppings. Peltops perch on tree branches and then sally out in swooping flights to catch insects on the wing.

Butcherbirds and currawongs build large, coarse, cup-shaped nests of sticks, lined with grass, high in tree forks. They tend to breed in monogamous pairs, but in many, the female builds the nest alone and incubates eggs. In the magpie, the female builds the nest, incubates, and feeds nestlings, but other adults in the group help feed fledglings. Butcherbirds have reputations as aggressive defenders of their nests; they will swoop at people who get too close and even hit them. None of the birds in this family are threatened.

MAGPIE-LARKS (sometimes called mudnest-builders) is the group name given to the Magpie-lark and Torrent-lark, the two species comprising family Grallinidae; the former occurs in Australia and New Guinea, the latter only in New Guinea. Both are black and white, 8 to 10 inches (20 to 25 cm) in length, long-legged and slender-billed. The Magpie-lark, in Australia an abundant bird of settled areas, parks and roadsides, forages on damp ground for insects, earthworms, and mollusks. Both species are monogamous breeders. They construct bowl nests of fibrous vegetation bound with mud, usually on tree branches. Neither is threatened.

MAGPIE-LARKS

Distribution:
Australia, New Guinea

No. of Living
Species: 2

No. of Species
Vulnerable,
Endangered: 0, 0

No. of Species Extinct
Since 1600: 0

COMMON IORA
Aegithina tiphia
5 in (13 cm)
Southern Asia

GOLDEN-FRONTED LEAFBIRD
Chloropsis aurifrons
7 in (18 cm)
Southern Asia

BLUE-WINGED LEAFBIRD
Chloropsis cochinchinensis
7 in (18 cm)
Southern Asia

ASIAN FAIRY-BLUEBIRD
Irena puella
10 in (26 cm)
Southern Asia

ORANGE-BELLIED LEAFBIRD
Chloropsis hardwickii
7.5 in (19 cm)
Southern Asia

Leafbirds, Ioras, and Fairy-bluebirds

LEAFBIRDS, IORAS, and FAIRY-BLUEBIRDS are brightly colored, small to midsize arboreal songbirds of forested regions of southern Asia. Leafbirds are mainly green, ioras green and yellow, and fairy-bluebirds blue and black. Some of them, such as the Golden-fronted Leafbird, Common Iora, and especially the brilliant turquoise blue Asian Fairy-bluebird, are so luminous and lovely that catching them with binoculars in just the right light is a bird-watcher's delight. Some of the leafbirds are accomplished mimics of other birds' songs, a single pair hidden high in a tree's dense foliage at times spewing forth sufficiently varied vocalizations to convince a person on the ground that the tree must contain a multispecies flock. The family, Irenidae, has fourteen species (eight leafbirds, four ioras, two fairy-bluebirds; sometimes Irenidae is split into two or three separate families), and is distributed from Pakistan and India eastward to the Philippines, with most species occurring in Southeast Asia. The group has the distinction of being the only bird family unique to the Oriental faunal, or zoogeographic, region of the world.

Leafbirds (6.25 to 8.75 inches [16 to 22.5 cm] long), ioras (4.75 to 6.5 inches [12 to 17 cm] long), and fairy-bluebirds (9.5 to 11 inches [24 to 28 cm] long) have slender, slightly down-curved bills, some being fairly long; fairy-bluebird bills are a bit heavier than in the ioras and leafbirds. Most in the family have short, often thick legs and small toes. Body plumage is thick and fluffy, and some in the group are known for easily shedding feathers when handled. Males of many of the species are more colorful than the females,

Distribution: Southern Asia

No. of Living Species: 14

No. of Species Vulnerable, Endangered: 1, 0

No. of Species Extinct Since 1600: 0

most dramatically in fairy-bluebirds. Male leafbirds tend to have glossy black and blue throat patches that females lack.

Birds in this group are found primarily in the midlevels and canopy of forests. Some of the ioras also range into other, more open habitats, such as forest edges, woodlands, agricultural sites, and even some scrub areas. Most of the food of leafbirds and ioras is fruit and seeds taken from large fruits, such as guavas and figs; fairy-bluebirds have a mostly fruit diet. Some insects are also taken from leaves and fruit, more so by ioras than leafbirds or fairy-bluebirds. Ioras apparently have a special fondness for caterpillars, and leafbirds and fairy-bluebirds commonly take nectar from flowers. All these birds search energetically for food among leaves in dense foliage, often hanging at odd angles to reach the best bits. These birds forage alone, in pairs, or in small family parties; larger groups may gather at fruit- or flower-laden trees. In the nonbreeding season, leafbirds and ioras often form flocks of their own species or participate in large mixed-species foraging flocks that move through forest canopies. Swift, straight flyers, leafbirds have a reputation for as aggressively defending good foraging sites, often trying to chase away other birds.

Leafbirds, ioras, and fairy-bluebirds are monogamous breeders. Ioras have elaborate courtship displays, males leaping, flying about, and falling through the air to impress females. Nests of leafbirds and ioras are cup-shaped, constructed of grass, rootlets, other plant materials, and lichens, bound with spider webbing and placed at the crux of two small branches high in a tree's foliage. Fairy-bluebirds build a larger, shallow platform nest of sticks, rootlets, and moss in a tree fork. Eggs are incubated by female leafbirds and fairy-bluebirds alone and by both parents in the ioras; both sexes feed offspring.

Most species in family Irenidae are abundant and many are widespread. Because they are forest birds, deforestation and forest degradation are the most severe threats they face. The Philippine Leafbird, endemic to some of the Philippine islands, is rare and now considered vulnerable; with its forest habitats being destroyed by logging and what remains being highly fragmented, its population has been declining. The stunning Asian Fairy-bluebird is sought after as a cage bird in some parts of its range and has disappeared from some areas, but with its broad range (India to the Philippines), it is not considered threatened.

Old World Orioles

OLD WORLD ORIOLES are medium-size forest and woodland songbirds of Eurasia, sub-Saharan Africa, and Australia renowned for their flashy yellow and black plumage and attractive, melodious songs. These attributes conspire to render orioles fairly conspicuous even though their preferred physical situation is to be sheltered in the dense and obscuring foliage of tree crowns. The family, Oriolidae, contains twenty-nine species (twenty-seven orioles and two figbirds, the latter confined to the Australia/New Guinea region). "Oriole" is from the Latin *aureolus*, which means "golden" or "yellow," the predominant color in many species, and the group is named for the common, widespread, bright yellow Eurasian Golden Oriole. (This oriole group is not to be confused with the very distantly related New World orioles, which are included in the New World blackbird family, Icteridae.)

Orioles range from 7 to 12 inches (18 to 31 cm) in length. All are very similar in shape—robust but fairly slender birds with an elongated look. They have long pointed wings, strong legs and feet, and straight, sturdy, slightly down-curved bills that, in males, are often pinkish or reddish orange (but dark in figbirds and non-yellow orioles). Male orioles are usually brightly turned out in yellow and black or green and black, often with lengthwise streaks; females typically are duller, greener or browner. In many species, males are bright yellow with black heads or eye stripes. Some species are mostly dull brown or brown and whitish. Figbirds are less brightly colored than most other orioles, often more olive green, and they have bare skin around their eyes, pinkish in males, gray in females.

Distribution:
Old World

No. of Living
Species: 29

No. of Species
Vulnerable,
Endangered: 2, 1

No. of Species Extinct
Since 1600: 0

IMM

BLACK-NAPED ORIOLE
Oriolus chinensis
10 in (26 cm)
Southeast Asia

♂

♀

BLACK-AND-CRIMSON ORIOLE
Oriolus cruentus
9.5 in (24 cm)
Southeast Asia

BLACK-WINGED ORIOLE
Oriolus nigripennis
8 in (20 cm)
Africa

AFRICAN GOLDEN ORIOLE
Oriolus auratus
8 in (20 cm)
Africa

♀

♂

OLIVE-BACKED ORIOLE
Oriolus sagittatus
10.5 in (27 cm)
Australia, New Guinea

GREEN FIGBIRD
Sphecotheres viridis
11 in (28 cm)
Australia, New Guinea

Almost exclusively arboreal and often staying in the higher reaches of tree foliage, orioles fly gracefully, with undulating flight, from tree to tree, looking among leaves for insects. They also take fruit when it is available, and some species, especially the figbirds, depend heavily on fruit, much of their diet consisting of figs and the like. Some species are known to come to the ground occasionally to feed on fallen fruit and even take insects from grassy areas. Orioles tend to forage solitarily or in small groups of two or three but will occasionally join large mixed-species foraging parties. Figbirds are strongly gregarious, usually seen in small flocks, and sometimes joining with groups of other fruit-eating birds. Orioles are fairly sedate in their foraging; figbirds more active and aggressive. Some orioles are year-round residents where they occur, but many are migratory; for instance, European-breeding populations of the Eurasian Golden Oriole, perhaps the best known oriole species, winter in tropical Africa. Some in the family, such as the Green Figbird in Australia, are nomadic when not breeding. During breeding seasons male orioles are highly vocal, producing the rich, liquid-noted songs for which the group is justifiably celebrated; these songs are frequent, characteristic sounds of many Old World forests. Figbirds often have harsher, chirpy, chattering songs.

Orioles are monogamous breeders, typically building rough but elaborate cup- or hammock-shaped nests woven of plant fibers, grass, and bark strips, which are suspended from forked tree branches and hidden by foliage. They frequently add moss and lichens to a nest for camouflage. In many species, the nest is constructed by the female alone, and she is largely in charge of incubating eggs. The male provides some food to the brooding female and helps feed the young after they are hatched. In figbirds, apparently both sexes perform nesting duties.

Most orioles are common and many have wide distributions. However, three species are currently considered threatened; all have small and declining populations, caused principally by habitat loss, mainly deforestation. Endangered is the Isabela Oriole, endemic to the northern Philippines, and vulnerable are the São Tomé Oriole, endemic to a small island off the western coast of Central Africa, and the Silver Oriole of southern China. Three other species, all of Southeast Asia, are near-threatened. Some orioles, migrating in flocks, become pests when they raid fruit orchards, and they are persecuted in parts of their ranges for these transgressions.

SCARLET MINIVET
Pericrocotus flammeus
7–8.5 in (18–22 cm)
Southern Asia

LARGE CUCKOO-SHRIKE
Coracina macei
11.5 in (29 cm)
Southern Asia

BAR-BELLIED CUCKOO-SHRIKE
Coracina striata
11.5 in (29 cm)
Southeast Asia

PURPLE-THROATED CUCKOO-SHRIKE
Campephaga quiscalina
8 in (20 cm)
Africa

WHITE-WINGED TRILLER
Lalage tricolor
6.5 in (17 cm)
Australia, New Guinea

BLACK-FACED CUCKOO-SHRIKE
Coracina novaehollandiae
12–14 in (30–36 cm)
Australia, New Guinea

WHITE-BELLIED CUCKOO-SHRIKE
Coracina papuensis
10.5 in (27 cm)
Australia, New Guinea

GROUND CUCKOO-SHRIKE
Coracina maxima
13–14 in (33–36 cm)
Australia

Cuckoo-shrikes, Trillers, and Minivets

CUCKOO-SHRIKES, TRILLERS, and MINIVETS, members of an Old World, mainly tropical family, are small to medium-size, slender, attractive birds that often conspicuously patrol the canopies of forests and woodlands in search of insects and fruit. While cuckoo-shrikes and trillers are undoubtedly handsome birds, generally shades of gray or brown and white, most minivets, on a small scale, are among the most resplendent of avian creatures, with males outfitted in bright red or orange, females in brilliant yellows. The family, Campephagidae (about eighty-two species spread over sub-Saharan Africa, southern Asia, and the Australian region), mainly includes multitudinous cuckoo-shrikes, but also twelve trillers (after their metallic chattering, trilling vocalizations), thirteen of the stunning little minivets (confined to Asia), two flycatcher-shrikes, and the Cicadabird (also named for its calls). People responsible for bird names obviously had a difficult time with this group: its common name, cuckoo-shrikes, is misleading, because the birds are neither cuckoos nor shrikes. The name may have arisen because the birds are gray and some are barred, as are many cuckoos, and the bill is somewhat shrikelike. Some in the family are often locally called caterpillar-shrikes or wood-shrikes; the name caterpillar birds, in fact, is sometimes applied to the entire family, a reference to the group's favorite dish.

Trillers and cuckoo-shrikes are fairly sleek, 6.5 to 14 inches (17 to 36 cm) long, with long tapered wings and tails, small feet, and short, broad, slightly down-curved bills. They tend to be gray with black or white markings;

Distribution:
Sub-Saharan Africa,
southern Asia, Australia

No. of Living
Species: 82

No. of Species
Vulnerable,
Endangered: 3, 1

No. of Species Extinct
Since 1600: 0

in some, females are browner and more streaked or barred. The generally smaller minivets, at 5.5 to 8.5 inches (14.5 to 22 cm), are chiefly flashy black and red (males) or gray/olive and yellow (females), although a few species are less gaudily attired in black, gray, and white. The male Cicadabird is dark blue gray, the female brown and barred. Many in the family have stiff feathers on the lower back that can be raised like porcupine quills.

Cuckoo-shrikes forage solitarily, in pairs, or in small parties, mostly high within tree foliage, taking large insects from branches and outer leaves. They will occasionally forage on the ground, and at least one species, Australia's Ground Cuckoo-shrike, regularly does so. Aside from insects, cuckoo-shrikes will take other small invertebrates and, usually, some berries and soft fruit. Caterpillars are preferred foods of some species. Foraging is often accomplished by perching on a branch, looking around, then pouncing on prey. Cuckoo-shrikes will also interrupt their graceful, undulating flights from tree to tree to chase and snatch flying insects from the air. Minivets often occur in large mixed-species feeding flocks high in the trees; some will also sometimes come to the ground in open areas to feed. They have an upright stance when perched, are usually lively and active, and eat mainly insects, but also some fruit. Many cuckoo-shrikes, especially the larger species, have an unusual, unexplained habit: after landing on a tree branch, they open and then quickly fold their wings, often several times. Males during courtship displays also repeatedly open and fold their wings.

Cuckoo-shrikes are monogamous, but many appear to be cooperative breeders. Helpers, presumably related to the breeding pair, assist with feeding young in the nest and after they fledge. Cuckoo-shrikes build small, shallow, cuplike nests of vegetation, bound with spider webbing, on branches or in forks high in a tall tree; sometimes they take over abandoned nests of other species. In most, the sexes share nesting duties, but in some, only the female incubates. Minivets build cup nests made of twigs, roots, and grass stems, usually placed on a horizontal tree branch; the builders cover the outside of the nest with lichens and spider webbing, perhaps for camouflage.

Most members of the cuckoo-shrike family are common birds with healthy populations. Four species are threatened: three are vulnerable and one, the Réunion Cuckoo-shrike, endemic to an island off Madagascar, is endangered. Deforestation is the primary threat to this group.

Vireos

VIREOS are mainly small, drably colored birds of New World forests and woodlands that flit about tree canopies looking for insects to eat. They are known for their persistent singing; in some areas, during early afternoon, for instance, vireos are often the only birds heard. Also, like other groups of small songbirds that contain multiple species that are confusingly similar in appearance and habits, vireos are notorious for presenting stiff challenges to bird-watchers trying to identify them to species. The family, Vireonidae, has about fifty-two species broadly distributed through the New World; included, with distributions limited to Mexico through South America, are about fifteen species known as greenlets, four shrike-vireos, and two peppershrikes. One, the Red-eyed Vireo, has one of the widest nesting ranges of any bird species in the Western Hemisphere, from northern Canada southward to parts of Argentina.

Varying from 4 to 7 inches (10 to 18 cm) long, vireos have short necks and stout legs. Bills are hooked somewhat like that of shrikes (appropriate, because shrikes are perhaps the vireos' nearest relatives). The family can be divided by its bills: typical vireos have medium-size bills slightly hooked at the tip; greenlets, which are the smallest in the family, have pointed bills; shrike-vireos have stout bills with hooked tips; and peppershrikes have heavy bills flattened sideways with a large hook at the tip. Most vireos are plainly colored in green, olive, or gray brown above, and yellow, grayish white, or buff below. Many have head stripes or eye rings; some have white or red eyes. Greenlets, small, warblerlike, and difficult sometimes to identify to species even for experts, are mainly olive above, yellowish or grayish below.

Distribution:
New World

No. of Living
Species: 52

No. of Species
Vulnerable,
Endangered: 1, 2

No. of Species Extinct
Since 1600: 0

THICK-BILLED VIREO
Vireo crassirostris
5 in (13 cm)
West Indies

SLATY-CAPPED SHRIKE-VIREO
Vireolanius leucotis
5.5 in (14 cm)
South America

RED-EYED VIREO
Vireo olivaceus
6 in (15 cm)
North America, South America,
West Indies

BLUE-HEADED VIREO
Vireo solitarius
5 in (13 cm)
North America, Central America,
West Indies

JAMAICAN VIREO
Vireo modestus
5 in (13 cm)
West Indies

WHITE-EYED VIREO
Vireo griseus
5 in (13 cm)
North America, Central America,
West Indies

YELLOW-THROATED VIREO
Vireo flavifrons
5.5 in (14 cm)
North America, South America, West Indies

The shrike-vireos and peppershrikes, chiefly limited to tropical areas and the largest members of the family, are somewhat more colorful, usually with head patches of chestnut or blue.

Vireos occur in forests, forest edges, woodlands, and, especially in the greenlets, in scrublands. They eat principally insects taken from tree leaves and branches; small fruits are also sometimes eaten. They move over branches and twigs more slowly, less actively than similar birds such as wood warblers. They also will hover near leaves to take food, and some will occasionally engage in flycatching, seizing flying insects in the air. Tropical species tend to join mixed-species foraging flocks that roam forest canopies. Many temperate-zone vireos are highly migratory, for instance, breeding in eastern North America and wintering in the Amazon region, or breeding in temperate areas of South America and migrating northward to the Amazon. A few species breed in the tropics in one place, such as Central America, and then migrate to another tropical area, such as northern South America, to spend the nonbreeding months. Vireos are territorial, some tropical species maintaining territories all year; temperate-zone species have breeding territories, but some also establish territories on their wintering grounds.

Vireos breed in monogamous pairs. They build small, hanging cup-shaped nests of grass, leaves, lichens, twigs, rootlets, and bark in the fork of a tall bush or tree, usually well out from the center of the plant. Both parents incubate eggs and feed insects to the young. Vireo nests are often "parasitized" by cowbirds, which lay their eggs in the nests of other species that then raise their young. Vireo nesting success is often much reduced by cowbirds, and many of their nests so affected end up without fledging any vireo young.

Most vireo species are abundant and widespread, but deforestation, forest fragmentation, and cowbird parasitism take their tolls. Three species are officially threatened. The Black-capped Vireo of Mexico and the United States is considered vulnerable (but classed as endangered on the U.S. Endangered Species List); it has been eliminated from large parts of its original range by habitat destruction, development, and, perhaps, cowbird parasitism. Colombia's Chocó Vireo, is endangered, and the San Andrés Vireo, endemic to Colombia's tiny San Andrés Island, is critically endangered. Two island-bound species, Jamaica's Blue Mountain Vireo and Brazil's Noronha Vireo, are near-threatened, and Bell's Vireo, another Mexican and U.S. species, is also in jeopardy.

NORTHERN SHRIKE
Lanius excubitor
10 in (25 cm)
Eurasia, North America

LONG-TAILED SHRIKE
Lanius schach
8–9.5 in (20–24 cm)
Southern Asia

MAGPIE SHRIKE
Corvinella melanoleuca
15.5–19.5 in (40–50 cm)
Africa

LONG-TAILED FISCAL
Lanius cabanisi
10–12 in (26–30 cm)
Africa

RED-BACKED SHRIKE
Lanius collurio
6.5 in (17 cm)
Eurasia, Africa

WHITE-RUMPED SHRIKE
Eurocephalus rueppelli
8 in (20 cm)
Africa

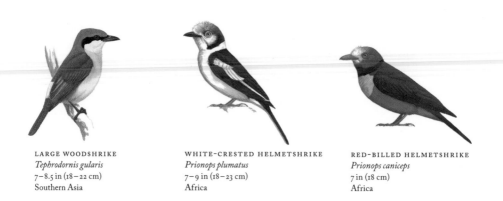

LARGE WOODSHRIKE
Tephrodornis gularis
7–8.5 in (18–22 cm)
Southern Asia

WHITE-CRESTED HELMETSHRIKE
Prionops plumatus
7–9 in (18–23 cm)
Africa

RED-BILLED HELMETSHRIKE
Prionops caniceps
7 in (18 cm)
Africa

Shrikes;
Helmetshrikes

SHRIKES are mostly medium-size, striking-looking songbirds, a charismatic bunch with reputations as minihawks; that is, they are carnivorous, taking animal prey by swooping down and grabbing it on the ground, then flying to a perch to eat it. Shrikes' most celebrated, if grisly, trait is that, after having eaten their fill, they continue to hunt, caching captured prey, for future consumption, by impaling it on thorns or other sharp plant parts, or these days, on spikes of barbed wire. They also impale larger prey to make it easier to dismember and eat. Because of these feeding habits, shrikes are often considered cruel; indeed, a common name for them is butcherbird, and in various regions of the world, other common names are variations on hangman and murderer. The name shrike is not much better: its derivation is the same as that of shriek, referring to the screeching harsh calls many of them make.

The family of true shrikes ("true" to distinguish it from helmetshrikes and bushshrikes), Laniidae, is comprised of thirty species that range over North America, Africa, Eurasia, and New Guinea. Most occur in Africa and Asia; only two inhabit the New World. They range in length mainly from 6 to 14 inches (15 to 36 cm), but one, Africa's Magpie Shrike, has a very long tail, the bird's total length reaching 17 inches (43 cm). Furthering their hawklike images, shrikes have heavy, hooked bills and strong legs and feet, with sharp claws. Most appear in handsome if somber combinations of black, white, and gray, although some add various shades of brown. Almost all shrikes have black heads or bold, black eye stripes. Six mainly black, white, and

SHRIKES

Distribution:
North America, Africa,
Eurasia, New Guinea

No. of Living
Species: 30

No. of Species
Vulnerable,
Endangered: 0, 1

No. of Species Extinct
Since 1600: 0

gray African shrikes are called fiscals, referring to a group of South African functionaries who traditionally wore black and white clothing. Male and female shrikes generally look alike or almost alike.

Shrikes are birds of open habitats, where they perch on tree branches or other elevated sites, swinging their tails and waiting for prey to make itself known. They take large insects, but also mice, small lizards, frogs, and birds. Upon spotting prey, a shrike drops to the ground, catches it, and kills it by repeatedly hitting it with its bill. Small prey may be eaten on the ground, but most are taken to a perch for consumption. Shrikes can kill and carry animals nearly up to their own body weight. They most often wedge prey into crevices so they can pick at it and tear it; or they impale it. Shrikes will also, like some hawks, hover over a spot where prey was spotted before diving down to take it, and some will catch insects in the air. Cached food may be eaten later on the day of capture or many days later. Most shrikes hunt alone, but some, more social, hunt in small groups. Shrikes are territorial, aggressively defending areas in which they feed and nest; they are noted for the large size of their territories, which are necessary to provide enough small birds, rodents, and lizards to satisfy their needs.

Monogamous breeders, shrikes build cup-shaped nests of leaves, twigs, and rootlets, often lined with feathers, grass, and spider webbing. Both sexes build the nest, mainly the female incubates eggs (sometimes being fed by her mate), and both feed young. Only one shrike is endangered, Newton's Fiscal, endemic to a small island off Africa's western coast. Populations of several other species have experienced declines, including North America's Loggerhead Shrike.

HELMETSHRIKES are a group of shrikes now usually placed in their own family, Prionopidae, which includes seven African species called helmetshrikes (because of stiff, forward-pointing head feathers that produce a helmeted appearance), and four in Southern Asia known as woodshrikes and philentomas. Most helmetshrikes, 6.5 to 10 inches (16.5 to 25 cm) long, live in groups of between three and twelve individuals that share a territory. They occur mainly in wooded habitats, feeding on insects taken from trees. Helmetshrikes breed cooperatively, the dominant pair in a group nesting, the others helping to care for young. Nests are cuplike, placed in a tree or thicket. Two African helmetshrikes, both with small populations and tiny ranges, are threatened.

HELMETSHRIKES

Distribution:
Africa, Southern Asia

No. of Living
Species: 11

No. of Species
Vulnerable,
Endangered: 1, 1

No. of Species Extinct
Since 1600: 0

Bushshrikes

BUSHSHRIKES are a group of shrikes known for their insect-eating ways, their occupation of densely vegetated habitats and frequent retiring, skulking lifestyles, and their vocal duets. Furthermore, some species of the group have incredibly beautiful plumage. Unlike the "true" shrikes (family Laniidae), bushshrikes (family Malaconotidae, which is restricted to Africa) are not generally known for impaling their prey. Among the 43 species in the group are 17 bushshrikes, 9 boubous, 6 puffbacks, 5 tchagras, and 4 gonoleks.

Bushshrikes are mostly stocky, medium-size songbirds, 5.5 to 10.5 inches (14 to 27 cm) long, with some of the typical bushshrikes being the largest in the family, and the puffbacks being the smallest. They all have strongly hooked bills that they use to dispatch their prey, and strong feet with sharp claws for handling struggling insects. Bird fanciers in North America and Europe think of shrikes as handsome but unimaginatively colored birds, because the shrikes distributed on those continents are mainly black, white, and gray. But Africa's bushshrikes include some spectacularly colored birds. In addition to the more colorful ones illustrated here are some multicolored gems, such as Doherty's Bushshrike of East Africa, with vivid green back, yellow belly, and red and black face and upper chest. Degree of brightness varies among the subgroups: boubous (named for some of their calls) are all black or black and white; puffbacks are black and white; tchagras are gray brown with reddish brown wings; gonoleks are black and red; and typical bushshrikes are either black, white, and chestnut, or brightly colored, often green above with yellow underparts. Puffbacks are so named because males

Distribution: Africa

No. of Living Species: 43

No. of Species Vulnerable, Endangered: 1, 5

No. of Species Extinct Since 1600: 0

have long, loose feathers on their lower backs that they puff out during courtship displays. In most species, males and females look alike, but in some, such as the puffbacks, sexes differ in coloring.

Bushshrikes are found in most terrestrial African habitats, from the leafy canopies of dense forests to the edges of deserts, but most species inhabit woodlands and savanna. Many skulk about in dense, impenetrable (to humans) thickets, and are more often heard than seen. Others frequent more open habitats such as forest midlevels or treetops. A good number, such as the tchagras, typically forage on or near the ground. All are primarily insect-eaters, but some other invertebrates, such as snails, are also taken. Larger bushshrikes will also take small vertebrates such as lizards, frogs, and tiny rodents. Some, especially the boubous, commonly raid other birds' nests to eat eggs and chicks. Most species, however, chiefly hunt insects, usually by moving methodically over tree or shrub trunks and limbs and through foliage. Some will occasionally fly after and grab flying insects. Most bushshrikes are seen foraging alone or in pairs, and many live year-round in pairs on defended territories; some, following breeding, join mixed-species foraging flocks. Many in the family are brilliant singers, with memorable calls or songs. The precisely timed duets produced in some species by mated individuals, the female instantaneously answering the male's notes, especially among the boubous, likely function in maintaining pair-bonds in the dense-vegetation habitats that these birds occupy and also in territory defense.

Most bushshrikes breed in monogamous pairs. Nests are usually open cups, constructed of twigs, stems, rootlets, grass, or bark strips, placed in trees or shrubs, or hidden in dense undergrowth. Some are camouflaged with an outer layer of lichens and spider webbing. In most, both sexes build the nest and feed the young, but sometimes only the female incubates and during this time she may be fed by her mate; there is much variation among species in gender contributions to nesting efforts.

Most birds in the bushshrike family are relatively secure, especially since some have adapted to modified habitats such as agricultural lands. Six species are considered threatened: one is vulnerable, four endangered, and one, Somalia's Bulo Burti Boubou, critically endangered; the latter may now have a tiny population or, indeed, already be extinct. Very small ranges combined with ongoing habitat loss are the main factors jeopardizing these species.

SOUTHERN BOUBOU
Lanarius ferrugineus
8.5 in (21 cm)
Southern Africa

BRUBRU
Nilaus afer
5 in (13 cm)
Sub-Saharan Africa

BLACK-BACKED PUFFBACK
Dryoscopus cubla
6.5 in (17 cm)
Africa

BLACK-CROWNED TCHAGRA
Tchagra senegala
8.5 in (21 cm)
Africa

GRAY-HEADED BUSH-SHRIKE
Malaconotus blanchoti
10 in (25 cm)
Africa

YELLOW FORM

CRIMSON FORM

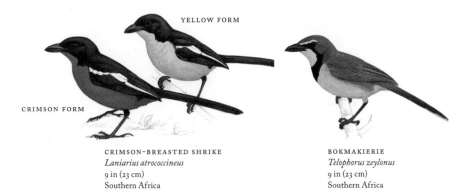

CRIMSON-BREASTED SHRIKE
Laniarius atrococcineus
9 in (23 cm)
Southern Africa

BOKMAKIERIE
Telophorus zeylonus
9 in (23 cm)
Southern Africa

RED-TAILED VANGA
Calicalicus madagascariensis
5.5 in (14 cm)
Madagascar

BLUE VANGA
Cyanolanius madagascarinus
6.5 in (16 cm)
Madagascar

NUTHATCH VANGA
Hypositta corallirostris
5.5 in (14 cm)
Madagascar

LAFRESNAYE'S VANGA
Xenopirostris xenopirostris
9.5 in (24 cm)
Madagascar

SICKLE-BILLED VANGA
Falculea palliata
12.5 in (32 cm)
Madagascar

HELMET VANGA
Euryceros prevostii
11.5 in (29 cm)
Madagascar

Vangas

VANGAS are small to midsize shrikelike birds mainly restricted to the island of Madagascar. Perhaps closely related to helmetshrikes (family Prionopidae) and bushshrikes (family Malaconotidae), and resembling these other groups in some aspects of their appearance and feeding behavior, vangas are sometimes called vanga shrikes. There are fourteen species (family Vangidae); all occur in Madagascar, but one, the Blue Vanga, also occurs in the nearby Comoro Islands.

The group is best known for being one of the few bird families endemic to islands and for its presumed evolutionary history: a single ancestor species may have successfully colonized Madagascar millions of years ago and developed into the multiple vanga species we see today. Sometimes when a colonizing species reaches an isolated island where there are few or no competitors, evolution is sped up and the single species rapidly diverges into several new ones as it fills previously empty habitats and ways of life, or ecological niches. The result (known as an adaptive radiation) is a diverse assemblage of closely related species that occupy an array of habitats but that often differ physically from one another only in small ways. For example, if each species in the group became highly specialized through evolution to pursue a particular food type or specific foraging technique, then many of them might resemble one another in body shape and coloring but have widely differing bills—such as found among the Hawaiian honeycreepers or Galápagos finches—to which the vangas are sometimes compared. Indeed, vanga bills vary widely, from smallish but wide and mostly straight (such as in the Red-tailed and Blue Vangas), to stout but compressed side to side

Distribution: Madagascar, Comoro Islands

No. of Living Species: 14

No. of Species Vulnerable, Endangered: 3, 1

No. of Species Extinct Since 1600: 0

(Lafresnaye's Vanga), to long, humped, and hooked (Helmet Vanga), to long, thin, and highly curved (Sickle-billed Vanga).

Ranging from 5 to 12.5 inches (13 to 32 cm) in length, vangas have long wings and moderately long tails. Coloring is variable, but many are mainly black, white, or gray, or a combination of these; some have brown added into the mix; and the Blue Vanga is blue and white. Bills in many are heavy and hooked at the tip.

Vangas inhabit forests, woodlands, and, in drier regions, even semidesert scrub areas. They are arboreal, mainly canopy birds that, outside breeding seasons, are usually in groups of four to twelve or more. Most species typically associate with mixed-species foraging flocks, and indeed, some of the vangas appear to be chief organizers of these flocks, playing some role in their formation. A few vangas stay in single-species groups. Vangas eat primarily insects but also spiders and small vertebrates (including frogs, lizards, mouse lemurs, and bird nestlings). Feeding methods vary considerably, as suggested by the variety of bill types. Large vangas with robust bills, such as the Helmet Vanga, use them to grab large insects (like beetles) and small vertebrates; others with strong bills, such as Lafresnaye's Vanga, rip bark on dead wood, looking for hiding insects; those with slighter bills seek insects on tree branches or in foliage, or they catch insects in flight; and the Sickle-billed Vanga uses its long bill to probe for food in cracks and crevices in tree bark. Some vangas are known for their melodic songs and, in some species, vocal duets.

Vangas are apparently monogamous. Males have been observed courting females by following them and then displaying to them by fanning their tails and spreading and drooping their wings. Both male and female contribute to nest-building, egg incubation, and feeding young. Nests, bowl-shaped, are made of leaves, moss, lichens, rootlets, bark, and twigs, sometimes bound together with spider webbing, and placed high or low in a tree, depending on the species.

Primarily birds of forests and woodlands, vangas are threatened generally because many of Madagascar's forest regions are under intense pressure from subsistence agricultural development, commercial logging, and increasing human population. Consequently, the populations of all fourteen species may be declining. Currently three vangas are considered vulnerable and one, Van Dam's Vanga, with a very small range in Madagascar's northwest, is endangered. Also, the Nuthatch Vanga (also called Coral-billed Nuthatch) may be endangered, but very little is known of this species.

Wattle-eyes, Batises, and Shrike-flycatchers

WATTLE-EYES, BATISES, and SHRIKE-FLYCATCHERS are a midsize group of small to medium-size, mostly black and white, flycatcher-like birds that occur only in Africa. In the past they were most often thought to be bona fide members of one or another of the formal flycatcher families, sometimes incorporated with the monarch flycatchers (Monarchidae), sometimes considered part of a large, broadly inclusive Old World flycatcher group (Muscicapidae). Now the thirty-one species considered here are usually thought to comprise a separate family, Platysteiridae. Aside from their insect-catching lifestyles, these sometimes inconspicuous birds are perhaps most noted for the obvious wattles some of them have around their eyes, rings of bare skin colored red, blue, greenish, or purplish.

Most wattle-eyes, batises, and shrike-flycatchers are 3 to 5 inches (7.5 to 13 cm) long, but the largest ones, somewhat resembling small shrikes, range up to 7 inches (18 cm). Like true flycatchers, they have short flat bills, broad at the base, surrounded by a fringe of small bristlelike feathers. Their bill shape and bristles, which funnel food to the mouth, help the birds snatch and hold insects. Bills are usually hooked at the tip. These birds have short tails, and wings that are either pointed or rounded; legs vary from longish in batises and some wattle-eyes, to short in shrike-flycatchers and other wattle-eyes. Black and white is the predominant color scheme in the family, especially among males. Many females are reddish brown where males have patches of black. Rump feathers in the group are often light colored, yellow or white, sometimes spotted, and quite fluffy, thus providing an alternative

Distribution:
Africa

No. of Living
Species: 31

No. of Species
Vulnerable,
Endangered: 0, 1

No. of Species Extinct
Since 1600: 0

CAPE BATIS
Batis capensis
5 in (13 cm)
Africa

PRIRIT BATIS
Batis pririt
4.5 in (12 cm)
Southern Africa

JAMESON'S WATTLE-EYE
Dyaphorophyia jamesoni
3.5 in (9 cm)
Africa

BLACK-AND-WHITE SHRIKE-FLYCATCHER
Bias musicus
5 in (13 cm)
Africa

name for the group (applied particularly to batises): puff-back flycatchers. Males and females within a species often differ in looks from each other; this is especially pronounced in shrike-flycatchers. The Chin-spot Batis is one of the few birds to be named after the female and not the male; only females possess a brown throat spot.

Species in this family, many of them common, chiefly inhabit forests and woodlands, including open savanna. Almost all are arboreal, foraging mainly in tree canopies. They are agile flyers, taking a lot of their insect prey on the wing. Some of these birds perch on a tree branch and scan their surroundings, looking for moving insects, then dash out in quick, powerful sallies to catch flying insects in the air or take moving prey from the surfaces of leaves. Many, however, make their living primarily as foliage gleaners; that is, they hop about from twig to twig in a tree's foliage, catching prey on or under leaves, sometimes making short, hovering flights to snap up escaping bugs. The quick, seemingly erratic movements of the birds as they make their way through foliage serve to flush insects from hiding places and make them easily detectable. Large insects or stinging ones are beaten on a perch until they are senseless, and then eaten. In addition to many kinds of insects and spiders, some of these birds will take small scorpions, small lizards, and occasional small fruits. One, the White-tailed Shrike, is largely terrestrial; it has long legs it uses to forage on the ground in scrub areas amid dense vegetation and in low bushes for insects. Most of the batises and wattle-eyes spend their days in pairs or small family parties. Some are common members of mixed-species feeding flocks, groups of up to fifteen or more species of insectivorous birds that, especially in winter, move around together in search of food. Batises and wattle-eyes are nonmigratory; most establish and live on permanent territories.

These birds breed monogamously, generally building shallow cup nests in trees; nests may be partly covered on the outside with spider webbing and bits of lichens, presumably for camouflage. Some species have helpers at the nest, offspring from previous years that assist the parents with feeding young. Both sexes construct the nest, but only the female incubates in most species; both sexes, however, feed young. A single member of this family is threatened: the Banded Wattle-eye, with a tiny range in western Cameroon, is endangered.

BOHEMIAN WAXWING
Bombycilla garrulus
8.5 in (21 cm)
Eurasia, North America

CEDAR WAXWING
Bombycilla cedrorum
7 in (18 cm)
North America, Central America

LONG-TAILED SILKY-FLYCATCHER
Ptilogonys caudatus
9.5 in (24 cm)
Central America

GRAY SILKY-FLYCATCHER
Ptilogonys cinereus
8 in (20 cm)
Mexico, Central America

PHAINOPEPLA
Phainopepla nitens
8 in (20 cm)
North America

PALMCHAT
Dulus dominicus
8 in (20 cm)
Hispaniola

Waxwings;
Silky-flycatchers;
Palmchat

WAXWINGS are smallish, soft-plumaged songbirds with sleek, pointed crests and wing-feather tips modified to look like drops of wax. They are known for eating sugary fruits and for their nomadic movements. One species, the Bohemian Waxwing, is named for its nomadic lifestyle. Three waxwings comprise family Bombycillidae: the Bohemian Waxwing occurs in northern Eurasia and North America, the Cedar Waxwing in North and Central America, and the Japanese Waxwing in East Asia. These pretty birds are 6 to 8 inches (15.5 to 20 cm) long, with slightly hooked, short bills, relatively long, pointed wings, and short legs. They are mainly brown and gray but have dark tails tipped with yellow or red, and the "wax" droplets on their wing feathers, which not all individuals possess and the functions of which are unknown, are red. The sexes look alike.

Waxwings are primarily fruit-eaters that form nomadic flocks during winter that roam the countryside searching for berry-laden trees and shrubs. In summer they still eat fruit, but also take insects from foliage or even catch them in the air, like flycatchers. Outside of the breeding season they are found in many habitats, wherever fruit and berries are available, but they breed in coniferous or mixed coniferous-deciduous forests. They are monogamous, both sexes helping to build the cup nest (the female doing most of the actual construction), made of twigs, grass, leaves, bark shreds, lichens, and moss, in a tree. Only the female incubates; the male brings her food. Both sexes feed the young.

Waxwings are mainly common birds, and with fruit orchards to raid,

WAXWINGS

Distribution:
North and Central
America, Eurasia

No. of Living
Species: 3

No. of Species
Vulnerable,
Endangered: 0

No. of Species Extinct
Since 1600: 0

SILKY-
FLYCATCHERS

*Distribution:
North and Central
America*

*No. of Living
Species: 4*

*No. of Species
Vulnerable,
Endangered: 0, 0*

*No. of Species Extinct
Since 1600: 0*

PALMCHAT

*Distribution:
West Indies*

*No. of Living
Species: 1*

*No. of Species
Vulnerable,
Endangered: 0, 0*

*No. of Species Extinct
Since 1600: 0*

the Cedar Waxwing is probably more abundant and widespread now than before the advent of agriculture. Some consider the Japanese Waxwing, which breeds only in eastern Russia, to be near-threatened.

SILKY-FLYCATCHERS are a group of four species of slender, mostly crested flycatchers with long tails and silky plumage that occur in Central and North America. The family, Ptilogonatidae, is closely related to waxwings (but only distantly related to other flycatchers); indeed, they are sometimes placed with waxwings in family Bombycillidae. Two species occur only in Costa Rica and Panama; one is endemic to Mexico; and one, the Phainopepla, occurs in Mexico and the southwestern United States. Silky-flycatchers, or silkies, are 6.5 to 9.5 inches (17 to 24 cm) long, with short, broad bills, and short wings and legs. They are mainly gray or blue gray but with patches of yellow and/or black and white; females are duller than males. The Phainopepla male is shiny black with white wing patches seen in flight.

Silky-flycatchers inhabit woodlands and forests, although the Phainopepla ranges into desert areas. They feed chiefly by catching insects on the wing, sitting high on exposed perches, such as bare tree branches, and then darting out repeatedly to capture flying insects. They also eat berries, the Phainopepla taking especially mistletoe berries. Most of the silkies form loose flocks during nonbreeding periods, but Central America's Black-and-yellow Silky-flycatcher stays in mated pairs that often join mixed-species feeding flocks. Silkies are monogamous, building shallow, open-cup nests of twigs and grass in trees or shrubs. In the Phainopepla, only the male builds the nest; both sexes incubate eggs and feed young. None of the silkies are threatened.

The PALMCHAT, closely related to silky-flycatchers and waxwings, is a medium-size songbird restricted to the Caribbean island of Hispaniola, and the only member of its family, Dulidae. Greenish brown above and streaked below, 7 to 8 inches (18 to 20 cm) long, Palmchats have slightly down-curved, robust bills, short, rounded wings, and longish tails; the sexes look alike. Palmchats are arboreal birds of open areas with scattered trees, including parks and gardens. They typically occur in noisy flocks, and feed on small fruits, berries, flowers, seeds, and, occasionally, insects. They breed communally, as many as thirty pairs collaborating on constructing a huge nest of sticks and twigs; each mated pair has a separate nesting chamber within the larger nest, with its own entrance. The Palmchat, common and widespread on Hispaniola, is the national bird of the Dominican Republic.

Thrushes; Dippers

THRUSHES, often drab and nondescript, as a group are tremendously successful birds, especially when they have adapted to living near humans and benefited from their environmental modifications. On five continents, thrushes are among the most common and recognizable park and garden birds, including North America's American Robin, Latin America's Rufous-bellied Thrush and Clay-colored Robin, and Europe's Eurasian Blackbird and Redwing. The family, Turdidae, with about 175 species, is distributed on all continents except Antarctica. The family includes birds called rock-thrushes, whistling-thrushes, ground-thrushes, nightingale-thrushes, bluebirds, shortwings, alethes, rock-jumpers, and solitaires, as well as robins and typical thrushes.

Thrushes vary in appearance but generally they are medium-size songbirds, slim or slightly plumpish, with round heads, slender bills, often square-ended tails, and strong legs. Most range in length from 5 to 12 inches (13 to 30 cm). Generally they are not brightly colored, dressed in browns, brown reds, grays, olive, and perhaps some black and white. The sexes often look similar; juveniles are usually spotted.

Many thrushes eat fruits; some primarily consume insects; and most are at least moderately omnivorous. Although arboreal, many thrushes frequently forage on the ground for insects, other arthropods, and a particular favorite, earthworms. Many species forage like the familiar thrushes of North America and Europe; they hop and walk along the ground, stopping at intervals and cocking their heads to peer downward. These birds occupy many habitats: forest edges and clearings and other open sites such as shrub areas, grasslands, gardens, parks, suburban lawns, and agricultural lands. Many are quite social, spending

THRUSHES

Distribution:
All continents except
Antarctica

No. of Living
Species: 173

No. of Species
Vulnerable,
Endangered: 11, 8

No. of Species Extinct
Since 1600: 5 or 6

RUFOUS-THROATED SOLITAIRE
Myadestes genibarbis
7.5 in (19 cm)
West Indies

CUBAN SOLITAIRE
Myadestes elisabeth
7.5 in (19 cm)
Cuba

BARE-EYED ROBIN
Turdus nudigenis
9 in (23 cm)
South America, West Indies

VARIED THRUSH
Ixoreus naevius
9.5 in (24 cm)
North America

MOUNTAIN BLUEBIRD
Sialia currucoides
7 in (18 cm)
North America

RED-LEGGED THRUSH
Turdus plumbeus
10.5 in (27 cm)
West Indies

RUFOUS-BELLIED THRUSH
Turdus rufiventris
9.5 in (24 cm)
South America

TOWNSEND'S SOLITAIRE
Myadestes townsendi
8.5 in (22 cm)
North America

RUSSET NIGHTINGALE-THRUSH
Catharus occidentalis
7 in (18 cm)
Mexico

AZTEC THRUSH
Zoothera pinicola
9 in (23 cm)
Mexico

IMM

OMAO
Myadestes obscurus
7 in (18 cm)
Hawaii

SIBERIAN RUBYTHROAT
Luscinia calliope
6.5 in (16 cm)
Asia

WHITE-RUMPED SHAMA
Copsychus malabaricus
8.5 in (22 cm)
Southern Asia

ORIENTAL MAGPIE-ROBIN
Copsychus saularis
8 in (20 cm)
Southern Asia

SUNDA WHISTLING-THRUSH
Myiophoneus glaucinus
10 in (25 cm)
Southeast Asia

BLUE ROCK-THRUSH
Monticola solitarius
8.5 in (22 cm)
Eurasia, Africa

CAPE ROCK THRUSH
Monticola rupestris
8 in (20 cm)
Southern Africa

GROUNDSCRAPER THRUSH
Psophocichla litsipsirupa
9 in (23 cm)
Africa

ORANGE-BREASTED ROCKJUMPER
Chaetops aurantius
9.5 in (24 cm)
Southern Africa

OLIVE-TAILED THRUSH
Zoothera lunulata
11 in (28 cm)
Australia, New Guinea

WHITE-CAPPED DIPPER
Cinclus leucocephalus
6 in (15 cm)
South America

AMERICAN DIPPER
Cinclus mexicanus
7.5 in (19 cm)
North America, Central America

their time during nonbreeding months in same-species flocks. Some tropical thrushes make seasonal migrations from higher to lower elevations, following abundant food supplies; some of the temperate-zone species are long-distance migrants, a few moving up to 6,000 miles (10,000 km) twice a year.

Thrushes are monogamous. Male and female defend an exclusive territory during the breeding season; pairs may associate year-round. Nests, usually built by the female in tree branches, shrubs, or crevices, are cup-shaped, made of grass, moss, and like materials, and often lined with mud. Only the female incubates eggs; young are fed by both parents.

A few European thrushes were intentionally spread to other continents by European settlers during the 1800s, for instance, the Song Thrush and Eurasian Blackbird are now naturalized citizens of Australia and New Zealand. With these human-assisted range expansions and their liking of human-altered landscapes, several thrush species are now more common and widespread than they were 500 years ago. However, about twenty species worldwide are threatened, with at least eleven considered to be vulnerable and eight endangered. A few species of thrushes endemic to the Hawaiian Islands became extinct during the last few decades, and another Hawaiian species, the Puaiohi, is now critically endangered.

DIPPERS are medium-size songbirds specialized for living along fast-flowing streams in mountainous areas; they feed by wading or by plunging from a rock or stream bank into rushing water, then using their wings to swim underwater searching for food. They are called dippers because they bounce up and down on their legs every few seconds while standing or foraging. The family, Cinclidae, has five species distributed over large swaths of the Americas and Eurasia and a small part of Northern Africa. Dippers are stocky and 6.5 to 8 inches (16 to 20 cm) long, all gray or brownish or with white or reddish brown areas below; the sexes look alike.

Foraging primarily for larvae of aquatic insects, dippers will also take small clams, crayfish, and fish. They walk into the water of rapidly moving streams and then forage among underwater rocks and on stream bottoms; they will also float on the surface like a duck and paddle with their unwebbed toes. Each dipper pair is territorial and defends a stretch of stream year-round. Monogamous breeders, they construct large roofed nests of moss and leaves, usually on vertical cliffs, sometimes behind waterfalls, and now often on the undersides of bridges. Both sexes participate in nest construction and feed the young; only the female incubates the eggs. One species, South America's Rufous-throated Dipper, is vulnerable.

DIPPERS

Distribution:
New World, Eurasia,
northern Africa

No. of Living
Species: 5

No. of Species
Vulnerable,
Endangered: 1, 0

No. of Species Extinct
Since 1600: 0

WHITE-CROWNED FORKTAIL
Enicurus leschenaulti
11 in (28 cm)
Southern Asia

VERDITER FLYCATCHER
Eumyias thalassina
6.5 in (16 cm)
Southern Asia

GRAY-HEADED CANARY-FLYCATCHER
Culicicapa ceylonensis
5 in (13 cm)
Southern Asia

RED-THROATED FLYCATCHER
Ficedula parva
5 in (13 cm)
Eurasia

HILL BLUE-FLYCATCHER
Cyornis banyumas
6 in (15 cm)
Southern Asia

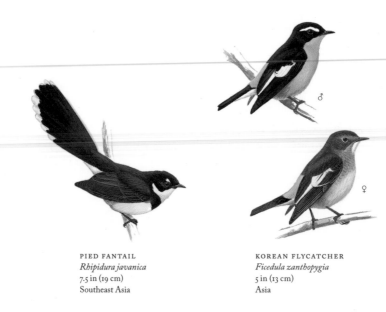

PIED FANTAIL
Rhipidura javanica
7.5 in (19 cm)
Southeast Asia

KOREAN FLYCATCHER
Ficedula zanthopygia
5 in (13 cm)
Asia

Old World Flycatchers

OLD WORLD FLYCATCHERS are a large group of mostly small, mostly forest and woodland songbirds that, in the main, make their livings by catching flying insects. Many are drab and fairly inconspicuous, but others are brighter and more easily noticed, and some, such as the European Robin, Pied Flycatcher, and Indian Robin, are common denizens of parks and gardens. Some have reputations as fine singers. Birds such as nightingales and shamas, with loud, melodic songs, claim top honors, and some are good mimics of other birds' songs. The family, Muscicapidae, has a checkered classification history. Now, with supportive DNA evidence, it is often considered to encompass about 270 species, including not only a multitude of flycatchers (and the Asian niltavas), but a subgroup that in the past was often considered part of the thrush family, including akalats, nightingales, rubythroats, shamas, redstarts, forktails, cochoas, chats, robin-chats, scrub-robins, stonechats, and wheatears. Old World flycatchers are widely distributed in Africa, Europe, and Asia; only 2, the Bluethroat and Northern Wheatear, occur in the New World (in Alaska and northern Canada).

Being such a large assemblage, Old World flycatchers, not surprisingly, vary considerably in shape, color, and size. Most are small, 4 to 6 inches (10 to 15 cm) long, but some in the group with long tails (shamas, forktails) range up to 10 inches (25 cm), and the chunky Asian cochoas range up to 12 inches (30 cm). Many come in dull browns or grays, but some are black and white, or black and yellow, or blue and white, or have reddish chests; forktails are striking black and white birds, and cochoas are blue, green,

Distribution:
Old World, northern
North America

No. of Living
Species: 270

No. of Species
Vulnerable,
Endangered: 22, 8

No. of Species Extinct
Since 1600: 1

MALAYSIAN BLUE-FLYCATCHER
Cyornis turcosus
5.5 in (14 cm)
Southeast Asia

ASIAN BROWN FLYCATCHER
Muscicapa dauurica
5 in (13 cm)
Asia

GREEN COCHOA
Cochoa viridis
11 in (28 cm)
Southern Asia

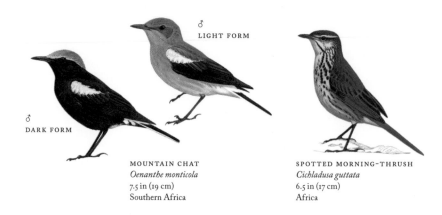

♂
LIGHT FORM

♂
DARK FORM

MOUNTAIN CHAT
Oenanthe monticola
7.5 in (19 cm)
Southern Africa

SPOTTED MORNING-THRUSH
Cichladusa guttata
6.5 in (17 cm)
Africa

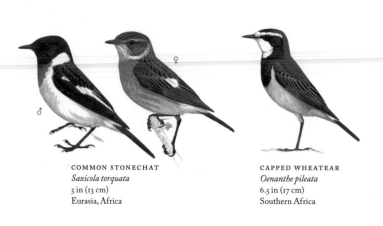

♂

♀

COMMON STONECHAT
Saxicola torquata
5 in (13 cm)
Eurasia, Africa

CAPPED WHEATEAR
Oenanthe pileata
6.5 in (17 cm)
Southern Africa

CLIFF CHAT
Myrmecocichla cinnamomeiventris
8.5 in (21 cm)
Africa

MARICO FLYCATCHER
Bradornis mariquensis
7 in (18 cm)
Southern Africa

SICKLE-WINGED CHAT
Cercomela sinuata
6 in (15 cm)
Southern Africa

CHORISTER ROBIN-CHAT
Cossypha dichroa
8 in (20 cm)
Africa

RED-CAPPED ROBIN-CHAT
Cossypha natalensis
6.5 in (17 cm)
Africa

EASTERN BEARDED SCRUB-ROBIN
Cercotrichas quadrivirgata
6 in (15 cm)
Africa

KALAHARI SCRUB-ROBIN
Cercotrichas paena
6 in (15 cm)
Southern Africa

FISCAL FLYCATCHER
Sigelus silens
7.5 in (19 cm)
Southern Africa

black, and purplish. The sexes are usually similar in appearance, although in some, males are more brightly colored. Most in the family have short, flattened bills, broad at the base, with bristlelike feathers sticking out from the bill base (which help funnel flying insects into the mouth and perhaps protect eyes from struggling bugs). Wings are often long, permitting the rapid, acrobatic flight necessary to seize flying insects; legs are short and weak; tails of many are long, rounded, or wedge-shaped. Many perched flycatchers compulsively flick their tails.

Flycatchers chiefly inhabit forests and woodlands, often staying in the canopy, but others are birds of forest edges, clearings, parks, and gardens. Some, such as many chats and wheatears, are open-country birds, found in meadows, tundra regions, and even deserts. A few species, such as redstarts and forktails, dwell among rocks of fast-moving streams and rivers. Many flycatchers feed by sitting on an elevated perch and waiting until insects fly by. They then launch themselves into the air and, with some chasing and maneuvering, snatch the insects out of the air with their bills, then return to the same perch to dine. Other species swoop to the ground to catch moving insects there, whereas still others hop about in the canopies of trees or forest undergrowth or in shrubs, taking insects from leaves and branches. Several species also eat fruits when they are available. Many are solitary during nonbreeding months, but some forest dwellers participate in mixed-species foraging flocks. In the tropics, many flycatchers remain in the same areas year-round; species that breed at high northern latitudes migrate south for winter, as far as southern Africa and Southeast Asia. Most species are territorial during breeding periods.

Old World flycatchers are monogamous or polygamous. The male Pied Flycatcher is famous for attracting one female to his first territory and then, soon after, establishing another territory and attempting to attract a second mate. Some flycatchers nest in tree holes, others in rock crevices, and some build cup nests in trees. Others in the family (some wheatears and chats) build nests in old rodent burrows or dig nest tunnels in earthen banks or termite mounds. Both parents usually help in nest construction; both or only the female incubates eggs; and both feed young. In at least a few species, young stay with parents to help feed young at subsequent nests. Many flycatchers are common birds, but at least thirty species in the family are threatened, principally by deforestation; twenty-two are vulnerable, and eight are endangered.

Starlings, Mynas, and Oxpeckers

STARLINGS, MYNAS, and OXPECKERS comprise a highly successful group of sturdily built, often dark, medium-size songbirds widespread in the Old World. They are renowned for their often highly iridescent plumage and their vocal abilities. Many have loud ringing or noisy voices, and some imitate sounds of other animals; some, such as the Hill Myna, can mimic human speech quite well. The two species of oxpeckers, restricted to Africa, are known for their habit of clinging to the hides of large, often hairy mammals and feeding on ticks and other parasites attached to giraffes, rhinos, buffalos, and zebras. But in North America and some other regions, the starling family is known best for the incredible competitiveness and opportunism of, not to mention agricultural and ecological damage caused by, some of its members, which have been spread by people to previously alien lands. The prime example is the European Starling, native to Eurasia, which was spread by people to the New World and to Australia and New Zealand during the nineteenth century and is now one of the world's most abundant birds. Likewise, southern Asia's Common Myna was introduced by people to places like Florida, South Africa, and Hawaii and other Pacific islands. Large flocks of these nonnative birds often cause serious damage to fruit and seed crops (as they sometimes do in their natural haunts), and they outcompete some native birds for breeding holes, driving down their populations.

The starling family, Sturnidae, contains 114 species, including mynas, oxpeckers, and glossy-starlings. They are 7 to 17.5 inches (18 to 45 cm) long, usually stocky, with strong, sharp, pointed bills and stout legs. Most species

Distribution:
Old World

No. of Living
Species: 114

No. of Species
Vulnerable,
Endangered: 5, 4

No. of Species Extinct
Since 1600: 4

SUPERB STARLING
Lamprotornis superbus
7.5 in (19 cm)
Africa

WATTLED STARLING
Creatophora cinerea
8 in (20 cm)
Africa, Arabia

AFRICAN PIED STARLING
Spreo bicolor
11 in (28 cm)
Southern Africa

BURCHELL'S GLOSSY-STARLING
Lamprotornis australis
13 in (33 cm)
Southern Africa

CAPE GLOSSY-STARLING
Lamprotornis nitens
10 in (25 cm)
Southern Africa

BLACK-BELLIED GLOSSY-STARLING
Lamprotornis corruscus
7 in (18 cm)
Africa

GREATER BLUE-EARED GLOSSY-STARLING
Lamprotornis chalybaeus
9 in (23 cm)
Africa

COMMON MYNA
Acridotheres tristis
10 in (26 cm)
Southern Asia

HILL MYNA
Gracula religiosa
11.5 in (29 cm)
Southern Asia

WHITE-VENTED MYNA
Acridotheres grandis
10.5 in (27 cm)
Southern Asia

GOLDEN-CRESTED MYNA
Ampeliceps coronatus
9 in (23 cm)
Southern Asia

YELLOW-BILLED OXPECKER
Buphagus africanus
8.5 in (22 cm)
Africa

BRISTLE-CROWNED STARLING
Onychognathus salvadorii
16.5 in (42 cm)
Africa

VIOLET-BACKED STARLING
Cinnyricinclus leucogaster
6.5 in (17 cm)
Africa, Arabia

BRD   NON-BRD

EUROPEAN STARLING
Sturnus vulgaris
8.5 in (22 cm)
Eurasia

ASIAN GLOSSY STARLING
Aplonis panayensis
8.5 in (21 cm)
Southeast Asia

♀

♂

WHITE-SHOULDERED STARLING
Sturnus sinensis
8 in (20 cm)
Southern Asia

ASIAN PIED STARLING
Sturnus contra
9.5 in (24 cm)
Southern Asia

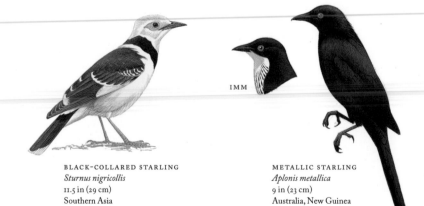

IMM

BLACK-COLLARED STARLING
Sturnus nigricollis
11.5 in (29 cm)
Southern Asia

METALLIC STARLING
Aplonis metallica
9 in (23 cm)
Australia, New Guinea

are predominantly black or brown, but many also have patches of white. Some, however, such as many African starlings, are stunning, clad in iridescent blue greens and orange, or violet. Several mynas are quite striking in black and yellow. Oxpeckers are gray brown, with sharp, curved claws to help them cling to their hairy food sources. Some in the family have crests, bare skin on their head, or fleshy, hanging wattles. Males and females generally look alike, although in some, males are more colorful.

Mynas and starlings are mainly arboreal, but some also function well on the ground. They inhabit forests, savanna, and grasslands. Most are opportunistic and omnivorous, although a few prefer insects or flower nectar, and some, particularly in the tropics, feed mainly on fruit. Many that forage on the ground have particularly strong head muscles that allow them to insert their closed bills into the ground or under stones or debris and then force them open (called gaping), moving soil or stones to reveal hiding bugs. Many also take advantage of insects disturbed by the movements of wild and domesticated mammals. Starlings tend to be gregarious, feeding and roosting in groups, and sometimes gathering into enormous flocks. In Africa large groups often form around herds of elephants and zebras, eating insects scared up by the moving mammals, and in North America, flocks of European Starlings are crop pests, as are groups of Common Mynas in southern Asia. The flocking and nomadic Wattled Starling moves over Africa, following its main food source, locusts.

Starling breeding varies from monogamy to cooperative nesting, as with the oxpeckers, in which young of previous years help a breeding pair raise young. Most nest in tree or cliff holes, but some, such as the Asian Pied Starling, construct massive domed nests made of twigs and other vegetation that are placed in isolated trees. Some nest in holes made by woodpeckers or barbets. Females only or both sexes incubate eggs; both feed the young.

Many starlings and mynas are common and widespread, and some have even expanded their natural ranges where people have transformed heavily forested habitats into open agricultural lands. Nonetheless, nine species, most of them island dwellers, are now threatened—five are considered to be vulnerable and four endangered. Indonesia's Bali Starling and Micronesia's Pohnpei Mountain Starling, both critically endangered, are some of the rarest birds on Earth. Populations of Hill Mynas in southern Asia are depleted in some regions because these pretty vocal mimics are captured for the pet trade.

BLUE MOCKINGBIRD
Melanotis caerulescens
10 in (26 cm)
Mexico

LONG-TAILED MOCKINGBIRD
Mimus longicaudatus
11.5 in (29 cm)
South America

NORTHERN MOCKINGBIRD
Mimus polyglottos
10 in (25 cm)
North America, West Indies

GRAY CATBIRD
Dumetella carolinensis
8.5 in (21 cm)
North America, Central America, West Indies

CRISSAL THRASHER
Toxostoma crissale
11.5 in (29 cm)
North America

PEARLY-EYED THRASHER
Margarops fuscatus
11.5 in (29 cm)
West Indies

SCALY-BREASTED THRASHER
Margarops fuscus
9 in (23 cm)
West Indies

BROWN TREMBLER
Cinclocerthia ruficauda
10 in (25 cm)
West Indies

Mockingbirds
and Thrashers

MOCKINGBIRDS and THRASHERS are handsome, long-tailed songbirds best known for the striking ability some of them possess to imitate the calls and songs of other birds, and even sounds of other types of animals. The United States' broadly distributed Northern Mockingbird is one of the most impressive of the group's mimics, often producing loud songs that include mixtures of mockingbird-specific and imitated vocalizations, sometimes incorporating the mimicked songs of ten or more different bird species. The reason for the vocal mimicry (in many mockingbirds as well as in some thrashers and catbirds) is not known for sure, but it probably relates to mate attraction: males with longer, more complex, and more varied songs may be more attractive to females. Some in the group are continuous singers, producing not the usual brief two- or three-second-long songs associated with many common songbirds, but long strings of vocalizations—incredible, virtuoso singing performances.

The mockingbird family, Mimidae, distributed from southern Canada to southern South America, consists of thirty-five species of mockingbirds (or mockers), thrashers, catbirds (named for their catlike mewing vocalizations), and tremblers (named for their habit of drooping their wings and trembling); most are tropical. Birds in the family, 8 to 13 inches (20 to 33 cm) long, are slender-looking, with fairly short, rounded wings, and long, sturdy legs for hopping about on the ground. Most have strong, moderately long bills that are often down-curved. They tend toward gray, brown, or reddish brown, often with lighter underparts and flashes of white in the tail and wings; some are streaked or spotted. Male and female generally look alike.

Distribution:
New World

No. of Living
Species: 35

No. of Species
Vulnerable,
Endangered: 1, 4

No. of Species Extinct
Since 1600: 0

With their loud and frequent vocalizations and long tails, and in the open habitats they prefer, mockingbirds are often highly conspicuous. They occur in scrub areas as well as forest edges and some forests; they are also common garden and park birds. Mockingbirds, catbirds, and most thrashers are mainly birds of the ground, shrubs, and low trees. They skulk about, using vegetation as cover, foraging for insects and other small invertebrates, also taking some fruit and berries. Many use their bills to dig for prey in soft soils and leaf litter. The two trembler species, restricted to the West Indies, are more arboreal, using their long bills to toss and tear vegetation, looking for insects. Birds in this family are usually seen solitarily or in pairs; the Galápagos mockingbirds occur in small family groups. (The Galápagos mockingbirds, now usually considered to number four species, have a specialized diet as well: in addition to insects, they eat bird and lizard eggs, small crabs, and ticks plucked from the skin of iguanas.) These birds, especially the mockers, are known for their aggressive territoriality during breeding seasons; many a person who wandered innocently across a mockingbird territory during nesting has been struck on the head by the swooping mockers. They defend their territories against not only other mockingbirds but also other birds as well. Species that breed in northern regions often migrate south for the winter; tropical species may remain on their territories year-round.

Mockingbirds and their kin are monogamous. Bulky, open-cup nests, sometimes lined with grass or hair, are built of twigs and leaves by both sexes or by the female alone, usually on the ground or in a bush. (Tremblers build domed nests or breed in tree cavities.) The female alone (mockers and catbirds) or both sexes (thrashers) incubate; both parents feed young. Galápagos mockingbirds raise their young in nests that are tended cooperatively by a group of helpers that are close relatives of the parents.

Five in the mockingbird/thrasher group are threatened: one of the Galápagos mockingbirds is vulnerable; another Galápagos mockingbird and a West Indies thrasher are endangered; and the Cozumel Thrasher (possibly extinct) and Socorro Mockingbird (endemic to one of Mexico's Pacific islands; only a few hundred individuals are left) are critically endangered. The Black Catbird, a slim, iridescent black beachcomber with red eyes, endemic to the Yucatán Peninsula, is apparently rare over much of its range and is considered near-threatened.

Nuthatches;
Sittellas

NUTHATCHES are mainly smaller forest birds of north temperate regions and also of Asian tropics. The twenty-five species, family Sittidae (including the Wallcreeper, which is sometimes placed in its own family, Tichodromadidae), excel at tree climbing, being the only birds that regularly climb not just up tree trunks, but down, headfirst, also. Another of the group's claims to fame is that the Brown-headed Nuthatch is one of the very few tool-using birds. Some individuals break off small pieces of bark and use them to pry up other pieces to find and dislodge insects that would otherwise be inaccessible. Four to 8 inches (10 to 20 cm) long, nuthatches have compact bodies, long, tapered bills, short necks, and short tails. They are blue gray or blue green above and paler below, with dark caps, dark eye lines, or both. Females are often duller than males.

Nuthatches use their long, slender bills to probe into cracks in tree bark and pry up loose pieces of bark. They also probe moss clumps and leaf masses on branches. They seek mainly insects and spiders, but also take some seeds, especially in winter, and, as their name suggests, nuts. Two Eurasian species, known as rock nuthatches, occupy rocky slopes, cliffs, and gorges, and forage for insects, spiders, snails, and seeds, mostly on rocks and the ground. Nuthatches cache seeds in the fall, and then visit their caches during winter, when food is scarce, seeming to remember many of their thousands of hiding spots. Usually seen singly or in pairs, they are strongly territorial; some species, however, form flocks during nonbreeding periods.

The Wallcreeper is considered by bird fanciers to be one of Eurasia's

NUTHATCHES

Distribution:
North America,
Eurasia, North Africa

No. of Living
Species: 25

No. of Species
Vulnerable,
Endangered: 2, 2

No. of Species Extinct
Since 1600: 0

RED-BREASTED NUTHATCH
Sitta canadensis
4.5 in (11 cm)
North America

BROWN-HEADED NUTHATCH
Sitta pusilla
4.5 in (11 cm)
North America, West Indies

VELVET-FRONTED NUTHATCH
Sitta frontalis
5 in (13 cm)
Southeast Asia

CHESTNUT-BELLIED NUTHATCH
Sitta castanea
5 in (13 cm)
Southern Asia

WALLCREEPER
Tichodroma muraria
6.5 in (17 cm)
Eurasia

VARIED SITTELLA
Daphoenositta chrysoptera
4.5 in (12 cm)
Australia, New Guinea

most spectacular small birds. Dull blackish and gray but with large red wing patches and a long, slender, down-curved bill, it occurs, somewhat uncommonly, over mountainous areas from Spain to China. As it forages for insects and spiders on cliffs and rock faces, it constantly flicks its wings open, revealing more of its red coloring.

Nuthatches are monogamous hole nesters. They excavate their own nests in dead trunks or branches or enlarge small natural cavities; a few nest in rock crevices. Some species narrow the nest opening with mud, for protection. Both sexes build nests (one sex often predominates), but only the female incubates eggs; both sexes feed young. Most nuthatches are common birds of forests and woodlands, and many live easily in close proximity to people. They are some of the most frequent and charming visitors to suburban bird feeders. Four Old World nuthatches are threatened (two are vulnerable, two are endangered).

SITTELLAS are small birds (5 inches, 12 cm, long) of tree trunks and branches that occur only in Australia and New Guinea. The family, Neosittidae, has two species, both compact birds with longish, slim bills, and short tails. New Guinea's Black Sittella is all black with a pink face and tail tip; the Varied Sittella of Australia and New Guinea is gray-brown, black, and white. Males and females show slight differences in color pattern. The Black Sittella is mainly confined to higher-elevation forests; Varied Sittellas occur in a range of habitat types, including forests, woodlands, shrublands, orchards, parks, and gardens.

Also known as treerunners, sittellas climb rapidly around a tree's trunk and branches, foraging for insects and spiders; they probe in cracks and crevices, and use their bills to pry off pieces of bark, looking for hidden bugs. They usually fly to near the top of a tree trunk and work downward, or onto branches and work toward the trunk. With their sharp claws they are able to cling to the undersides of limbs. Sittellas usually traverse woodlands in communal groups of three to twelve or more; the groups are very social, the members huddling together when perched on tree branches and often preening each other. Sittellas breed communally, with the dominant male and female reproducing, the others in the group, presumably relatives, helping construct nests and feed the young. The breeding female alone incubates eggs in the deep-cup nest placed in a tree fork. Neither of the sittellas is threatened, but the Black Sittella is uncommon.

SITTELLAS

Distribution:
Australia, New Guinea

No. of Living
Species: 2

No. of Species
Vulnerable,
Endangered: 0, 0

No. of Species Extinct
Since 1600: 0

BLACK-CAPPED DONACOBIUS
Donacobius atricapillus
8.5 in (22 cm)
South America

INCA WREN
Thryothorus eisenmanni
6.5 in (16 cm)
South America

SOUTHERN NIGHTINGALE-WREN
Microcerculus marginatus
4.5 in (11 cm)
Central America, South America

WINTER WREN
Troglodytes troglodytes
4 in (10 cm)
Eurasia, North America

SPOTTED WREN
Campylorhynchus gularis
6.5 in (17 cm)
Mexico

CAROLINA WREN
Thryothorus ludovicianus
5.5 in (14 cm)
North America, Central America

HOUSE WREN
Troglodytes aedon
4.5 in (12 cm)
North America, South America

Wrens

Distribution:
New World, Eurasia,
North Africa

No. of Living
Species: 80

No. of Species
Vulnerable,
Endangered: 3, 3

No. of Species Extinct
Since 1600: 0

WRENS are mostly small brownish songbirds with an active, snappish manner and characteristically upraised tails. They are recognized for a variety of traits: their vocal abilities (long, complex songs and duetting), breeding habits (some are cooperative nesters; some are polygamous), capacity to live in association with people (the common House Wren, ranging from Canada to southern Chile, tends to forage around and nest in human-crafted structures), and geographic distribution. The eighty wren species, comprising family Troglodytidae, are confined to the Western Hemisphere—all, that is, except one. The single exception, known as the Winter Wren or Northern Wren, in addition to its wide range in North America, occurs over a broad swath from northern Africa through Eurasia (where, understandably, it is usually known simply as Wren). All in the family are generally called wrens except one marsh-inhabiting South and Central American species, previously believed to be part of the mockingbird family, called the Black-capped Donacobius.

Ranging in length from 4 to 8.5 inches (10 to 22 cm), many toward the smaller extreme, wrens appear usually in shades of brown or reddish brown, with smaller bits of gray, tan, black, and white. Some are mainly black and white, and some are heavily spotted or streaked; wings and tails are frequently embellished with fine bars. Bills in the group are slender and slightly down-curved; eye stripes are common. Wrens have rather broad, short wings and owing to this, are considered poor flyers. The sexes look alike. Some of these birds are tiny, weighing less than half an ounce (15 grams).

Most wrens skulk about in thick ground vegetation, but a few are arboreal, staying in trees in more open areas. Wrens' cryptic coloration is advantageous as they flip, flutter, hop, and poke around low levels of forests and through thickets, grasslands, and marshes, foraging for insects and other small arthropods. Some, like the Cactus, Canyon, and Rock Wrens, inhabit a variety of barren rocky and arid areas, such as those of the southwestern United States and northern Mexico. Many wrens live year-round in pairs, defending territories in which during the breeding season they nest. Some larger species spend their days in small family flocks and, owing to their size, are a bit bolder in their movements. After using their nests for breeding, wrens will use them as overnight roosting sites. Those that breed in northern climes are migratory, those in Canada and the northern United States, for instance, moving to the U.S. coasts or southern states for the winter. Wren vocalizations have been studied extensively. A pair will call back and forth as they lose sight of each other while foraging in thickets, keeping in contact. In certain species, mated pairs sing some of the world's most complex avian duets, male and female alternating song phrases so rapidly and expertly that it sounds as if one individual utters the entire sequence. Such duets probably function as keep-out signals, warning away from the pair's territory other members of the species, and in maintaining the pair-bond between mated birds. Other wrens, such as the Winter Wren, have amazingly complex songs, trains of notes in varied sequences up to 10 seconds long.

Most wrens, especially the tropical ones, are monogamous, with a single pair carrying out all nesting duties, but some breed cooperatively, with members of the small family group helping out at the single nest of the parents. Several species, such as North America's Marsh Wren, are polygamous. Nests, generally of woven grass, are placed in vegetation or in tree cavities. (The family name, Troglodytidae, arose because many species nest in cavities, both natural and artificial; troglodytes are cave dwellers.) Nests are small but elaborate, roofed, with inconspicuous side entrances. In some species, the male builds many more nests on his territory than his mate (or mates, in polygamous species) can use, apparently as a courtship signal, perhaps as an inducement for a female to stay and mate. Only the female incubates, sometimes being fed by the male; both parents feed young. Most wrens are fairly or very abundant. However, six species (half of them island dwellers) are threatened: three are vulnerable, three are endangered.

Gnatcatchers;
Kinglets

GNATCATCHERS are small, slender songbirds that flit about tree foliage, seeking insect prey. The fifteen species, family Polioptilidae, include three called gnatwrens. All are confined to the New World, where they range from extreme southern Canada to northern Argentina. In the past, the gnatcatchers were usually included in the very large family Sylviidae, the Old World Warblers, but recent studies suggest they are not closely related to that group. Gnatcatchers and gnatwrens are known among bird fanciers as active, agile little birds that often constantly wave or twitch their tails, which are usually held in an upright, or cocked, posture.

Four to 5 inches (10 to 12.5 cm) in length, gnatcatchers are mainly bluish gray, with long, narrow, black and white tails and, usually, some black on their heads. Gnatwrens, 4 to 4.75 inches (10 to 12 cm) long, as you might expect, are more wrenlike in appearance, being predominantly brown. Two of the gnatwrens have very stubby tails, and one, the Long-billed Gnatwren, has a long tail. All in the family have long, slender bills, longer in gnatwrens. Males and females look alike or almost so.

Gnatcatchers, chiefly arboreal, move quickly about forests, forest edges, woodlands, mangroves, and even some semiopen scrub areas. Most usually stay fairly high in trees, but gnatwrens prefer undergrowth regions of forests and woodlands. They all take small insects and spiders from foliage and other vegetation. The birds' rapid movements and constant tail-waving may help flush bugs from hiding spots. These birds occur solitarily, in pairs, or small groups; some typically join mixed-species feeding flocks. The most northerly species, the Blue-gray Gnatcatcher, which occurs over a good portion of the

GNATCATCHERS

Distribution:
New World

No. of Living
Species: 15

No. of Species
Vulnerable,
Endangered: 0, 0

No. of Species Extinct
Since 1600: 0

MASKED GNATCATCHER
Polioptila dumicola
5 in (13 cm)
South America

BLUE-GRAY GNATCATCHER
Polioptila caerulea
4.5 in (11 cm)
North America, West Indies

LONG-BILLED GNATWREN
Ramphocaenus melanurus
4.5 in (12 cm)
Mexico, Central America, South America

BLACK-TAILED GNATCATCHER
Polioptila melanura
4 in (10 cm)
North America

CUBAN GNATCATCHER
Polioptila lembeyei
4 in (10 cm)
Cuba

GOLDEN-CROWNED KINGLET
Regulus satrapa
4 in (10 cm)
North America, Central America

RUBY-CROWNED KINGLET
Regulus calendula
4.5 in (11 cm)
North America, West Indies

United States, is migratory; its northernmost breeding populations move seasonally over long distances, including to Mexico and the Caribbean.

Monogamous breeders, gnatcatchers build cup nests in trees, well out from the trunk, weaving together vegetation, bark strips, moss, and spider webbing. Both sexes build the nest, although the male sometimes predominates; both sexes incubate eggs and feed young. Some species in the family are fairly rare, but only one, South America's Creamy-bellied Gnatcatcher, is considered near-threatened. The California Gnatcatcher is not globally threatened (it occurs in southern California and Baja California), but the coastal subspecies is listed by the United States as threatened; and this tiny bird, broadly protected by the U.S. Endangered Species Act, has a reputation of being able to halt coastal development projects in southern California.

KINGLETS are tiny, hyperactive, mostly dully colored songbirds that have patches of bright color on their heads. There are six species in the family, Regulidae. Two, Ruby-crowned and Golden-crowned Kinglets, are widely distributed in North America; two, the Goldcrest and Firecrest, occur over broad swaths of Eurasia and bits of northern Africa; one occurs on Taiwan and one in the Canary Islands. Three to 4.5 inches (8 to 11 cm) long, kinglets are olive green or light brownish above and lighter below, with slender, pointed bills and pale wing bars. Their crowns have bright red, orange, and/or yellow stripes or patches. Not all these color patches are always noticeable, however, because sometimes the brightly colored feathers are only erected during breeding behaviors such as singing and courtship. Females are drabber than males and may lack bright head coloring.

Kinglets occur mainly in coniferous forests, but also in some deciduous and mixed forests, woodlands, and parks. They move quickly and actively about the canopy, searching for insects (they also take spiders and small amounts of seeds). They usually travel over small outside branches, probing bark and leaves, and often hang and flutter under leaves, to search for prey there. Northerly breeding kinglet populations are migratory. Kinglets are monogamous, pairs defending territories during the breeding season. Both sexes or only the female builds small elongated or cup-shaped nests of various plant materials that hang high in trees; only the female incubates, sometimes surprisingly large clutches of ten or more eggs; both sexes feed the young. None of the kinglets are threatened; some are incredibly numerous forest birds, and the Golden-crowned Kinglet has even been expanding its breeding range.

KINGLETS

Distribution:
North and Central
America; Eurasia

No. of Living
Species: 6

No. of Species
Vulnerable,
Endangered: 0, 0

No. of Species Extinct
Since 1600: 0

GRAY-HEADED PARROTBILL
Paradoxornis gularis
7 in (18 cm)
Southern Asia

BEARDED REEDLING
Panurus biarmicus
6.5 in (17 cm)
Eurasia

GREAT PARROTBILL
Conostoma oemodium
11 in (28 cm)
Southern Asia

VERDIN
Auriparus flaviceps
4.5 in (11 cm)
North America

AFRICAN PENDULINE-TIT
Anthoscopus caroli
3.5 in (9 cm)
Africa

Parrotbills; Penduline Tits

PARROTBILLS are small to medium-size, mostly brownish Eurasian birds with unusual, somewhat parrotlike bills. Often yellow, the bills are stubby but broad, slightly bulging, and powerful—sometimes even used to strip bamboo stalks. There are a total of twenty parrotbill species, family Paradoxornithidae. Most occur in China, Nepal, and eastern India; one, named the Bearded Reedling, with a non-parrotlike bill, occurs in Europe. Parrotbills are 3.5 to 11 inches (9 to 28 cm) long, with short, rounded wings, generally longish tails, and strong legs and feet. Plumage, soft and loose-looking, is mostly brown, buff, and gray. Many species have black markings on head and/or throat. The sexes look alike.

Birds mainly of reeds, tall grasses, dense scrub, and, especially, bamboo thickets, some parrotbills also inhabit forests, although usually the lower parts. Many occur in mountainous regions, some at up to 12,000 feet (3,700 m) in the Himalayas. They feed chiefly on insects, but also on berries, seeds, and some other plant materials such as buds and shoots. They are gregarious, foraging in small to largish groups. Parrotbill breeding is not well known, but many may be monogamous; both sexes incubate eggs and feed young. Their nests, situated in reeds or low vegetation, are usually cups of grass with strips of bamboo leaves, bound with spider webbing. Three parrotbills, one Indian and two Chinese, are considered vulnerable, owing to habitat loss and small ranges. Although little is known about them, they apparently have very small and declining populations. One of them, China's Rusty-throated Parrotbill, is seen infrequently, and may actually be endangered. A fourth species,

PARROTBILLS

Distribution:
Eurasia

No. of Living
Species: 20

No. of Species
Vulnerable,
Endangered: 3, 0

No. of Species Extinct
Since 1600: 0

eastern Asia's Reed Parrotbill, is near-threatened. (The Maui Parrotbill, endemic to Maui in the Hawaiian Islands, and a threatened species, is not a member of the parrotbill family; it is in the Hawaiian honeycreeper family, Drepanididae.)

PENDULINE TITS are very small, sprightly insect-eaters restricted mainly to Asia and Africa, but with single representatives in Europe (the Eurasian Penduline Tit) and North America (the Verdin). Their reputation is as active foragers that acrobatically pursue bugs in trees and bushes. The thirteen species in the family, Remizidae, range in length from 3 to 5 inches (8 to 13 cm) and are typically brown, olive, and/or grayish above and lighter below; many have black facial masks. They have short, rounded wings and short legs. Bills are short, conical, and finely pointed. The sexes look alike.

PENDULINE TITS

Distribution: Eurasia, Africa, North America

No. of Living Species: 13

No. of Species Vulnerable, Endangered: 0, 0

No. of Species Extinct Since 1600: 0

Many penduline tits inhabit open scrubby areas, where they move around trees and bushes, seeking food. However, the African species are predominantly woodland/savanna birds, while others inhabit marsh reed areas. They often forage around a plant's outer branches and leaves, and will hang upside down to reach a food item or to climb along the bottom side of a branch. They eat mainly insects, but also spiders and some seeds, buds, and fruit. Many will use one foot to hold a larger food item, such as a large insect, tightly against a perch, and then use their bill to attack and eat it. Another of their feeding methods is to use their bill to pry open plant matter or insect cocoons to expose hidden insects. These birds occur in pairs or small family parties, usually of three to eight individuals, occasionally up to twenty or more; some typically join mixed-species foraging flocks. Most species are sedentary, but some in colder regions migrate, such as the Eurasian Penduline Tit in the northern parts of its European range. The breeding habits of most species are not known, but many are thought to be monogamous. The hanging, pendulous, nests after which the group is named, are spherical, oval, or pear-shaped and are usually suspended from a tree fork. They are often made of woolly plant down finely woven into a feltlike material; some are made of twigs, leaves and grass, or include animal hair. Some of the felted nests were once used in Central Europe as slippers for small children. Both the male and female or only the male builds the nest; both sexes or only the female incubates eggs; and both feed young. None of the penduline tits are threatened, although several are rare in various parts of their ranges.

Long-tailed Tits

LONG-TAILED TITS constitute a small group of diminutive, fluffy-looking songbirds that acrobatically navigate trees and shrubs in pursuit of insect prey. They are known for their energetic and agile foraging behavior and for their gregariousness. The group occurs chiefly in Eurasia, with some members having very large ranges, such as the Long-tailed Tit (distributed from England to Japan) and Black-throated Tit (ranging from Pakistan to Southeast Asia). Others have small ranges limited to mountainous regions of Central Asia. Only one, the Bushtit, occurs in the New World, from the southwestern corner of Canada southward to Guatemala. The family, Aegithalidae, contains only eight species. In the past, they were usually included with the "true" tits and chickadees in family Paridae, but recent studies, including DNA evidence, suggest they are not closely related to that group and warrant separate family status.

Most of the long-tailed tits are 4 to 5 inches (10 to 13 cm) in length, but Eurasia's Long-tailed Tit ranges up to 5.75 inches (16 cm) and Indonesia's tiny Pygmy Tit measures only about 3.5 inches (8.5 cm). All have compact bodies, very short wings, and narrow, disproportionately long tails. Their legs are longish and the feet strong. The bills are stubby and conical, and somewhat compressed from side to side. Body feathers are rather loose, providing these birds a "fluffball" appearance. Most are gray or brownish above, lighter below; many have black masks or eyelines. The sexes look alike.

Long-tailed tits are chiefly arboreal inhabitants of forests, woodlands, parks, and some scrub areas, although some frequent shrubs and forest

*Distribution:
Eurasia and North
America*

*No. of Living
Species: 8*

*No. of Species
Vulnerable,
Endangered: 0, 0*

*No. of Species Extinct
Since 1600: 0*

BUSHTIT
Psaltriparus minimus
4.5 in (11 cm)
North America, Central America

LONG-TAILED TIT
Aegithalos caudatus
5.5 in (14 cm)
Eurasia

BLACK-THROATED TIT
Aegithalos concinnus
4.5 in (11 cm)
Southern Asia

RUFOUS-FRONTED TIT
Aegithalos iouschistos
4.5 in (11 cm)
Southern Asia

PYGMY TIT
Psaltria exilis
3 in (8 cm)
Indonesia

undergrowth. They feed on insects, spiders, and other small invertebrates, which they take from leaves and twigs; small seeds are occasionally eaten. They move systematically about the outer reaches of trees and shrubs, searching for prey, and will frequently hang upside down, with one or two feet anchored to a twig to reach an escaping bug or probe under a leaf or into a bark crevice. Movement from tree to tree is with a characteristic skipping, undulating flight. Long-tailed tits live during nonbreeding months in groups of three to twenty or more, continually roaming over a defended group territory. At night, group members perch together in a tree, huddling for warmth. Often these groups join mixed-species feeding flocks of insect-eating songbirds. Occasionally several groups of tits join together into flocks of up to two hundred or more individuals.

These little birds are monogamous, with some species, such as the Black-throated Tit, breaking into mated pairs during breeding seasons. In such cases males and females share nesting duties. Some other species are usually cooperative breeders, with helpers (up to eight other individuals) assisting a mated pair with feeding young and perhaps, in some cases, with nest building. Helpers are other members of the social group, presumably close relatives of one or another of the mated individuals; some helpers may begin helping only after their own nesting attempts end in failure. Nests are complex hanging domed structures with side entrances that are suspended high in bushes or trees. The nests, constructed by both sexes, are made of combinations of moss, grass, twigs, and rootlets that are bound together with spider webbing. Sometimes the nests are covered on the outside with bits of lichen, perhaps for camouflage; they are often lined with feathers and hair. Depending on species, the female only or both sexes incubate eggs, and both sexes feed the young.

Many of the long-tailed tits are common or fairly common over most of their ranges, and the broadly distributed Long-tailed Tit and Bushtit are very successful. The latter species has been expanding its North American range, moving in historic times into the northwestern corner of the United States and southwestern corner of Canada. This species first bred on Vancouver Island in 1937, but is now common there. Certain of the Old World species with limited ranges are considered by some authorities to be near-threatened, particularly southern Asia's White-throated Tit and central China's Sooty Tit; and the Pygmy Tit, endemic to higher-elevation forests of Java, will almost certainly be threatened in the future by deforestation.

BLACK-CAPPED CHICKADEE
Poecile atricapillus
5 in (13 cm)
North America

CHESTNUT-BACKED CHICKADEE
Poecile rufescens
4.5 in (12 cm)
North America

TUFTED TITMOUSE
Baeolophus bicolor
6.5 in (17 cm)
North America

BRIDLED TITMOUSE
Baeolophus wollweberi
5 in (13 cm)
North America

YELLOW-CHEEKED TIT
Parus spilonotus
6 in (15 cm)
Southern Asia

SULTAN TIT
Melanochlora sultanea
8 in (20 cm)
Southern Asia

RED-THROATED TIT
Melaniparus fringillinus
4.5 in (11 cm)
Africa

WHITE-WINGED BLACK-TIT
Melaniparus leucomelas
5.5 in (14 cm)
Africa

Tits and Chickadees

TITS and CHICKADEES are small, quite social, arboreal birds found mainly across the colder, northern portions of the world. They are known for their often plump silhouettes and the lively, busy way they move through the canopies of trees and bushes, hopping through branches, poking into corners, and hanging upside down, all in search of insects. The tit family, Paridae, has about fifty-five species distributed through much of North America, Europe, Asia, and the forested regions of Africa. Birds in the family are often called "true" tits to distinguish them from members of two other, smaller families of somewhat similar birds, the penduline tits (family Remizidae) and long-tailed tits (family Aegithalidae). The name tit is shortened from the original name for some of these birds, titmouse; both mean, generally, "small bird." Seven of the eleven tits that occur in North America are known as chickadees, after the "chick-a-dee-dee-dee" calls of some of them; the others are called titmice. Old World species are all called tits.

Most of these birds range from 4 to 5.5 inches (10 to 14 cm) long (but Asia's showy Sultan Tit is about 8 inches [20 cm]). They have short, strong, conical bills, short, rounded wings, square or slightly notched tails, and short, strong feet for hanging from twigs. Some tits, including North America's titmice, have jaunty crests. Most are brown, gray, or bluish gray above and whitish, gray, or yellow below, but some are mainly black and white, and many have black and white heads. White wing bars and white outer tail feathers are common. Chickadees are gray, brown, or reddish brown with dark caps and throats and light cheek patches. Sexes in the family usually look alike, although females are often duller than males.

Distribution: Eurasia, Africa, North America

No. of Living Species: 55

No. of Species Vulnerable, Endangered: 1, 0

No. of Species Extinct Since 1600: 0

Chickadees and tits mainly inhabit forests and woodlands, but also some more open habitats, especially those modified by people, such as parks, gardens, and roadsides. All eat insects and other small invertebrates but also seeds and some buds, berries, and other fruit. When foraging they move almost constantly through the foliage of trees and shrubs, taking insects from small twigs and leaves, sometimes hanging acrobatically or hovering briefly to reach a fleeing bug. Fruits and seeds may be held against a perch with both feet or wedged into a crevice, then hammered open with their bills; larger insects are dismembered piece by piece in a similar manner. Many in the group, including chickadees, store seeds (as well as insects and spiders) in fall, using their bills to tuck them into knotholes, bark nooks and crannies, amid dead leaves, or in soil, for later use in winter, when food is less available.

Tits are gregarious—they are often found in family groups, calling to one another continually as they move around their territory and roosting together at night. In nonbreeding seasons, many tits form flocks that roam forests and some commonly join mixed-species foraging flocks. During breeding seasons, pairs of tits rigorously defend their territories against other members of their species. Most in the family are not migratory but some that breed in far northern regions do make seasonal movements, and those that breed at high elevation in summer move to lower elevations in fall and winter.

Tits nest in cavities, usually in trees (in natural holes or old woodpecker holes), but sometimes among tree roots, in stone walls, or even in holes in the ground. A few excavate their own cavities or enlarge natural ones in rotted, soft wood. Some will breed in human-provided nest boxes, and Eurasia's Great Tit, the most widespread and well-known member of the family, actually prefers this situation. Usually the cavity is lined with soft material, such as moss, grass, fur, and feathers. Most species are probably monogamous. Both sexes may work on the nest but only the female incubates (sometimes being fed on the nest by her mate), and both sexes feed young.

Many tits are abundant birds with broad ranges. Only one is threatened: India's White-naped Tit is considered vulnerable. It has a small, highly fragmented population in a limited tropical thorn-scrub habitat that is increasingly cleared or degraded. Three other Asian tits are considered near-threatened.

Swallows

SWALLOWS are small, streamlined birds that catch insects on the wing during long periods of sustained flight. Their pointed wings and often forked tails, which enable them to sail through the air with high maneuverability, and their conspicuousness in a variety of habitats as they fly back and forth during their foraging flights, make swallows familiar to many; and owing to this familiarity, they have a long history of association with people. The ancient Greeks revered swallows as sacred birds, probably because they nested in and flew around the great temples. In the New World, owing to their insect-eating ways, swallows have been popular with people going back to the ancient Mayans. Today, arrival of the first migratory Barn Swallows in Europe is considered a sign of approaching spring, as is the arrival of Cliff Swallows at Capistrano, an old Spanish mission in California.

The approximately ninety species of swallows (family Hirundinidae), including birds called martins and sawwings, have a worldwide distribution. Ranging in length from 4.5 to 8 inches (11.5 to 20 cm), they have short necks, bills, and legs. Some are shades of blue, green, or violet, but many are gray or brown; in most, the sexes look alike.

Swallows occur predominantly in open or semiopen habitats, often near freshwater, over and around lakes, rivers, marshes, cliffs, grassland, savanna, and forest edges and clearings. They sometimes seem to fly all day, circling low over land or water or in patterns high overhead, snatching insects from the air. But they do land, usually resting during the hottest part of the day. Directly after dawn, however, and at dusk, swallows are always airborne.

Distribution:
All continents except
Antarctica

No. of Living
Species: 89

No. of Species
Vulnerable,
Endangered: 4, 0

No. of Species Extinct
Since 1600: 1

BARN SWALLOW
Hirundo rustica
6.5 in (17 cm)
Worldwide

MANGROVE SWALLOW
Tachycineta albilinea
4.5 in (12 cm)
North America, Central America

WHITE-BANDED SWALLOW
Atticora fasciata
5.5 in (14 cm)
South America

BAHAMA SWALLOW
Tachycineta cyaneoviridis
6 in (15 cm)
West Indies

PURPLE MARTIN
Progne subis
8 in (20 cm)
North America, South America

SOUTH AFRICAN SWALLOW
Petrochelidon spilodera
6 in (15 cm)
Southern Africa

BLACK SAWWING
Psalidoprocne holomelas
6 in (15 cm)
Africa

WHITE-BACKED SWALLOW
Cheramoeca leucosternus
6 in (15 cm)
Australia

Some species have relatively short, unforked tails and tend to forage high in the air, where a high degree of maneuverability is unnecessary; others, with very long, deeply forked tails, forage near the ground, where they must be able to turn suddenly to avoid obstacles and catch low-flying insects. Because they depend each day on capturing enough bugs, their daily habits are largely tied to prevailing weather. Flying insects are thick in the atmosphere on warm, sunny days, but relatively scarce on cold, wet ones. Therefore, on nice days, swallows can catch their fill of insects in only a few hours of flying, virtually anywhere. But on cool, wet days, they may need to forage all day to find enough food, and they tend to do so over water or low to the ground, where under such conditions insects are more available. Swallows mostly consume flying insects of the beetle, fly, true bug, and ant/bee/wasp groups; spiders drifting in the wind and caterpillars suspended from vegetation on silk threads are also taken; and some swallows also eat berries. Tropical species are usually fairly sedentary, but those in temperate-zone areas are often strongly migratory. For instance, Barn Swallows in Europe migrate to winter in sub-Saharan Africa; those in North America winter in South and Central America.

All swallows are monogamous, and many species breed in dense colonies of several to several thousand nesting pairs. (Some, such as Cliff Swallows, locate their colonies near to or actually surrounding the cliff-situated nests of large hawks, basking in the protection afforded by having a nest close to a fearsome predator.) Nests are constructed of plant pieces placed in a tree cavity, burrow, or building, or, alternatively, consist of a mud cup attached to a vertical surface such as a cliff. Both sexes build the nest and both or the female alone incubate eggs; both feed young.

Many swallows are widespread and abundant. Four species, two in the West Indies and two in Africa, all with small ranges, are considered vulnerable; little is known of the population sizes or vulnerabilities of three or four other species. Another, Thailand's White-eyed River Martin, discovered only in 1968, may be extinct. People's alterations of natural habitats, harmful to so many birds, are often helpful to swallows, which adopt buildings, bridges, road culverts, and quarry walls as nesting areas. Barn Swallows, common in many parts of the world, have for the most part given up nesting in anything other than human-crafted structures.

RED-VENTED BULBUL
Pycnonotus cafer
9 in (23 cm)
Southern Asia

RED-WHISKERED BULBUL
Pycnonotus jocosus
8 in (20 cm)
Southern Asia

BLACK-CRESTED BULBUL
Pycnonotus melanicterus
7.5 in (19 cm)
Southern Asia

SOOTY-HEADED BULBUL
Pycnonotus aurigaster
8 in (20 cm)
Southern Asia

BLACK BULBUL
Hypsipetes leucocephalus
10 in (25 cm)
Southern Asia

STREAK-EARED BULBUL
Pycnonotus blanfordi
7.5 in (19 cm)
Southeast Asia

YELLOW-VENTED BULBUL
Pycnonotus goiavier
8 in (20 cm)
Southeast Asia

Bulbuls; Hypocolius

BULBULS are a large group of small to midsize, mainly tropical, mainly forest-dwelling songbirds, usually with conspicuous crests, which are widely distributed throughout much of Africa and southern Asia. They are known generally as highly successful, adaptable birds, as manifested by their diversity (the family, Pycnonotidae, contains about 130 species of bulbuls, greenbuls, and brownbuls, as well as birds called finchbills, bristlebills, and nicators) and their abilities to conform to new realms when people have transported them from their native regions to alien lands. For anyone who bird-watches in, say, Southeast Asia, bulbul diversity and abundance quickly become clear: some days most of the birds one sees are bulbuls, often many varieties within the same forest habitat. As for their adaptability, the best example is the Red-whiskered Bulbul, a native of southern Asia that now, through human transportation, is a thriving inhabitant of such far-flung sites as Australia, Florida, Hawaii and other Pacific islands, and various Indian Ocean islands. Because of their singing abilities, bulbuls are also popular cage birds in many parts of their range; their singing is a prime reason people introduced them to various parts of the world.

Bulbuls, 5 to 9 inches (13 to 23 cm) in length and slender, have narrow, moderately long, often slightly down-curved bills and longish tails. They tend to have hairlike feathers on the back of the head that often form into crests. Most are dully turned out in subdued grays, browns, or greens; the sexes look alike. In many regions several bulbul species occur that are difficult to distinguish, differing only slightly in eye, throat, or tail color.

BULBULS

Distribution:
Africa, southern Asia

No. of Living
Species: 130

No. of Species
Vulnerable,
Endangered: 10, 3

No. of Species Extinct
Since 1600: 0

STRIPE-THROATED BULBUL
Pycnonotus finlaysoni
8 in (20 cm)
Southeast Asia

SCALY-BREASTED BULBUL
Pycnonotus squamatus
6.5 in (16 cm)
Southeast Asia

STRAW-HEADED BULBUL
Pycnonotus zeylanicus
11.5 in (29 cm)
Southeast Asia

Hemixos flavala
8.5 in (21 cm)
Southern Asia

CHESTNUT BULBUL
Hemixos castanonotus
8.5 in (21 cm)
Southern Asia

STRIATED BULBUL
Pycnonotus striatus
9 in (23 cm)
Southern Asia

BLACK-AND-WHITE BULBUL
Pycnonotus melanoleucus
7 in (18 cm)
Southeast Asia

CRESTED FINCHBILL
Spizixos canifrons
8.5 in (21 cm)
Southeast Asia

COMMON BULBUL
Pycnonotus barbatus
7 in (18 cm)
Africa

YELLOW-BELLIED GREENBUL
Chlorocichla flaviventris
8.5 in (22 cm)
Africa

SOMBRE GREENBUL
Andropadus importunus
7 in (18 cm)
Africa

CAPE BULBUL
Pycnonotus capensis
8 in (20 cm)
Southern Africa

BLACK-FRONTED BULBUL
Pycnonotus nigricans
8 in (20 cm)
Southern Africa

HYPOCOLIUS
Hypocolius ampelinus
9 in (23 cm)
Iraq, Arabia

Bulbuls are chiefly forest birds, but some favor forest edges, woodlands, thickets, or even more open sites, including parks and gardens. The widely introduced species Red-whiskered and Red-vented Bulbuls often stick to open, scrubby sites, cultivated regions, parks, and settled areas in their new homes. Bulbuls principally eat fruits, including berries, but some also regularly take insects or probe flowers for nectar; some, such as the African brownbuls, eat mainly insects. A few will pursue flying insects in the air, but most search tree and shrub leaves for food. Many forest bulbuls are shy and skulking, but in habitats near people, bulbuls often become quite bold. Most are gregarious, traveling in noisy family groups (although some are typically seen solitarily), and sometimes joining mixed-species feeding flocks. Many, particularly following breeding seasons, form fairly large flocks that can cause harm to fruit crops. Most bulbuls are not migratory. Many have loud, striking, whistling or chattering songs, and some of them quite melodious.

HYPOCOLIUS

Distribution: Southwestern Asia

No. of Living Species: 1

No. of Species Vulnerable, Endangered: 0, 0

No. of Species Extinct Since 1600: 0

Mating systems among bulbuls are quite variable. Some, perhaps most, are monogamous, with both parents usually sharing in nest building, incubation and feeding young. Others are probably cooperative breeders, with family members helping a mated pair feed young. Nests, fairly bulky, consist of loosely woven twigs, leaves, and stems formed into a cup, lined with grass, and placed usually in the fork of a tree or bush (but some nest low in undergrowth).

Many bulbuls are common to abundant, with broad distributions. However, being predominantly forest dwellers at a time when forests in Africa and Asia are increasingly cleared or degraded, some are threatened. Additionally, some species are trapped in large numbers for the pet industry, both local and international. Currently ten species are considered vulnerable and three endangered. Most of these are from Indonesia, Africa, or Madagascar; about half are restricted to islands.

The HYPOCOLIUS (single-species family Hypocoliidae) is a midsize grayish songbird restricted to the Middle East, chiefly Iraq, Jordan, and Arabia. About 9 inches (23 cm) long, Hypocoliuses have short, wide, slightly hooked bills, short, rounded wings, long tails, short legs, and strong feet. Males are pale gray with black markings; females are duller, browner. The species inhabits open areas such as scrublands, shrub areas, and gardens. Usually in small groups, they forage for small fruits and berries and occasional insects. Mating is monogamous, male and female building a bulky cup nest of twigs, lined with vegetation, in a shrub or small tree. The species is not threatened.

White-eyes

WHITE-EYES are a large group of mostly very small, generally nondescript, greenish birds recognized for their rather consistent looks and often conspicuous white eye rings. They are also noted for their many successful invasions and occupations of small oceanic islands—often where few or no other songbirds thrive. Many of them are fairly sedentary (nonmigratory). As a group, however, white-eyes have a propensity for congregating in flocks that sometimes disperse over long distances, including over large expanses of open water. In this way these hardy little birds have reached many islands and succeeded in colonizing them. In some cases, such as Heron Island in Australia's Great Barrier Reef, the islands are tiny, with very limited habitat—such as small groves of trees—that might support terrestrial, arboreal birds. Yet the Silver-eye maintains a healthy population on Heron Island. This same species first reached New Zealand by crossing more than 1,000 miles (1,600+ km) of the Tasman Sea, from Tasmania, during the mid-1800s; today it is one of New Zealand's most abundant land birds. The Japanese White-eye, introduced from Asia to Hawaii in the 1930s, is now probably Hawaii's most common land bird.

The white-eye family, Zosteropidae, has ninety-four species distributed over sub-Saharan Africa, southern Asia, Australasia, and the southwestern Pacific. White-eyes are slender, 3.5 to 6 inches (9 to 15 cm) long (most being 4 to 5 inches, 10 to 13 cm), with small, delicate, slightly down-curved, sharply pointed bills, fairly rounded wings, and sometimes relatively short tails. Their eye rings, occasionally subtle and indistinct, are composed of fine

Distribution: Sub-Saharan Africa, southern Asia, Australasia, southwestern Pacific

No. of Living Species: 94

No. of Species Vulnerable, Endangered: 10, 11

No. of Species Extinct Since 1600: 1

JAPANESE WHITE-EYE
Zosterops japonicus
4.5 in (11 cm)
Asia

ORIENTAL WHITE-EYE
Zosterops palpebrosus
4.5 in (11 cm)
Asia

BLACK-CAPPED WHITE-EYE
Zosterops atricapilla
4.5 in (11 cm)
Southeast Asia

SILVE-EYE
Zosterops lateralis
4.5 in (11 cm)
Australia, Tropical Pacific islands

AUSTRALIAN YELLOW WHITE-EYE
Zosterops luteus
4.5 in (11 cm)
Australia

MOUNTAIN BLACK-EYE
Chlorocharis emiliae
5.5 in (14 cm)
Borneo

AFRICAN YELLOW WHITE-EYE
Zosterops senegalensis
4.5 in (11 cm)
Africa

CAPE WHITE-EYE
Zosterops pallidus
4.5 in (12 cm)
Southern Africa

silvery white feathers. Almost all species have white rings; one, Borneo's Mountain Black-eye, has black rings. White-eyes are mainly greenish above and yellowish below, but often with patches of gray, white, or brown. Four species, known as speirops, endemic to islands off the coast of West Africa, are darker birds, sporting black, brown, and gray. Male and female white-eyes look alike, although females are sometimes a bit smaller and/or duller.

Birds of forests, forest edges, woodlands, orchards, and gardens, white-eyes eat insects and spiders, fruit such as berries, and sometimes, nectar from flowers. They have brush-tipped tongues that they use to mop up nectar, as well as pulp and juice from fruits. They are lively and fast-moving, landing in a tree or shrub, searching leaves and bark for bugs, probing into crevices for spiders, then quickly moving on to another tree or shrub. They will also occasionally take flying insects in the air. White-eyes tend to forage through the mid and upper levels of tree canopies, but sometimes come lower, and some are common shrub birds. During nonbreeding periods, white-eyes are often in small flocks, and sometimes join mixed-species foraging flocks. Many are territorial for all or part of the year. Some species at higher latitudes migrate to warmer regions to spend nonbreeding periods; for instance, white-eyes that breed in the northern parts of their Asian range migrate to southern Asia, and some of the white-eyes that breed in Tasmania migrate northward to the Australian mainland.

Monogamous breeders, white-eyes build loosely woven, cup-shaped nests of grasses, other vegetation, and spider webbing, suspended by the rim in the fork of a small branch in a tree or shrub. Both sexes build the nest, incubate eggs, and feed young. Some white-eyes remain in long-term mated pairs, perhaps for life.

Many white-eyes are widespread, common birds—too common at times. Flocks of Silver-eyes are so abundant and destructive of fruit in parts of Australia that they are considered orchard and vineyard pests. Many other white-eyes, however, are in trouble. Currently ten species are considered vulnerable and eleven endangered. Fully eighteen of these twenty-one species are restricted to small islands; main threats they face are extremely small populations, typhoons, deforestation, and island development. One, the Robust White-eye, endemic to Australia's Lord Howe Island, became extinct in the 1920s, probably because the rats that were introduced to the island preyed on the birds and their nests.

YELLOW-BELLIED PRINIA
Prinia flaviventris
5 in (13 cm)
Southern Asia

GOLDEN-HEADED CISTICOLA
Cisticola exilis
4.5 in (11 cm)
Southern Asia, Australia, New Guinea

BLACK-CAPPED APALIS
Apalis nigriceps
4.5 in (11 cm)
Africa

YELLOW-BREASTED APALIS
Apalis flavida
5 in (13 cm)
Africa

GREEN-BACKED CAMAROPTERA
Cameroptera brachyura
4 in (10 cm)
Africa

NON-BRD

BRD

GRAY-BACKED CISTICOLA
Cisticola subruficapilla
5 in (13 cm)
Southern Africa

BLACK-CHESTED PRINIA
Prinia flavicans
5.5 in (14 cm)
Southern Africa

SPOTTED PRINIA
Prinia maculosa
5 in (13 cm)
Southern Africa

Cisticolas, Prinias, and Apalises

CISTICOLAS, PRINIAS, and APALISES constitute the majority of a large group of small, plainly marked, often brownish birds that flit about in tall grass, shrubs, or tree foliage mainly in Africa but also, to a much lesser extent, in Eurasia and Australia. Being often tiny, visually inconspicuous birds, they are usually noticed only by bird-watchers, and they are notorious for being bewilderingly similar in appearance. Typically there are at least three or four similar-looking species present in, for example, any African grassland, shrubland, or savanna. Identification of particular species, therefore, is extremely challenging, usually accomplished by using some combination of a particular bird's specific habitat choice, its tail length, and its vocalizations. Sometimes identification must be left until the birds' breeding seasons, when their species-specific songs and courtship displays are evident. There are approximately 110 species in the family, Cisticolidae, including a few called longtails, camaropteras, and warblers. The group in the past was considered part of the huge assemblage of small birds known as the Old World warblers (family Sylviidae), but recent research, including DNA comparisons, suggest these African warblers, as they are sometimes known, warrant separate family status.

Cisticolas and their relatives are 3.5 to 6 inches (9 to 15 cm) long, usually slim-bodied, with slender, pointed bills, some of which are slightly or moderately down-curved. Many have short, rounded wings, and most are rather short-tailed. Many are unobtrusively colored in brown, buff, gray, or green, often with streaks, but some tropical species, especially among the apalises, are brighter and more striking. Males and females usually look alike, although males may

Distribution: Africa, Eurasia, Australia

No. of Living Species: 111

No. of Species Vulnerable, Endangered: 6, 4

No. of Species Extinct Since 1600: 0

be brighter during breeding and/or a bit larger. Prinias, often brownish like cisticolas, are more compact-looking and have longer tails that they often keep raised and moving over their backs. Apalises are distinctive because they typically have dark chest bands or large patches of yellow.

All birds in this family eat primarily insects (some also take spiders, snails, seeds, or some combination of these), which they pull off foliage. The main habitats they occupy vary. The cisticolas (about forty-five species, forty-two in Africa) and prinias (twenty-eight species), known collectively as grass warblers, occur in grassy habitats: grassland, savanna, woodland clearings, scrub areas, and marshes. Cisticolas are mainly terrestrial, foraging on or near the ground, creeping or running along. Prinias also tend to forage low down, in shrubs or low in trees, but some are more arboreal. Apalises (about twenty-two species, all in Africa), which make up a sizable fraction of an assemblage called African tree warblers, occur predominantly in forest and woodland canopies, but some also inhabit forest edges, savanna, or thickets. Most of these birds are territorial, living year-round in pairs; forest-dwelling apalises, however, occur in pairs or small family parties, occasionally joining mixed-species foraging flocks. The majority of these birds are nonmigratory, although some in cooler regions, such as some populations in Australia, sometimes make seasonal movements.

These birds are chiefly monogamous, although some cisticolas are at least occasionally polygamous and some may breed cooperatively. Some, especially among the cisticolas, are known for aerial courtship displays in which, to attract a female, a male flies up, utters songs in flight, then dives or circles, sometimes making snapping noises with his wings. Nests, most often oval-, ball-, or bag-shaped, and woven into small branches and living leaves of a tree or bush, are made mostly of grass but often with other materials added, such as rootlets, leaves, or lichens, or a combination of these. Some nests are placed in grass clumps. Depending on species, one or both sexes build the nest; the female only or both sexes incubate eggs, and usually both feed the young.

Cisticolas and their relatives are often very common and some have enormous distributions (such as the Golden-headed Cisticola of Asia and Australia and the Zitting Cisticola of Eurasia, Africa, and Australia). Still, ten species, almost all endemic to Africa, are considered threatened: six are vulnerable and four are endangered (one of the latter, Kenya's Taita Apalis, critically so). Restricted ranges, tiny populations, and continued habitat loss are the main threats to these small birds.

Old World Warblers

OLD WORLD WARBLERS comprise a huge, diverse group of small, mostly plain, inconspicuous songbirds that make their livings chiefly by moving actively through dense foliage, seeking insects. That so many of them are undistinguished-looking brown birds naturally leads most North American bird-watchers who first see them in, say, Europe, to conclude quickly that they are the unidentifiable LBJs (little brown jobs) of the Old World (LBJ being the despairing epithet Americans often use for sparrows and other difficult-to-identify, small, dark, flitting birds). Aside from their diversity, usually drab plumage, and insect-catching ways, the group is probably most recognized for its songs, which, as the name warbler connotes, are often sweet, musical, and at least sometimes warbling, or trilling. The family Sylviidae, with about 279 species, is distributed widely, mainly across Africa and Eurasia, but also in Australia and New Zealand. A single species, the Arctic Warbler, occurs in the New World; it crossed the Bering Sea from Siberia and now breeds in Alaska. In addition to warblers, the group includes birds called tailorbirds, crombecs, eremomelas, longbills, grassbirds, whitethroats, and chiffchaffs, among others; it is not closely related to the New World warblers (family Parulidae).

These slim-bodied birds are mostly from 3.5 to 6.25 inches (9 to 16 cm) long (but Asian grassbirds range up to 10 inches [25 cm]). Chiefly brown, olive, or yellow, often with darker streaks, they typically have small, slender, pointed bills, strong feet, and some have longish tails. In most, the sexes are similar, but in some, such as many of the Asian tailorbirds, there are slight to moderate differences in coloring.

Distribution: Africa, Eurasia, Australia, New Zealand, Alaska

No. of Living Species: 279

No. of Species Vulnerable, Endangered: 23, 9

No. of Species Extinct Since 1600: 2

ARCTIC WARBLER
Phylloscopus borealis
5 in (13 cm)
Eurasia, North America

JAPANESE BUSH-WARBLER
Cettia diphone
5.5 in (14 cm)
Asia

DARK-NECKED TAILORBIRD
Orthotomus atrogularis
4.5 in (12 cm)
Southern Asia

INORNATE WARBLER
Phylloscopus inornatus
4.5 in (11 cm)
Asia

CHESTNUT-CROWNED WARBLER
Seicercus castaniceps
4 in (10 cm)
Southern Asia

MOUNTAIN WARBLER
Phylloscopus trivirgatus
4.5 in (12 cm)
Southeast Asia

YELLOW-BREASTED WARBLER
Seicercus montis
4 in (10 cm)
Southeast Asia

YELLOW-BELLIED HYLIOTA
Hyliota flavigaster
4.5 in (11 cm)
Africa

WILLOW WARBLER
Phylloscopus trochilus
4.5 in (12 cm)
Eurasia, Africa

AFRICAN YELLOW WARBLER
Chloropeta natalensis
5.5 in (14 cm)
Africa

BANDED PARISOMA
Parisoma boehmi
5 in (13 cm)
Africa

GREEN-BACKED EREMOMELA
Eremomela canescens
4.5 in (11 cm)
Africa

RUFOUS-VENTED WARBLER
Parisoma subcaeruleum
5.5 in (14 cm)
Southern Africa

LESSER SWAMP-WARBLER
Acrocephalus gracilirostris
6 in (15 cm)
Africa

WHITE-WINGED SCRUB-WARBLER
Bradypterus carpalis
6.5 in (17 cm)
Africa

VICTORIN'S SCRUB-WARBLER
Bradypterus victorini
6.5 in (16 cm)
Southern Africa

RED-FACED CROMBEC
Sylvietta whytii
3.5 in (9 cm)
Africa

AFRICAN MOUSTACHED WARBLER
Melocichla mentalis
7.5 in (19 cm)
Africa

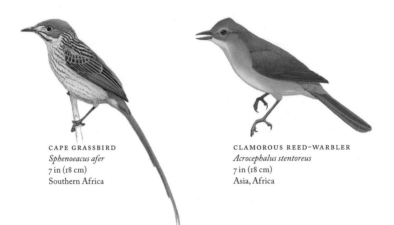

CAPE GRASSBIRD
Sphenoeacus afer
7 in (18 cm)
Southern Africa

CLAMOROUS REED-WARBLER
Acrocephalus stentoreus
7 in (18 cm)
Asia, Africa

RUFOUS SONGLARK
Cincloramphus mathewsi
6.5 in (16 cm)
Australia

BROWN SONGLARK
Cincloramphus cruralis
7 in (18 cm)
Australia

Old World warblers occur in a wide variety of habitats, from forests and woodlands to grasslands, thickets, shrublands, croplands, marshes, and mangroves. They are mainly arboreal insect-eaters, pulling bugs off foliage, flowers, and bark, sometimes hovering for a few seconds to do so, but also take some seeds and other plant materials. Spiders, other small invertebrates, and some fruits and berries are also consumed. The crombecs also take nectar. Many species stay within tree canopies, but others forage at various levels of a forest, some even partly on the ground. They prefer feeding within dense foliage, and some are skulking. They tend to feed alone or in pairs. Most are territorial, a male and female defending a piece of real estate from other members of their species. During nonbreeding periods some join into small flocks, and some forest dwellers join mixed-species foraging flocks. The majority of these warblers are migratory. Many of the European ones breed in spring/summer, and then migrate to Africa for the nonbreeding winter months. Some of these journeys are remarkable: the tiny, widespread Willow Warbler, for example, flies up to 5,500 miles (9,000 km) twice a year between breeding and wintering sites. The ones that breed in Siberia may fly even longer distances to winter in sub-Saharan Africa.

Old World warblers are usually monogamous. Both sexes generally help in nest construction (the female often predominating), egg incubation (female-only incubation in some), and feeding young. Nests are often elaborately woven from grasses, spider webbing, mosses, and hair. Most species construct cup-shaped nests suspended from a fork in branches of a bush or tree. But some build spherical nests on or near the ground, usually in dense vegetation or grass clumps. Tailorbirds construct the group's most elaborate nests by sewing together edges of large leaves. The tailorbird's bill is used as a sewing needle, vegetable fiber as thread. The nest is placed in the cone formed by the stitched leaves and consists of plant down, spider webbing, and bark fiber. Many Old World warblers are common hosts for brood-parasitic cuckoos, which lay their eggs in nests of other species, the host parents then raising cuckoo young instead of their own.

Most Old World warblers are common to abundant birds, but some are confined to very specific habitats or narrowly restricted ranges. Currently twenty-three species are vulnerable and nine are endangered, two of the latter critically. Two, New Zealand's Chatham Island Fernbird and the Seychelles' Aldabra Warbler, became extinct in historic times.

PUFF-THROATED BABBLER
Pellorneum ruficeps
6.5 in (17 cm)
Southern Asia

WHITE-BROWED SHRIKE-BABBLER
Pteruthius flaviscapis
6.5 in (17 cm)
Southern Asia

PYGMY WREN-BABBLER
Pnoepyga pusilla
3.5 in (9 cm)
Southern Asia

CUTIA
Cutia nipalensis
7 in (18 cm)
Southern Asia

STRIPED TIT-BABBLER
Macronous gularis
5.5 in (14 cm)
Southeast Asia

RUFOUS-CROWNED BABBLER
Malacopteron magnum
7 in (18 cm)
Southeast Asia

MOUNTAIN FULVETTA
Alcippe peracensis
6 in (15 cm)
Southeast Asia

Babblers; Rockfowl

BABBLERS form an immense assemblage of small to large, often nondescript songbirds, recognized for their ecological diversity, gregarious behavior, and vocalizations. In some regions, such as India and Southeast Asia, babblers are extremely diverse and abundant, and are common park and garden birds. They are generally noisy, often engaging in long, boisterous bouts of babbling, chattering, or chuckling songs, sometimes in group choruses; in addition to babblers, some in the group are called laughingthrushes or chatterers (others, not named for vocalizations, include barwings, fulvettas, minlas, illadopsises, and yuhinas). Because many stick to dense undergrowth, babblers tend to be more often heard than seen. The family, Timaliidae, contains 265 species distributed mainly through Eurasia and Africa; most occur in tropical southern Asia; a single North American species, the Wrentit, inhabits the United States' western coast.

Babblers, 3.5 to 12.5 inches (9 to 32 cm) long, typically are rather robust-looking, with short, rounded wings and sturdy legs (long in the more terrestrial species). They usually have narrow bills that, depending on subgroup, are slightly to extremely down-curved. Most are brown or olive above and lighter, often with dark streaks, below, but a few Asian forest dwellers, such as the Red-billed Leiothrix, are more boldly colored. Babbler sexes mostly look alike.

Although most common in forests and dense brushy areas, babblers occupy a variety of habitats, from rainforests to scrubby deserts and treeless, high-elevation areas. Many are arboreal but some are semi- or mainly terrestrial, foraging by hopping about lower levels of forests, bushes, or on the ground. Most consume insects and other small invertebrates, foraging by using their

BABBLERS

Distribution:
Africa, Southern Asia,
North America

No. of Living
Species: 265

No. of Species
Vulnerable,
Endangered: 18, 6

No. of Species Extinct
Since 1600: 0

GREATER NECKLACED LAUGHINGTHRUSH
Garrulax pectoralis
10.5 in (27 cm)
Southern Asia

MELODIOUS LAUGHINGTHRUSH
Garrulax canorus
10 in (25 cm)
Southern Asia

RED-BILLED LEIOTHRIX
Leiothrix lutea
5.5 in (14 cm)
Southern Asia

WHITE-BELLIED YUHINA
Yuhina zantholeuca
5 in (13 cm)
Southern Asia

CHESTNUT-CRESTED YUHINA
Yuhina everetti
5.5 in (14 cm)
Borneo

LESSER NECKLACED LAUGHINGTHRUSH
Garrulax monileger
10 in (26 cm)
Southern Asia

WHITE-CRESTED LAUGHINGTHRUSH
Garrulax leucolophus
10 in (26 cm)
Southern Asia

WHITE-BROWED SCIMITAR-BABBLER
Pomatorhinus schisticeps
8.5 in (22 cm)
Southern Asia

GOLDEN BABBLER
Stachyris chrysaea
4.5 in (12 cm)
Southern Asia

SILVER-EARED MESIA
Leiothrix argentauris
6.5 in (17 cm)
Southern Asia

BLUE-WINGED MINLA
Minla cyanouroptera
6 in (15 cm)
Southern Asia

LONG-TAILED SIBIA
Heterophasia picaoides
11 in (28 cm)
Southern Asia

CHESTNUT-CAPPED LAUGHINGTHRUSH
Garrulax mitratus
9 in (23 cm)
Southeast Asia

IMM

RUFOUS CHATTERER
Turdoides rubiginosus
7.5 in (19 cm)
Africa

SOUTHERN PIED-BABBLER
Turdoides bicolor
9 in (23 cm)
Southern Africa

GRAY-NECKED ROCKFOWL
Picathartes oreas
13 in (33 cm)
Africa

bills to turn over leaves on the ground and to probe into nooks and crannies; arboreal species pick insects off leaves and other vegetation. Many also eat fruits; some also take seeds, nectar, or are omnivorous. Babblers, usually quite social, are often found together in family groups of three to thirty that huddle together when resting or roosting; when not breeding, some join mixed-species foraging flocks. Many species defend group territories, in which, seasonally, the group breeds. Generally babblers are not migratory.

Babblers employ a variety of mating systems, and cooperative breeding is common. In cooperative breeding, only one or perhaps two pairs in a family group breed, other members of the group participating as helpers, assisting with feeding young and sometimes with incubation. Helpers may be young of previous years that stay around, gaining nesting experience, before breeding on their own. Many species build domed nests of dead leaves and moss, often hidden in dense shrubbery on or near the ground. Others build cuplike nests of coarse plant materials high in bushes or trees. Both sexes incubate eggs and feed young.

ROCKFOWL

*Distribution:
West Africa*

*No. of Living
Species: 2*

*No. of Species
Vulnerable,
Endangered: 2, 0*

*No. of Species Extinct
Since 1600: 0*

Currently eighteen babbler species are considered vulnerable, and six are endangered; many are in China or Southeast Asia. Main threats are habitat destruction and alteration, especially for forest species. Also, some Asian babblers are trapped in large numbers for the pet trade; caged, vocal laughingthrushes are popular in China and other parts of Asia.

The two species of ROCKFOWL (family Picathartidae) are large, shy, semiterrestrial rainforest birds of West Africa. Their checkered classification history includes periods when they were considered members of the babbler, starling, and crow families, but now the group is usually considered a separate family. Rockfowl are also called bald crows, after their bare-skin heads, and picathartes, after their genus name. They range from 13 to 16 inches (33 to 41 cm) long, and are mainly gray or blackish above, and light below. The White-necked Rockfowl's head is yellow and black, the Gray-necked Rockfowl's is red, blue, and black. Both have large eyes, robust bills, and long legs and tails. The sexes look alike. Rockfowl are confined to forests, where they stay in low vegetation or on the ground, hunting insects, worms, snails, centipedes, millipedes, frogs, and lizards. Apparently monogamous, rockfowl build cup nests, of mud mixed with vegetation, in a cave or on a cliff wall or boulder; both sexes incubate the eggs and feed the young. Both species are considered vulnerable because they have relatively small populations and fragmented distributions in regions undergoing extensive deforestation.

Larks

LARKS are small or medium-size, ground-dwelling, open-country songbirds, typically outfitted in dull brown, often streaked plumage that allows them to meld well into their grassland habitats. They are known widely for their rich, melodious songs, which are delivered sometimes while in flight (during breeding-season aerial territorial displays) or while perched atop a bush. The musical vocalizations of the broadly distributed Eurasian Skylark, for instance, are considered by many to be among the bird world's most beautiful songs. Among bird-watchers where larks are plentiful, however, they are perhaps most noted for their cryptic plumage (larks are often essentially invisible until flushed) and for the fact that so many of them are similar in appearance and therefore difficult to tell apart. Field guides sometimes provide complex charts that (supposedly) help bird-watchers identify larks by comparing minute differences in structural features, bills types, and plumage markings. The ninety-one larks, family Alaudidae, constitute an almost entirely Old World group, with drier regions of Africa having the most representatives. A single species, the Horned Lark, long ago crossed the Bering Strait and now, in addition to its Eurasian and African range, occurs across much of North America (with isolated populations in Colombia).

Larks, 4.5 to 9 inches (11 to 23 cm) in length, have relatively long, pointed wings and, often, shortish tails. Their strong bills vary, depending on feeding specialization, from slender and fairly straight or slightly down-curved to short and conical. Lark feet are strong, each having a long, straight hind claw that furnishes good support while walking. Because they spend a lot

Distribution: Eurasia, Africa, Australia, North America

No. of Living Species: 91

No. of Species Vulnerable, Endangered: 4, 4

No. of Species Extinct Since 1600: 0

EURASIAN SKYLARK
Alauda arvensis
7 in (18 cm)
Eurasia, Africa

CLAPPER LARK
Mirafra apiata
6 in (15 cm)
Southern Africa

SABOTA LARK
Mirafra sabota
6 in (15 cm)
Southern Africa

CHESTNUT-HEADED SPARROW-LARK
Eremopterix signata
4.5 in (11 cm)
Africa

LARGE-BILLED LARK
Galerida magnirostris
7 in (18 cm)
Southern Africa

GRAY-BACKED SPARROW-LARK
Eremopterix verticalis
5 in (13 cm)
Southern Africa

HORNED LARK
Eremophila alpestris
7.5 in (19 cm)
Eurasia, Africa, North America

AUSTRALASIAN BUSHLARK
Mirafra javanica
5.5 in (14 cm)
Australia, Southeast Asia

of time on the ground (foraging, nesting), these birds are colored mostly in various shades of brown, usually darker above and lighter brown, buff, or whitish below, effectively camouflaging them from aerial predators. Many larks match, with their back coloring, the predominant soil color in the area in which they live; many also are streaked above and/or below, furthering the cryptic effect. Some have black and white head markings, and some are crested. Male and female larks usually look alike.

Open habitats in which larks occur include grasslands, savanna, woodland edges, tundra, deserts, farmland, and roadsides. They tend to stay on the ground as they forage, walking (or running), rather than hopping, along, but some also search low-lying vegetation for food. They seek mainly insects (but also some other small invertebrates, including mollusks) and seeds (but also some flowers, buds, and occasional leaves). Many larks form flocks during nonbreeding seasons, sometimes coming together in large numbers to feed in, among other places, agricultural stubble. Larks that live in colder, northern regions generally are migratory; some that live in desert areas are fairly nomadic, moving frequently in search of food.

Most larks breed monogamously, pairs defending territories together during the breeding season. Song flights, in which male larks fly up, circle, sing (sometimes for minutes at a time), and then fly away or spiral or dive back to the ground, likely function in mate attraction or territory maintenance, or both. In some species, such as Africa's Flappet and Clapper Larks, males fly very high and clap their wings together loudly as they climb. All larks nest on the ground, building in a depression a shallow cup-shaped or domed nest. The nest may be lined with fine grass, rootlets, hair, or some combination of these. In hot deserts, nests are often hidden under low vegetation, providing critical shade protection. Both sexes or the female only builds the nest; the female alone often incubates eggs. Both sexes feed chicks.

Larks are birds of open habitats and this feature of their ecology both benefits and harms them. Such habitats, including deserts and grasslands, often cover huge areas, so provide resources for large overall lark populations. On the other hand, natural grasslands are among Earth's most threatened environments because of their obvious suitability for cultivation of grain crops. Eight lark species are currently threatened, four vulnerable and four endangered (two critically so); all are African. But many larks are still widespread, abundant, highly successful birds, and some, including the Horned Lark, are thought to be among the world's most numerous birds.

CHESTNUT SPARROW
Passer eminibey
4.5 in (11 cm)
Africa

GRAY-HEADED SPARROW
Passer griseus
6 in (15 cm)
Africa

YELLOW-SPOTTED PETRONIA
Petronia pyrgita
6 in (15 cm)
Africa

CAPE SPARROW
Passer melanurus
6 in (15 cm)
Southern Africa

HOUSE SPARROW
Passer domesticus
5.5 in (14 cm)
Eurasia, Africa

EURASIAN TREE SPARROW
Passer montanus
5.5 in (14 cm)
Eurasia

PLAIN-BACKED SPARROW
Passer flaveolus
5.5 in (14 cm)
Southeast Asia

Old World Sparrows

OLD WORLD SPARROWS are small, stubby-billed, mostly brown and gray, open-country birds that forage on the ground for seeds. They are renowned because several of them (and one in particular) have adjusted their ecologies and now "make their livings" in association with people. That is, they live in parks, villages, towns, and cities, nest in these places, sometimes in buildings, and eat a variety of foods, including what many refer to euphemistically as table scraps (and more directly as garbage). The House Sparrow, having been introduced from its native Eurasia and North Africa to many other regions, now flourishes over vast stretches of the terrestrial Earth, essentially wherever there are people, excepting rainforests, deserts, and arctic tundra areas. Living in and near cities and towns at elevations from sea level to 14,700 feet (4,500 m), it is one of the globe's most common, widespread, and well-known birds. The House Sparrow probably began associating with people when towns and cities were first developed, and transport animals such as horses filled streets with their seed-rich droppings. Europeans brought House Sparrows with them when they colonized great sections of the planet during the seventeenth through nineteenth centuries, and these birds adapted to new realms swiftly. The species' first successful North American introduction occurred during the mid-1850s in New York. It then took only 50 years for the bird to reach the continent's Pacific coast, essentially occupying all suitable habitat on the U.S. mainland. It now occurs over much of subarctic Canada, Mexico, and Central and South America, and is still spreading; it also thrives in Australia, New Zealand, and

Distribution:
Africa, Eurasia

No. of Living
Species: 35

No. of Species
Vulnerable,
Endangered: 0, 0

No. of Species Extinct
Since 1600: 0

Hawaii. The Eurasian Tree Sparrow, likewise, often nests in buildings and is now a naturalized citizen of such far-flung outposts as Australia and the midwestern United States, and several additional members of the family nest at least occasionally in buildings.

The Old World sparrow family, Passeridae, consists of thirty-five species distributed in Europe, Asia, and, especially, Africa. In the past the group was often considered part of the finch family (Fringillidae), but now is thought more closely related to weavers (family Ploceidae). The family consists of birds known as "true" sparrows as well as some called petronias and snowfinches. Generally chunky and from 4 to 8 inches (10 to 20 cm) long, almost all are drably clad in brown, buff, and gray (males in two species known as golden sparrows are yellow). Many have streaked plumage, and black markings are common. They have short, stout, conical bills, strong skulls, and large jaw muscles, all associated with crushing and eating seeds. Tails are short, legs are fairly thick, and feet are relatively large, used to scratch the ground in search of food. Petronias, or rock sparrows, birds of semidesert scrub and rocky outcrops, are generally dull gray or brown and have a yellow spot on their throat. Snowfinches, sparrows of the Eurasian mountains, are relatively light in color and have white patches on wings and tail. Male and female sparrows sometimes look alike, sometimes not.

Sparrows occupy open environments such as scrub and rocky areas, lightly wooded and bush habitats, as well as agricultural and urban sites. They eat mostly seeds, but some, especially Eurasian Tree and House Sparrows, will eat a great variety of things, including insects, plant buds and leaves, and nectar. Many are very social, living and breeding in small colonies. After breeding, sparrows tend to form into large foraging flocks, sometimes with other seed-eating birds. Sparrows are not great singers, their vocalizations centering on various chatters, chirps, and buzzes.

Sparrows tend to be monogamous, mated pairs breeding in loose colonies, sometimes hundreds of nests together in one group of trees or bushes. Both sexes share nest building, both or the female only incubates eggs, and both feed young, which are reared on insects. Nests are messy-looking, ball- or pad-shaped affairs of dry grass, often lined with feathers. Some species nest in crevices, including tree, rock, or rodent holes. None of these sparrows are currently threatened. In their large winter flocks they sometimes become significant crop pests on cereals and grains.

Weavers

WEAVERS comprise a large group of mainly small songbirds that primarily inhabit savanna, grasslands, and other open habitats of Africa and, to a lesser extent, southern Asia. They are most celebrated, as their name suggests, for their nest-building skills: they construct elaborate, woven, roofed grass nests. Because they typically breed in colonies, trees festooned with several hanging weaver nests are a common sight through many regions of Africa. The weaver family, Ploceidae, contains 114 species, variously named weaver, sparrow-weaver, buffalo-weaver, social-weaver, fody, malimbe, quelea, bishop, and widowbird. In addition to their nests, these birds are noted for their gregariousness and for their numbers. They sometimes gather in enormous groups, flocks so huge that, as they move, they resemble clouds of dark smoke. Some in the family—the widowbirds—are renowned for the males' very long tails.

Weavers, chiefly 4.5 to 10 inches (11.5 to 26 cm) long (widowbirds with long tails range up to 28 inches [71 cm]), have short, stout, conical bills, rounded wings, and strong legs and feet. In many species, male and female differ markedly in appearance, especially during breeding, when males become more colorful or elaborate, sometimes spectacularly so, but females retain their year-round, subdued, streaky brown, olive, buff, and black coloring. The largest subgroup in the family is the yellow weavers, such as the Village and Spectacled Weavers, wherein many males have black heads or face masks and many species look rather alike. Sparrow-weavers are brown and white, generally larger than yellow weavers, and their sexes are similar. Buffalo-weavers, also fairly large, are black or black and white, with heavy bills. Social-weavers are small, brown or brown and white birds.

Distribution:
Africa, southern Asia

No. of Living
Species: 114

No. of Species
Vulnerable,
Endangered: 7, 7

No. of Species Extinct
Since 1600: 0

PARASITIC WEAVER
Anomalospiza imberbis
4.5 in (11 cm)
Africa

WHITE-BROWED SPARROW-WEAVER
Plocepasser mahali
6.5 in (17 cm)
Africa

BLACK-CAPPED SOCIAL-WEAVER
Pseudonigrita cabanisi
5 in (13 cm)
Africa

RED-BILLED BUFFALO-WEAVER
Bulbalornis niger
8.5 in (22 cm)
Africa

VILLAGE WEAVER
Ploceus cucullatus
6.5 in (17 cm)
Africa

CAPE WEAVER
Ploceus capensis
6.5 in (17 cm)
Southern Africa

CHESTNUT WEAVER
Ploceus rubiginosus
6 in (15 cm)
Africa

♂
BRD

GROSBEAK WEAVER
Amblyospiza albifrons
6.5 in (17 cm)
Africa

RED-HEADED WEAVER
Anaplectes rubriceps
5.5 in (14 cm)
Africa

SOCIAL WEAVER
Philetairus socius
5.5 in (14 cm)
Southern Africa

RED-BILLED QUELEA
Quelea quelea
5 in (13 cm)
Africa

RED-HEADED QUELEA
Quelea erythrops
5 in (13 cm)
Africa

ORANGE BISHOP
Euplectes franciscanus
4.5 in (11 cm)
Africa

♂
BRD

♀

♂
NON-BRD

LONG-TAILED WIDOWBIRD
Euplectes progne
6 in (15 cm)
Africa

RED-COLLARED WIDOWBIRD
Euplectes ardens
5 in (13 cm)
Africa

♂

♀

♂

♀

YELLOW BISHOP
Euplectes capensis
6 in (15 cm)
Africa

FIRE-FRONTED BISHOP
Euplectes diadematus
4 in (10 cm)
Africa

♂
BRD

♂
NON-BRD

YELLOW-CROWNED BISHOP
Euplectes afer
4 in (10 cm)
Africa

MARSH WIDOWBIRD
Euplectes hartlaubi
6 in (15 cm)
Africa

Fodies, of the Madagascar region, are red, brown, and black. Malimbes are red and black forest weavers. Male queleas, mostly brown and streaked, during breeding have varying amounts of red on the head, chest, and bill. Male bishops during breeding adopt black and red or black and yellow plumage, and male widowbirds grow long, elaborate tail feathers and bright shoulder patches.

Although most common in open, grassy habitats, weavers occupy a variety of environments, from dry shrubland and agricultural areas to marshes and forests. They are typically seed-eaters but some species, particularly forest dwellers, eat a lot of insects, and most provision their nestlings with insects. While grassland and open-country species are often social, staying in groups, forest weavers are usually more solitary; they tend to form pairs that defend territories year-round, while the social, colonial-nesting species only defend a small area around their nests during the breeding season. After breeding, weavers often form mixed-species flocks that are sometimes nomadic, moving long distances to find goovd supplies of food.

Some weavers are monogamous, breeding in solitary pairs or colonies. Others are polygynous: a male builds a nest, displays from it, attracts a female to it, then starts a second nest, attracting to it a second female, and so on. Some are intensely colonial, with southern Africa's Social Weaver at the extreme: it lives in groups of up to three hundred, and each group constructs and maintains a single, huge, domed grass nest—the largest nest structure built by any bird. Most weaver nests, round or onion-shaped, are woven of grasses, palm-leaf strips, and other plant materials, and suspended from the outer branches of trees and shrubs. The sizes of weaver colonies vary from about a dozen nests to several hundred. Buffalo-weavers build large, unruly, communal nests of twigs and grass in large trees; each pair has a separate nest chamber within the larger nest. Most widowbirds and bishops nest solitarily; they have oval-shaped, simpler nests made of coarse grass, usually placed in grassland or a marsh. Depending on species, the male weaver, the female, or both sexes build nests; the female only or both sexes incubate eggs and feed young.

Many weavers are extremely abundant, some, such as the Red-billed Quelea (likely one of the world's most numerous land birds), forming seasonally into vast flocks that are serious grain pests. Fourteen weavers are threatened (seven are vulnerable, seven are endangered), thirteen African and one Indian.

DOUBLE-BARRED FINCH
Taeniopygia bichenovii
4.5 in (11 cm)
Australia

CHESTNUT-BREASTED MANNIKIN
Lonchura castaneothorax
4.5 in (12 cm)
Australia, New Guinea

CRIMSON FINCH
Neochmia phaeton
5 in (13 cm)
Australia, New Guinea

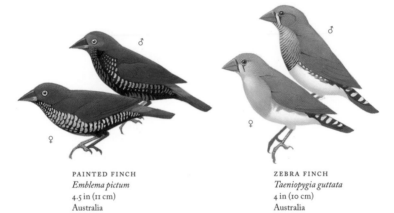

PAINTED FINCH
Emblema pictum
4.5 in (11 cm)
Australia

ZEBRA FINCH
Taeniopygia guttata
4 in (10 cm)
Australia

RED-EARED FIRETAIL
Stagonopleura oculata
4.5 in (12 cm)
Australia

RED-HEADED
FORM

♀

BLACK-HEADED
FORM

♂

GOULDIAN FINCH
Erythrura gouldiae
5 in (13 cm)
Australia

Waxbills

WAXBILLS, also called estrildid finches and grassfinches, are small, usually abundant Old World birds that mainly occupy grassy, brushy, open-country lands. Their family, Estrildidae, contains about 140 species of seed-eaters of southern Asia, sub-Saharan Africa, and Australasia. Included are birds called antpeckers, pytilias, crimson-wings, seedcrackers, bluebills, twinspots, firefinches, firetails, cordonbleus, avadavats, parrotfinches, and waxbills (because they have red, waxy-looking bills), among others. The group is recognized for several features of its members' biology, including coloring, sociality, and suitability for captivity. Many are drably clad in shades of brown, often with barring patterns, but some have large bold patches of black and/or white and others are quite colorful. Northern Australia's Gouldian Finch, for instance, is one of the globe's most visually arresting birds, with a green back, purple chest, yellow belly, and red, blue, and black head. Many species are extremely gregarious, sometimes coming together into huge flocks that, in agricultural districts, can cause significant crop damage. The Nutmeg Mannikin, for example, native to Asia but introduced to Hawaii during the 1860s, was largely responsible for eliminating rice farming in Hawaii and is now something of an agricultural pest in eastern Australia. Some waxbill species, such as Australia's pretty Zebra Finch, live and breed well in captivity and have become popular cage birds; the Zebra Finch especially has become a common lab study subject for animal behavior research. That these birds thrive in captivity also makes them good candidates for successful human transport, the result being that many waxbills now occur in the wild outside their native ranges, in such locations as the West Indies, many Pacific and

Distribution:
Sub-Saharan Africa,
southern Asia,
Australasia

No. of Living
Species: 140

No. of Species
Vulnerable,
Endangered: 8, 2

No. of Species Extinct
Since 1600: 0

GREEN-WINGED PYTILIA
Pytilia melba
5.5 in (14 cm)
Africa

PURPLE GRENADIER
Uraeginthus ianthinogaster
5 in (13 cm)
Africa

COMMON WAXBILL
Estrilda astrild
4 in (10 cm)
Africa

AFRICAN QUAILFINCH
Ortygospiza atricollis
3.5 in (9 cm)
Africa

SHELLEY'S CRIMSON-WING
Cryptospiza shelleyi
5 in (13 cm)
Africa

VIOLET-EARED WAXBILL
Uraeginthus granatina
6 in (15 cm)
Southern Africa

CUT-THROAT FINCH
Amadina fasciata
4 in (10 cm)
Africa

RED-HEADED FINCH
Amadina erythrocephala
5 in (13 cm)
Southern Africa

RED-BROWED FINCH
Neochmia temporalis
4.5 in (11 cm)
Australia

LAVENDER WAXBILL
Estrilda caerulescens
4.5 in (11 cm)
Africa

ORANGE-CHEEKED WAXBILL
Estrilda melpoda
4 in (10 cm)
Africa

RED AVADAVAT
Amandava amandava
4.5 in (11 cm)
Southern Asia

JAVA SPARROW
Padda oryzivora
6.5 in (16 cm)
Indonesia

NUTMEG MANNIKIN
Lonchura punctulata
4 in (10 cm)
Southern Asia

RED-CHEEKED CORDONBLEU
Uraeginthus bengalus
5 in (13 cm)
Africa

Indian Ocean islands, Australia, Southeast Asia, the Philippines, and even Los Angeles (the Nutmeg Mannikin).

Waxbills are 3.5 to 6 inches (6 to 16 cm) long (most 4 to 5.5 inches [10 to 14 cm]) and have short, chunky, pointed bills, and short, rounded wings. The bills of some species appear disproportionately large for such small birds, but they are required to handle and crush seeds. Patches of red are common in the group, particularly on head, rump, or upper tail; many species have white dots on their sides. Members of one subgroup, usually slightly heavier looking than others and mainly dully colored, are called munias (typically the Asian species) or mannikins (African and Australian species; not to be confused with the South American manakins). Waxbill sexes look alike in some species, different in others.

Principally occupying tropical regions, waxbills are predominantly birds of open and semiopen habitats, including grasslands, brush and scrublands, forest edges and clearings, reedbeds, and agricultural areas. But some dwell in woodlands and even dense forests. Most are seed-eaters (but some also take insects; two African species specialize on ants), with grass seed, collected from the ground or pulled from grass stalks, being the primary diet item; many species switch to an at least partial insect diet during breeding. A high degree of sociality is usual in these birds: various species flock all year (often in groups of five to fifty or more individuals), nest in colonies during breeding seasons, or flock together after pairs nest solitarily. Large numbers of individuals spend the entire day together, feeding, drinking, bathing, and roosting in flocks; during rests, they often preen each other.

Waxbills are usually monogamous, a male and a female maintaining a stable pair-bond for one or more breeding seasons; some may mate for life. Most build untidy domed nests of grass and other vegetation placed in shrubs, trees, or sometimes on the ground in tall, dense grass. Some nest in holes. Both sexes construct the nest (the male usually doing more finding and delivering materials, the female doing more actual building); the sexes take turns incubating eggs; both feed the young.

Waxbills are some of the world's most popular cage birds and some areas in which they are heavily hunted and trapped for the pet trade have experienced significant declines in these birds. The Java Sparrow, for instance, introduced to various parts of the world, including Hawaii, is now scarce over parts of its native range in Bali and Java. It, and seven other waxbills, are considered vulnerable; two others, including the gorgeous Gouldian Finch, are endangered.

Indigobirds and Whydahs

INDIGOBIRDS and WHYDAHS constitute a small group of diminutive seed-eating birds of Africa. The family, Viduidae, contains nineteen species and is most known for its unusual breeding. All its members are brood parasites, like many cuckoos and cowbirds: they do not build nests, they do not incubate eggs, and they do not feed young. Rather, the "parasitic" females, after mating, deposit their eggs in the nests of other species, the host parents then raising the indigobird or whydah young as their own. The species that are parasitized in this case are all waxbills, members of the family Estrildidae, which are closely related to the indigobirds and whydahs (in fact, the indigobirds and whydahs are sometimes placed within family Estrildidae). The parasitism is highly specialized and species-specific: in most cases, each species of indigobird or whydah places its eggs solely in the nests of a single waxbill species. Further, presumably to avoid the host recognizing the parasitic nestling for what it is and not feeding it or ejecting it from the nest, indigobird and whydah chicks have evolved mouth markings and colors that match those of the host's own young. For example, the Village Indigobird, the most widespread of the ten indigobirds, only parasitizes the Red-billed Firefinch; and Village Indigobird nestlings, when they open their mouths to beg to be fed by firefinch adults, have the same internal mouth color and spot patterns as the firefinches' own offspring. As an added layer of deceit, the young indigobirds also imitate the host species' begging calls, which stimulate parents to feed their young. Whydahs are also noted for the males' elaborate, elongated tails when in their breeding

Distribution: Africa

No. of Living Species: 19

No. of Species Vulnerable, Endangered: 0, 0

No. of Species Extinct Since 1600: 0

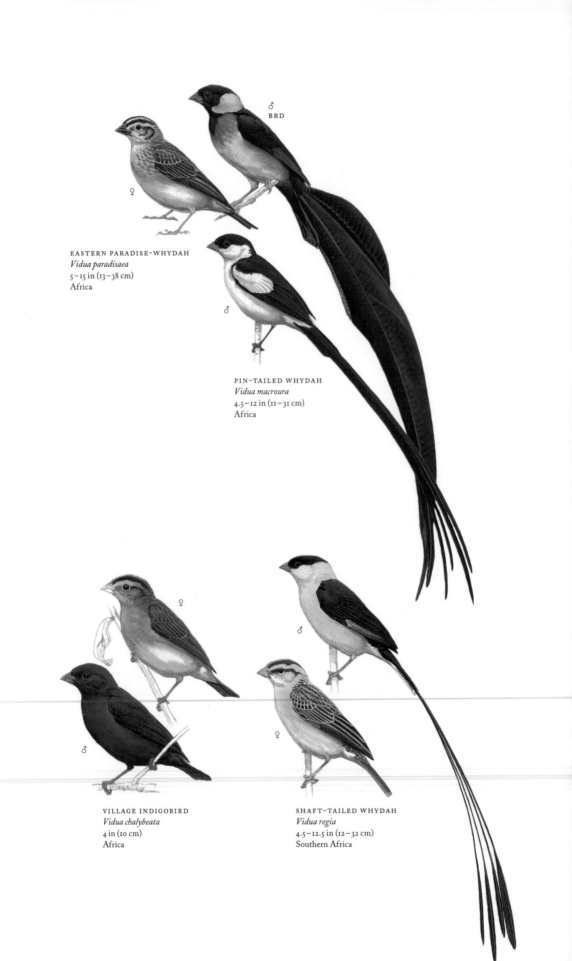

♂ BRD

♀

EASTERN PARADISE-WHYDAH
Vidua paradisaea
5–15 in (13–38 cm)
Africa

♂

PIN-TAILED WHYDAH
Vidua macroura
4.5–12 in (11–31 cm)
Africa

♀

♂

♂

♀

VILLAGE INDIGOBIRD
Vidua chalybeata
4 in (10 cm)
Africa

SHAFT-TAILED WHYDAH
Vidua regia
4.5–12.5 in (12–32 cm)
Southern Africa

plumage. And indigobirds are notorious among bird-watchers for their similar appearances—many are little blackish (male) or streaked brownish (female) birds with small white bills, identifiable as to particular species only by the color of the gloss on the males' dark plumage, by leg coloring, and, sometimes, only by vocalizations and mating behavior.

These birds, occasionally called widows, widow-finches, indigo-finches, or steel-finches (all referring to the males' dark coloring), are only 4 to 5 inches (10 to 13 cm) long, but breeding male whydahs have long central tail feathers that add another 6 to 10 inches (15 to 25 cm) to their total length (the four whydahs have long, slender tails; the five paradise-whydahs have long, wide, elaborate tails). Male indigobirds when breeding are mostly black (with blue, purple, or greenish gloss); females and nonbreeding males are brown, streaked, sparrowlike. Breeding male whydahs are combinations of black, white, buff, and reddish brown; females and nonbreeding males are brown and streaked. Bills in the family, generally reddish or white, are stout and conical, very short; wings are short and rounded; legs are short; feet are small.

Whydahs and indigobirds inhabit mainly woodlands and savannas but also areas around human settlements. They perch in trees and shrubs but generally feed on the ground. Small grass seeds are their main diet items, but some insects are also taken. Most species eat seeds that they either pick up from the ground or pull directly from grass stems. Although fairly solitary during breeding seasons, many whydahs and indigobirds form flocks afterward, sometimes with other species. These flocks will move considerable distances searching for wet weather and, consequently, growing grasses with seeds.

The whydahs and indigobirds have lek-type mating systems in which males display to attract females and then, after copulation takes place, females go off by themselves to lay their eggs in appropriate host-species nests. No long-term pair-bonds are formed. Males space themselves in conspicuous spots at intervals of several hundred yards, often on branches or wires, and sing for long periods. A female arriving in the lek area visits a number of males, which court them with aerial displays. After inspecting a number of males, the female chooses one and mates. Often one male is particularly popular (presumably because his looks and/or displays communicate his potential superiority as a genetic father for females' offspring) and copulates with many of the females in an area. None of the indigobirds or whydahs are threatened.

AMERICAN PIPIT
Anthus rubescens
6.5 in (17 cm)
North America, Asia

WHITE WAGTAIL
Motalcilla alba
7.5 in (19 cm)
Eurasia, Africa

PADDYFIELD PIPIT
Anthus rufulus
6.5 in (16 cm)
Southern Asia

FOREST WAGTAIL
Dendronanthus indicus
7 in (18 cm)
Asia

YELLOW WAGTAIL
Motacilla flava
7 in (18 cm)
Eurasia, Africa

GOLDEN PIPIT
Tmetothylacus tenellus
6.5 in (16 cm)
Africa

ORANGE-THROATED LONGCLAW
Macronyx capensis
8 in (20 cm)
Southern Africa

Wagtails and Pipits

WAGTAILS and PIPITS are small to midsize, slender, ground-dwelling songbirds that are probably most recognized for their incessant tail-wagging: as they walk or run along they constantly move their tails up and down; most continue bobbing their tails even when perched—they never seem to stop. The family, Motacillidae, with about sixty-two species, includes seven in Africa known as longclaws, and is noted for its worldwide distribution (excepting Antarctica, the high Arctic, parts of South America, and some oceanic islands). In addition, pipits are notorious among bird-watchers because so many of them look alike and thus are difficult to identify to species—clues to their identification in the field sometimes lie in minor differences in their displays and flight calls. Of special interest to those familiar with North America's birds, some of Africa's longclaws look eerily similar to Eastern and Western Meadowlarks, streaked brown above, yellow below, with black chest markings. It is a clear case of convergent evolution: two groups, not closely related (meadowlarks are in the New World blackbird family) and in different hemispheres, coming to look like each other, presumably because they have adapted to blend in to similar grassland environments.

Birds in this family have long toes and especially long hind claws (up to 1.5 inches [4 cm] long, in longclaws), that furnish support as they walk, and, generally, long tails with white edges. Bills are slender and pointed. Wagtails, which are mainly limited to the Old World (two species crossed the Bering Strait and now breed in Alaska), are conspicuous, brightly marked birds, either black and white or with mixtures of black, white, gray and yellow;

Distribution:
All continents except
Antarctica

No. of Living
Species: 62

No. of Species
Vulnerable,
Endangered: 3, 2

No. of Species Extinct
Since 1600: 0

many have yellow underparts. They are 6 to 8.25 inches (15 to 21 cm) long and have dark legs; the sexes sometimes differ slightly in appearance. Pipits (4.5 to 8 inches [12 to 20 cm]), among the most widely distributed songbirds, are brownish and streaked, with narrow tails somewhat shorter than in wagtails; the sexes look alike. Longclaws (6.5 to 8.5 inches [12 to 22 cm]) are streaked brown above and yellow, orange, or reddish below, with a black chest band.

Pipits and wagtails spend most of their time on the ground feeding, pursuing insects and other small arthropod prey with short sprints; they will sometimes fly up a short distance to capture an escaping bug, and will perch on bushes, wires, or buildings. Wagtails mainly occur along rivers, streams, and pond edges, or in wet meadows. Some associate with large mammals, such as cows, and capture insects that flush when the mammals move. Pipits tend to be in drier grasslands, meadows, pastures, and grassy woodlands, but several species commonly occur in wet meadows, bogs, or wet cultivations. Pipits eat insects but also some seeds and other plant materials. Cryptically colored, they are often invisible to people until flushed, at which point they ascend, circle, and either fly away or drop back to the ground. Wagtails, when flushed, typically fly in large arcs, calling loudly. Both wagtails and pipits fly with a strongly undulating flight. All these birds tend to form flocks in the nonbreeding season, and many are migratory.

Most of the birds in this group breed monogamously, with pairs defending territories together in the breeding season. Male pipits give advertising displays in which, while singing, they ascend quickly into the air and then either drop fast or flutter down slowly. Wagtail courtship displays stay more on the ground and involve tail- and wing-spreading. All nest on the ground or in rock crevices. The well-concealed nest, sometimes sheltered by a rock or overhanging grass, and built in a depression, is cuplike and constructed of grasses, plant fibers, moss, and hair. Typically the female alone incubates; both parents usually feed young.

Pipits and longclaws, primarily grassland species, are jeopardized where their habitats are widely converted to crop agriculture. Currently five species are threatened (three vulnerable, two endangered); three are African, one is South American. The fifth, Sprague's Pipit, which breeds in the north-central part of North America, is vulnerable because its prairie breeding habitat continues to shrink, the land taken for farming and degraded by livestock grazing.

Accentors

ACCENTORS are small, drably colored, sparrowlike songbirds of Europe, Asia, and northern Africa. There are thirteen species in the family, Prunellidae. All are called accentors except one: the common Dunnock (sometimes called European Hedge Sparrow or Hedge Accentor), which is distributed broadly in Europe and parts of western Asia. Accentors are recognized for their unusual Old World, temperate-zone-only distribution, their almost universal confinement to mountainous environments, and for their variety of mating systems. The Dunnock is the only species that, in addition to inhabiting higher-elevation sites (in the Alps, Pyrenees, and the Caucasus, for example), occurs in lowland areas also (in European forests, woodlands, shrublands, hedgerows, and gardens). The Dunnock is also widely known for its variable breeding system, in which monogamy, polygyny (one male mating with more than one female), and polyandry (one female mating with more than one male) are all observed.

Accentors range in length from 5.5 to 7 inches (14 to 18 cm). They have slender, pointed bills, rounded wings, and short, strong legs. Their thick plumage is brown, reddish brown, and/or gray above, often streaked, and grayish or tawny below. Some have black or black and white spotted or barred throat patches, and/or white eye lines. The sexes look mostly alike although males are a bit larger and sometimes brighter during breeding.

Birds mainly of high-altitude scrub and brush, mountain slopes, and alpine meadows, some accentors, such as the Alpine and Himalayan Accentors, routinely breed at up to 16,500 feet (5,000 m). The former species,

Distribution:
Eurasia, northern
Africa

No. of Living
Species: 13

No. of Species
Vulnerable,
Endangered: 0, 0

No. of Species Extinct
Since 1600: 0

HIMALAYAN ACCENTOR
Prunella himalayana
6 in (15 cm)
Asia

ROBIN ACCENTOR
Prunella ruberculoides
6.5 in (16 cm)
Asia

SIBERIAN ACCENTOR
Prunella montanella
6 in (15 cm)
Asia

MAROON-BACKED ACCENTOR
Prunella immaculata
6.5 in (16 cm)
Asia

RUFOUS-BREASTED ACCENTOR
Prunella strophiata
6 in (15 cm)
Asia

DUNNOCK
Prunella modularis
6 in (15 cm)
Eurasia, Africa

which usually breeds above treeline and up to snowline, has actually been seen near the top of Mount Everest, around 26,000 feet (8,000 m). The Siberian Accentor inhabits the most northerly forests on Earth. Members of this family feed mainly on the ground, but also in undergrowth and shrubs, and they can cling to rock faces. On the ground they quietly creep, or shuffle, along, sometimes hopping, generally keeping their body close to the ground, often flicking their wings and tail. When in more open areas, they tend to stay close to shrubs and boulders. They use their bill to turn over leaves and other vegetation as they forage, seeking insects and other small invertebrates (worms, snails, spiders), as well as seeds. Typically these birds switch to a seed and berry diet during winter. Accentors are not strong fliers, and usually fly only short distances when flushed. Some are migratory but others remain all year in the same region. The Dunnock is migratory only in the colder, northern parts of its range; these individuals move in autumn to such locations as Gibraltar, Mediterranean islands, and Iraq. Species that breed at high elevations regularly move to lower elevations for the winter, where food is more available. Accentors are usually in pairs during breeding seasons; most species form flocks in winter.

Little is known about accentor breeding except in Dunnocks. During a long-term study in England, biologists found that although many Dunnocks bred in monogamous pairs, others bred in various other combinations of the sexes, the second-most common, after one female/one male, being one female/two males. In these cases, the second, subordinate, male stays with the primary pair and, if he is able to mate with the female, contributes to feeding young and defending the territory. Nests placed on the ground, in a rock crevice, or in a shrub or low tree branch, are neatly woven open cups of plant pieces and moss, lined with feathers and hair. The female builds the nest; the female only or both sexes incubate eggs; both sexes feed the young.

Some accentors, such as the Dunnock and Alpine Accentor, are abundant and widespread, the latter species ranging from the mountains of Europe to the mountains of China and Taiwan. Others have more limited distributions in various Asian mountain ranges. None are currently threatened, but the Yemen Accentor, restricted to high-elevation western mountains of Yemen, is near-threatened. It has limited breeding habitat and occurs at fairly low densities. If its grassy/shrubby habitats are degraded (for instance, by increased livestock grazing) or reduced, the species could quickly become threatened.

GURNEY'S SUGARBIRD
Promerops gurneyi
9–11.5 in (23–29 cm)
Southern Africa

FIRE-BREASTED FLOWERPECKER
Dicaeum ignipectus
3.5 in (9 cm)
Southern Asia

SCARLET-BACKED FLOWERPECKER
Dicaeum cruentatum
3.5 in (9 cm)
Southern Asia

ORANGE-BELLIED FLOWERPECKER
Dicaeum trigonostigma
3.5 in (9 cm)
Southern Asia

YELLOW-VENTED FLOWERPECKER
Dicaeum chrysorrheum
4 in (10 cm)
Southern Asia

FAN-TAILED BERRYPECKER
Melanocharis versteri
6 in (15 cm)
New Guinea

CRESTED BERRYPECKER
Paramythia montium
8.5 in (21 cm)
New Guinea

Sugarbirds; Flowerpeckers; Berrypeckers

SUGARBIRDS are medium-size, long-billed, long-tailed, nectar-eating African songbirds. Because they are endemic to southern Africa, sugarbirds are sometimes used as symbols of that region's biodiversity. The family, Promeropidae, contains only two species, the Cape Sugarbird and Gurney's Sugarbird. Formerly thought to be close kin of the Australasian honeyeaters, sugarbirds are now thought be more closely related to sunbirds (family Nectariniidae). The two sugarbirds are much alike, being dull brown above and whitish below, with brown or reddish brown on the chest and yellow under the tail. They have slender, down-curved bills, short rounded wings, and strong legs and feet. The sexes look alike, although males have longer tails. Males range from 11.5 to 17 inches (29 to 44 cm) in length, females from 9 to 11.5 inches (23 to 29 cm).

Arboreal birds chiefly of shrublands, sugarbirds are particularly dependent on protea shrubs for their nectar food. They insert their long bills into protea flowers and their specialized tongues help them suck in nectar. They also take insects and spiders they find in or around the flowers. Breeding is timed to coincide with the flowering periods of proteas, and sugarbirds are territorial at this time. Once the proteas finish flowering, sugarbirds move around according to nectar availability and are often gregarious then. Many sugarbirds occur on mountain slopes, and are altitudinal migrants, moving to lower elevations during colder, nonbreeding periods. Sugarbirds are monogamous, the female building an open-cup nest of grass and twigs in a bush; only the female incubates eggs but both sexes feed young. Neither sugarbird is considered threatened.

SUGARBIRDS

Distribution:
Southern Africa

No. of Living
Species: 2

No. of Species
Vulnerable,
Endangered: 0, 0

No. of Species Extinct
Since 1600: 0

FLOWERPECKERS

Distribution:
Southern Asia,
Australasia

No. of Living
Species: 44

No. of Species
Vulnerable,
Endangered: 2, 1

No. of Species Extinct
Since 1600: 0

BERRYPECKERS
AND LONGBILLS

Distribution:
New Guinea

No. of Living
Species: 10

No. of Species
Vulnerable,
Endangered: 0, 0

No. of Species Extinct
Since 1600: 0

PAINTED
BERRYPECKERS

Distribution:
New Guinea

No. of Living
Species: 2

No. of Species
Vulnerable,
Endangered: 0, 0

No. of Species Extinct
Since 1600: 0

FLOWERPECKERS are very small, short-tailed, bullet-shaped arboreal birds of southern Asia and Australasia. There are forty-four species in the family, Dicaeidae. They are known for the rapid way they move through tree foliage when foraging, and for their sometimes brilliant plumage. Flowerpeckers are 3 to 5 inches (8 to 13 cm) long, with short, pointed, usually down-curved bills, pointed wings, and stumpy tails. Coloring ranges from some with rather plain, dull olive plumage, similar in both sexes, to others in which males are clad in bright reds, oranges, yellows, blues, or greens, but females are drably colored, olive to dull yellow.

Flowerpeckers utilize an array of habitats, from rainforests and mountain forests to bamboo groves, agricultural areas, and gardens, but most are forest, forest edge, or woodland birds. They feed on berries, soft fruits, and insects, and have specialized bills and tongues for nectar-eating. Flowerpeckers are usually alone or in pairs, but occasionally, during nonbreeding periods, small flocks form around good nectar or fruit sources. Flowerpeckers are monogamous, building oval, baglike nests suspended from small twigs high in trees. The nest, constructed by the female, is made of vegetable down, grass, rootlets, moss, and spider webbing. Females incubate alone; both parents feed young. Many flowerpeckers are common birds, but three species restricted to the Philippines are threatened (two are vulnerable, and the Cebu Flowerpecker, with a population below one hundred, is critically endangered).

BERRYPECKERS are forest songbirds of New Guinea. They are little known and their classification is controversial. In the past some were considered flowerpeckers, and some, honeyeaters. Now the twelve species are often separated out into two families: Melanocharitidae, containing six species of berrypeckers and four called longbills, and Paramythiidae, containing the Crested and Tit Berrypeckers. They range in length from the tiny Pygmy Longbill, at 2.8 inches (7.3 cm; New Guinea's smallest bird), to the Crested Berrypecker, at 8 inches (21 cm). Most have fairly small bills but those of longbills, adapted for feeding on flower nectar, are long, slender, and down-curved. Most are plainly marked, but the Crested and Tit Berrypeckers, sometimes called painted berrypeckers, are brightly colored. Berrypeckers occupy various forested habitats, from lowlands to high altitudes at treeline. Most flit about the forest understory or favor forest edges. They eat small fruits, spiders, and insects, many hovering at foliage to pull off food items; longbills take nectar at flowering shrubs and trees, but also some insects. None of the berrypeckers are officially threatened but there is scant information about them, and one, the Obscure Berrypecker, has been seen only a few times.

Sunbirds and Spiderhunters

SUNBIRDS are tiny to midsize, often very pretty, arboreal birds with long down-curved bills designed to probe and penetrate into flower parts to get nectar. Distributed mainly through sub-Saharan Africa, southern Asia, and northeastern Australia, they are noted for the rapid, acrobatic way they flit in and out of flowering trees and shrubs as they search for flowers and insects. Because of their iridescent colors, bill shape, nectar-feeding behavior, and aggressiveness in defending feeding territories, sunbirds are considered by many to be the Old World ecological equivalent of the New World hummingbirds. The two groups are not closely related; rather, the similarities are superficial, brought about by parallel evolutionary responses to similar habitats and ecological needs—in other words, the great similarity in form and behavior between sunbirds and hummingbirds is a good case of convergent evolution. The sunbird family, Nectariniidae, contains about 130 species, including 10 in Asia called SPIDERHUNTERS. Most are African or Asian; a single species, the Olive-backed Sunbird, occurs in Australia.

Ranging from 3.5 to 8.5 inches (9 to 22 cm) long, sunbirds have relatively thin, down-curved bills that vary in length among subgroups; some have shorter, sharper bills suitable for a more insect-centered diet; spiderhunters, essentially large sunbirds, have extremely long, curved bills. They all have short, rounded wings and strong legs; some, such as Africa's Black-bellied Sunbird, have very long central tail feathers. The often spectacularly colorful sunbird males typically have patches of red, blue, and/or yellow, with many having metallic green or blue iridescence that sparkles in the sun. Females are much duller, usually

Distribution:
Sub-Saharan Africa,
southern Asia,
Australia

No. of Living
Species: 130

No. of Species
Vulnerable,
Endangered: 4, 2

No. of Species Extinct
Since 1600: 0

BROWN-THROATED SUNBIRD
Anthreptes malacensis
5.5 in (14 cm)
Southeast Asia

RUBY-CHEEKED SUNBIRD
Anthreptes singalensis
4.5 in (11 cm)
Southern Asia

OLIVE-BACKED SUNBIRD
Nectarinia jugularis
4.5 in (11 cm)
Southeast Asia, Australia

GOULD'S SUNBIRD
Aethopyga gouldiae
4.5–6.5 in (11–16 cm)
Southern Asia

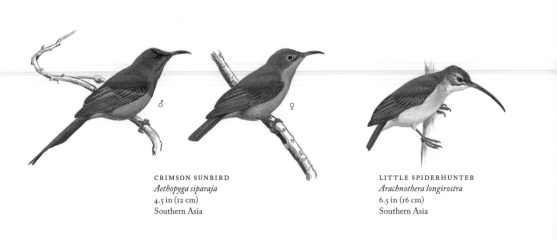

CRIMSON SUNBIRD
Aethopyga siparaja
4.5 in (12 cm)
Southern Asia

LITTLE SPIDERHUNTER
Arachnothera longirostra
6.5 in (16 cm)
Southern Asia

PURPLE-NAPED SUNBIRD
Hypogramma hypogrammicum
5.5 in (14 cm)
Southeast Asia

EASTERN VIOLET-BACKED SUNBIRD
Anthreptes orientalis
4.5 in (12 cm)
Africa

COLLARED SUNBIRD
Hedydipna collaris
4 in (10 cm)
Africa

AMETHYST SUNBIRD
Chalcomitra amethystina
5.5 in (14 cm)
Africa

MARIQUA SUNBIRD
Cinnyris mariquensis
4.5 in (12 cm)
Africa

WHITE-BELLIED SUNBIRD
Cinnyris talatala
4.5 in (11 cm)
Africa

BLACK-BELLIED SUNBIRD
Cinnyris nectarinioides
5 in (13 cm)
Africa

MALACHITE SUNBIRD
Nectarinia famosa
5.5–9.5 in (14–24 cm)
Africa

ORANGE-BREASTED SUNBIRD
Anthobaphes violacea
4.5–6 in (12–15 cm)
Southern Africa

SOUTHERN DOUBLE-COLLARED SUNBIRD
Cinnyris chalybea
4.5 in (12 cm)
Southern Africa

SUPERB SUNBIRD
Cinnyris superba
6.5 in (17 cm)
Africa

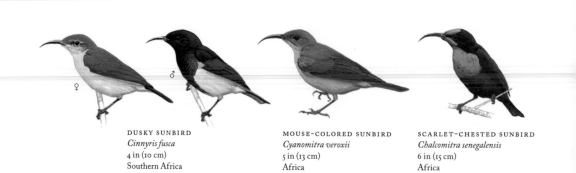

DUSKY SUNBIRD
Cinnyris fusca
4 in (10 cm)
Southern Africa

MOUSE-COLORED SUNBIRD
Cyanomitra veroxii
5 in (13 cm)
Africa

SCARLET-CHESTED SUNBIRD
Chalcomitra senegalensis
6 in (15 cm)
Africa

shades of olive or yellow. Spiderhunters have dull olive or yellow plumage, and the sexes look alike.

Sunbirds occupy a wide array of habitat types, including forests, forest edges, woodlands, mangroves, parkland, gardens, and more open areas such as scrublands and agricultural sites. Even though they are small nectar feeders, sunbirds rarely hover at flowers like hummingbirds; rather, they flit about foliage, stopping and perching each time they feed. They fit their long, slender bills into tubular flowers to get at nectar at the bottom; their long tongue, with a frayed, featherlike tip, flicked back and forth rapidly, is used to pull nectar into the mouth by a combination of suction and capillary action. With large flowers, they tear petals away or use their bill to rip through the bottoms of petals, to get at nectar. They also use the fine, pointed bill to probe into tight corners for spiders and insects (to add protein to their nectar diet), and they pluck spiders from their webs—which must be done carefully because these small birds can accidentally become entangled in webs (indeed, a group of large spider species in southeastern Asia, called bird-eating spiders, make their livings capturing birds). Many species switch to catching insects when they are rearing young. Sunbird bills have fine, saw-tooth edges at their tips that help hold struggling insects and spiders. Sunbirds are often seen foraging alone, and seem to lead fairly solitary lives; when they meet at popular feeding places, they are usually aggressive toward each other. Sometimes they will defend a group of flowers. They can be seen hovering conspicuously over the plants, perhaps giving their buzzy little songs, and chasing away any sunbird, butterfly, or even bee that tries to get near the nectar. Many defend territories throughout the year, often centered around favored flowering trees or shrubs.

A sunbird pair breeds monogamously on an exclusive territory that the male defends. The female builds the domed oval or purselike nest of plant fibers, grass, and spider webbing; the nest hangs from a tree branch or leaf. The female also incubates eggs, but both sexes feed the young. Nesting is timed to coincide with rains and the consequent blossoming of nectar-bearing flowers and abundance of insects.

Many sunbirds are very common and some have broad ranges. However, six species are currently threatened (four are vulnerable, two are endangered), five of them African. All are mainly forest dwellers and all six have tiny ranges and very small populations; the prime threat they face is continued forest loss.

HOODED SISKIN
Carduelis magellanica
4.5 in (12 cm)
South America

RED CROSSBILL
Loxia curvirostra
6.5 in (16 cm)
Eurasia, Africa, North America

WHITE-WINGED CROSSBILL
Loxia leucoptera
6.5 in (17 cm)
Eurasia, North America

COMMON REDPOLL
Carduelis flammea
5 in (13 cm)
Eurasia, North America

PINE SISKIN
Carduelis pinus
5 in (13 cm)
North America, Central America

EVENING GROSBEAK
Coccothraustes vespertinus
8 in (20 cm)
North America

Finches, Siskins, Crossbills, and Canaries

FINCHES, SISKINS, CROSSBILLS, and CANARIES, as well as serins, linnets, redpolls, and some grosbeaks and seedeaters, comprise a large group of mostly small songbirds widely distributed on all continents but Australia and Antarctica. The family, Fringillidae, contains about 134 species, most of which occur in the Old World (more in the north temperate zone than tropics). The group is commonly referred to simply as finches (and, indeed, includes birds called finches, chaffinches, goldfinches, rosefinches, rosy-finches, mountain-finches, and bullfinches). But *finch* is also a term used by many biologists and bird-watchers to describe seed-eating songbirds in general, birds usually with short, robust bills specialized to handle and crack open seeds. Furthermore, several other families (Emberizidae, Cardinalidae, Estrildidae, Ploceidae, Viduidae, Passeridae) are also sometimes referred to generally as finches. To avoid confusion, the term used for the group described here, the "true" finches, is fringillid finches, or fringillids.

Several fringillids are widely recognized as common park, garden, or agricultural-area birds, or have great familiarity as pets. The House Finch, which often breeds in urban areas, was originally a bird of Mexico and the southwestern United States, but it spread as forests were turned into farmland and cities, and eventually moved northward as far as southern Canada. After some were released in New York, it also spread into most of eastern North America. The Chaffinch, which breeds in parks and gardens, is one of Europe's most abundant and familiar land birds. Other noted fringillids are crossbills, famous for their bills that are "crossed," so that the birds can lift

Distribution: Worldwide except Australia and Antarctica

No. of Living Species: 134

No. of Species Vulnerable, Endangered: 3, 6

No. of Species Extinct Since 1600: 1

HOODED GROSBEAK
Coccothraustes albeillei
7 in (18 cm)
Mexico, Central America

PURPLE FINCH
Carpodacus purpureus
6 in (15 cm)
North America

HOUSE FINCH
Carpodacus mexicanus
6 in (15 cm)
North America

AMERICAN GOLDFINCH
Carduelis tristis
5 in (13 cm)
North America

BLACK-HEADED SISKIN
Carduelis notata
4.5 in (11 cm)
Mexico, Central America

COMMON ROSEFINCH
Carpodacus erythrinus
5.5 in (14 cm)
Eurasia

YELLOW-FRONTED CANARY
Serinus mozambicus
4.5 in (11 cm)
Africa

THICK-BILLED SEEDEATER
Serinus burtoni
6 in (15 cm)
Africa

STREAKY SEEDEATER
Serinus striolatus
5.5 in (14 cm)
Africa

BLACK-THROATED CANARY
Serinus atrogularis
4.5 in (11 cm)
Africa

CAPE CANARY
Serinus canicollis
4.5 in (12 cm)
Africa

YELLOW CANARY
Serinus flaviventris
5 in (13 cm)
Southern Africa

WHITE-THROATED CANARY
Serinus albogularis
6 in (15 cm)
Southern Africa

CAPE SISKIN
Serinus totta
4.5 in (12 cm)
Southern Africa

ISLAND CANARY
Serinus canaria
6 in (15 cm)
Canary Islands and Azores

scales of hard closed pinecones to get at seeds, and canaries, which are Old World, often yellow fringillids. Familiar examples include the Island Canary (of the Canary Islands and Azores) and Yellow-fronted Canary (of Africa), which are among the most common of the developed world's cage birds.

Fringillids range from 4.25 to 9 inches (11 to 23 cm) long, most being 5 to 6 inches (13 to 15 cm). They have short, stout, conical bills used to crush seeds (and large jaw muscles to operate the strong bills), longish, pointed wings, short to midlength, often forked, tails, and fairly short legs. Coloring varies from brownish or olive and highly cryptic, to quite colorful, especially among grosbeaks, goldfinches, and siskins (striking yellows and black), and bullfinches, rosefinches, and crossbills (reds). Many are heavily streaked; conspicuous wing and tail markings are common. Males are brighter than females in most species.

Fringillids occupy a variety of habitats, from forests and grasslands to deserts and arctic tundra. They are generally gregarious flocking birds during nonbreeding periods, roaming widely in search of seeds, which are usually taken from on or near the ground (some, such as canaries, feed in trees and shrubs as well). Fringillids are specially adapted for eating seeds: they actually crush a small seed by wedging it into a groove on the roof of their mouth, then raising the lower jaw into it; the broken husk is then peeled off with the tongue and discarded, the bare seed swallowed. Other foods, such as buds, blossoms, and some fruit, are taken from small trees; certain species also take some insects. House Finches consume mostly grass and weed seeds, but also some tree buds and fruit. A few fringillids, such as the Chaffinch, defend large territories during breeding seasons, in which they nest and locate all their food, but most nest in loose colonies, defend only small areas around their nests, and feed in groups away from the nesting area.

Fringillids breed in monogamous pairs, the female alone usually building the nest and incubating eggs, her mate sometimes feeding her as she incubates. Open, cup-shaped nests are constructed typically of twigs, grass, fine roots, moss, and lichens, and placed in trees or shrubs; some species, such as redpolls and rosy-finches, nest on the ground or in rock crevices. Both male and female tend nestlings, feeding them insects, insects and seeds, or, in crossbills, siskins, redpolls, and linnets, seeds alone. Many fringillids are common, widespread birds; nine are threatened (three are vulnerable, six are endangered). The prime threat they face is destruction and alteration of their natural habitats, mostly for farming and ranching.

Hawaiian Honeycreepers

HAWAIIAN HONEYCREEPERS are stunning little forest songbirds with a remarkable story: A sizable group of birds, perhaps fifty species or more, developed in the Hawaiian Islands from a single finchlike species that likely colonized 4 to 5 million years ago. These birds existed only in Hawaii, thriving there, protected by their midocean isolation. But from the time people arrived in the islands 1,600 years ago, these small birds have been persecuted: hunted for their feathers, their forest homes cleared for agriculture, their eggs and young eaten by introduced nest predators such as rats and mongooses, and their immune systems overwhelmed by introduced diseases. Only a few remain, most of them threatened, confined now mainly to patches of protected forests on the islands of Kauai, Maui, and Hawaii. They are perhaps Hawaii's most famous animal residents, and some of them are among the world's most critically endangered birds.

Distribution: Hawaiian Islands

No. of Living Species: 18

No. of Species Vulnerable, Endangered: 7, 7

No. of Species Extinct Since 1600: at least 15

Also called Hawaiian finches (because they evolved from a finchlike ancestor), Hawaiian honeycreepers now number about eighteen species. Their classification is controversial, but they are usually considered to comprise a separate family, Drepanididae, closely related to the finch/siskin/canary family, Fringillidae. Bird-watchers often call these birds dreps, after the family name. Most are from 4 to 5 inches (10 to 13 cm) long, but a few range up to about 7 inches (18 cm). Most are greenish or green and yellow, but some are bright red or orangish, or brownish, and one, the Akohekohe, is black. Within a species, male and female usually look alike, although females are often a bit duller. As a group, the birds' most distinctive physical trait is the bill, which varies from short and straight to very long and highly down-curved.

APAPANE
Himatione sanguinea
5 in (13 cm)
Hawaii

HAWAII AMAKIHI
Hemignathus virens
4.5 in (11 cm)
Hawaii

ANIANIAU
Hemignathus parvus
4 in (10 cm)
Hawaii

AKIAPOLAAU
Hemignathus munroi
5.5 in (14 cm)
Hawaii

MAUI PARROTBILL
Pseudonestor xanthophrys
5.5 in (14 cm)
Hawaii

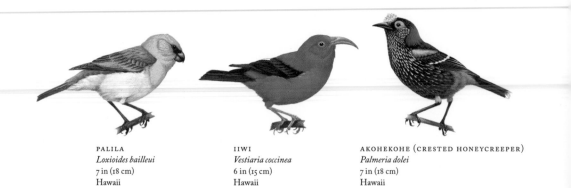

PALILA
Loxioides bailleui
7 in (18 cm)
Hawaii

IIWI
Vestiaria coccinea
6 in (15 cm)
Hawaii

AKOHEKOHE (CRESTED HONEYCREEPER)
Palmeria dolei
7 in (18 cm)
Hawaii

Honeycreepers primarily occupy Hawaii's native forests, although a few occasionally move into shrublands. Most are limited to higher-elevation sites, usually those dominated by ohia or koa trees, generally above 4,000 feet (1,200 m). Their low-elevation wooded habitats were mostly cleared for agriculture, and with the introduction of mosquitoes that carry bird diseases such as avian malaria, the birds can now survive only where mosquitoes cannot—in high, and therefore cool, places.

The most celebrated aspect of honeycreeper ecology is the relationship between the birds' bills and their respective feeding habits. Their bills have been adapted through evolution to closely reflect feeding niches. Some of the birds, seed-eaters, have short, stout, strong, finchlike bills to crush seeds, such as the Palila. Some, such as the Anianiau, have short, fairly narrow, sharply pointed, slightly down-curved, warblerlike bills that can quickly aim at and grab insects from tree trunks and branches. Other insect-eaters, such as the amakihis (which also feed on nectar), have midsize, slim, down-curved bills to probe for insects in mosses and tree crevices and under bark. Some, such as the Apapane, eat mostly nectar (and some insects), and have midsize, slightly down-curved bills specialized to take nectar from flowers. The Iiwi has a very long, strongly down-curved bill that permits it to take nectar from long, tubular flowers. And some have highly specialized bills in which the top and bottom parts differ significantly. For instance, the bottom part of the Akiapolaau's bill is short, thick, and straight and is used to dig small holes into soft tree wood and bark, like woodpeckers do; the top part, fine, down-curved, and nearly twice as long as the bottom, is used to probe into the holes and extract insect larvae.

Most or all honeycreepers are monogamous, male and female building shallow cup nests in trees or cavities, of twigs, grasses, mosses, and leaves. Usually only the female incubates; both sexes feed young. The current status of the honeycreepers is precarious and highly threatened. These birds occupy more domestic slots on the United States' endangered species list than members of any other bird family. A few have been reduced to tiny populations of fifty or fewer individuals. Many species became extinct during the past 150 years, several of them recently. Conservation efforts are underway to attempt to save some of the surviving species (seven are now considered vulnerable, seven endangered).

ADELAIDE'S WARBLER
Dendroica adelaidae
5 in (13 cm)
West Indies

ARROW-HEADED WARBLER
Dendroica pharetra
5 in (13 cm)
West Indies

BAHAMA YELLOWTHROAT
Geothlypis rostrata
6 in (15 cm)
West Indies

LOUISIANA WATERTHRUSH
Seiurus motacilla
6 in (15 cm)
North America, South America, West Indies

BLACKBURNIAN WARBLER
Dendroica fusca
5 in (13 cm)
North America, South America

GRAY-AND-GOLD WARBLER
Basileuterus fraseri
5.5 in (14 cm)
South America

SPECTACLED REDSTART
Myioborus melanocephalus
5 in (13 cm)
South America

BANANAQUIT
Coereba flaveola
4.5 in (12 cm)
West Indies, Central America,
South America

Wood Warblers

WOOD WARBLERS are sprightly, often dazzling, mostly small songbirds that flit about forests and woodlands of the Americas, taking insects from foliage. They are widely acclaimed among bird-watchers for their beauty and diversity. Unfortunately, many warblers look much alike, their myriad plumage color combinations of yellow, olive, gray, black and white differing only subtly. This, added to their sizes and agile natures, sometimes causes even experienced birders to despair of trying to identify the various species in the wild. Warblers are also recognized for their ecological importance especially in North American forests, where they are often so diverse and numerous that, as a group, they make up more of the birdlife than all other birds combined. The wood warbler group (also known as New World warblers), family Parulidae, contains approximately 117 species (including parulas, yellowthroats, redstarts, chats, waterthrushes, and the Ovenbird), about 50 of which breed in the United States/Canada, the remainder ranging from Mexico to southern South America and the Caribbean. One, the Olive Warbler, which occurs from the southwestern United States to Central America, is sometimes classed in its own single-species family, Peucedramidae. And the common Bananaquit, distributed from southern Mexico and the Caribbean to northern Argentina, has been variously lumped into the warbler group, considered with the tanagers, or made the sole member of its own family, Coerebidae.

Warblers are predominantly yellow or greenish, often mixed with varying amounts of gray, black and white; a few have patches of red, orange or blue, and a few are brown. Many have conspicuous whitish wing and tail

Distribution:
New World

No. of Living
Species: 117

No. of Species
Vulnerable,
Endangered: 6, 7

No. of Species Extinct
Since 1600: 1

SLATE-THROATED REDSTART
Myioborus miniatus
5 in (13 cm)
Mexico, Central America, South America

AMERICAN REDSTART
Setophaga ruticilla
5 in (13 cm)
North America, South America, West Indies

PAINTED REDSTART
Myioborus pictus
6 in (15 cm)
North America, Central America

MAGNOLIA WARBLER
Dendroica magnolia
5 in (13 cm)
North America, Central America,
West Indies

GOLDEN-BROWED WARBLER
Basileuterus belli
5 in (13 cm)
Mexico, Central America

RED-FACED WARBLER
Cardellina rubrifrons
5.5 in (14 cm)
North America, Central America

BLACK-AND-WHITE WARBLER
Mniotilta varia
5 in (13 cm)
North America, Central America, West Indies

PROTHONOTARY WARBLER
Protonotaria citrea
5.5 in (14 cm)
North America, Central America, West Indies

CRESCENT-CHESTED WARBLER
Parula superciliosa
4.5 in (11 cm)
North America, Central America

OVENBIRD
Seiurus aurocapillus
6 in (15 cm)
North America, South America,
West Indies

RED WARBLER
Ergaticus ruber
5 in (13 cm)
Mexico

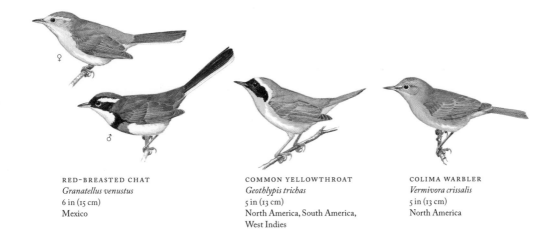

RED-BREASTED CHAT
Granatellus venustus
6 in (15 cm)
Mexico

COMMON YELLOWTHROAT
Geothlypis trichas
5 in (13 cm)
North America, South America,
West Indies

COLIMA WARBLER
Vermivora crissalis
5 in (13 cm)
North America

MACGILLIVRAY'S WARBLER
Oporornis tolmiei
5 in (13 cm)
North America, Central America

OLIVE WARBLER
Peucedramus taeniatus
5 in (13 cm)
North America, Central America

markings. Male and female in nonmigratory species tend to look alike, but in migratory species, males acquire brighter plumage for breeding. Warblers have narrow, pointed bills and slender legs. They range from 4 to 7.5 inches (10 to 19 cm) long, most being about 5 inches (13 cm); the Bananaquit, yellow and olive/grayish, is quite small at 4 to 4.25 inches (10 to 11 cm).

Although forests and woodlands are their main habitats, warblers also occupy rainforests, shrublands, brushy sites, and marshes. In North America, warblers are mainly arboreal, but in South America many skulk about in dense undergrowth. Most warblers move jauntily around trees, shrubs, and gardens, searching for insects and spiders that they tend to take from undersides of leaves and twigs; some use their thin bills to probe flowers and buds for hidden bugs. Some also take small fruits and berries, and one subgroup also consumes a lot of flower nectar and pollen. Redstarts sally out like flycatchers to snatch flying insects in the air; the Black-and-white Warbler moves over tree trunks and large branches, picking insects from bark; and waterthrushes and the Ovenbird walk along the ground, foraging among dead leaves. Bananaquits eat nectar and fruit. Many warblers typically join mixed-species feeding flocks with other small songbirds. Many temperate-zone species migrate long distances to tropical wintering grounds. The Black-and-white Warbler, for instance, nests as far north as northern Canada but winters as far south as Peru. Warblers generally are territorial: either during the breeding season (in migratory species) or year-round (in nonmigratory, tropical species) a male and female defend a piece of real estate from other members of the species. Some tropical warblers remain paired throughout the year. Consistent with the name warbler, many songs of these birds are melodic, but they range in quality from loud, clear notes to insectlike trills and buzzes.

Mainly monogamous breeders, warblers build open-cup or domed nests in trees or shrubs, but sometimes on the ground or in cavities. Often the female builds all or most of the nest, and incubates eggs; the male may feed his incubating mate, and both parents feed young. Both Bananaquit sexes build the round, domed, breeding nest; they also build lighter, domed "dormitory" nests, which they sleep in individually.

Six warblers are considered vulnerable; seven are endangered. This includes the Golden-cheeked Warbler (endangered), which breeds in central Texas, Kirkland's Warbler (vulnerable), which breeds in Michigan, and Semper's Warbler (critically endangered), which occurs only on the Caribbean island of St. Lucia. Bachman's Warbler, a swamp breeder of the southeastern United States, likely became extinct in the 1960s.

Flowerpiercers; Conebills

FLOWERPIERCERS and CONEBILLS are distinct groups of pretty little songbirds, mainly South American, that are now often considered members of the tanager family, Thraupidae. But their position in bird classification schemes is controversial. In the recent past they were sometimes formed into a separate family (Coerebidae) with honeycreepers, dacnises (both now also considered tanagers), and the Bananaquit; or conebills were included with the New World warblers (Parulidae). Both flowerpiercers and conebills are associated with flowers, and both are predominantly higher-elevation birds of the Andes Mountains. Flowerpiercers are noted, and named for, their peculiar feeding method: they use their highly specialized bill to make holes in the bottoms of flowers to get at nectar. The upper part of the bill, which is a bit longer than the bottom part and has a sharply hooked tip, is used to hold part of the flower steady; the lower part of the bill, which is slightly upturned and ends in a sharp point, then makes a hole through which the long tongue is inserted to suck up nectar. In parts of the Andes, most large flowers at mid- to high altitudes eventually have tiny holes in their bottoms, indicating the ubiquity of these small birds. Conebills, named for their mostly short, pointed bills, also forage at flowers, but they are normally there not for nectar but to search for insects. The eighteen flowerpiercer species occur from Central Mexico southward to northern Argentina, and the eleven conebills range from Panama to northern Argentina.

Varying from 4.5 to 6.5 inches (11.5 to 16.5 cm) in length, flowerpiercers are chiefly black or shiny blue, and some have white or reddish brown markings.

FLOWERPIERCERS

Distribution:
Neotropics

No. of Living
Species: 18

No. of Species
Vulnerable,
Endangered: 0, 2

No. of Species Extinct
Since 1600: 0

MASKED FLOWERPIERCER
Diglossopis cyanea
6 in (15 cm)
South America

GLOSSY FLOWERPIERCER
Diglossa lafresnayii
5.5 in (14 cm)
South America

MOUSTACHED FLOWERPIERCER
Diglossa mysticalis
5.5 in (14 cm)
South America

GIANT CONEBILL
Oreomanes fraseri
6.5 in (16 cm)
South America

CAPPED CONEBILL
Conirostrum albifrons
5 in (13 cm)
South America

CHESTNUT-VENTED CONEBILL
Conirostrum speciosum
4.5 in (11 cm)
South America

The sexes are similar in most but females are sometimes drabber than males, usually some shade of olive or brown. Conebills, mainly 4 to 5.5 inches (9.5 to 14 cm) long (but the Giant Conebill is 6.5 inches [16.5 cm]), show a wide range of plumage patterns, but many are blue or bluish gray above, paler below, often with reddish brown patches. Conebill sexes typically look alike, but females are duller in some. The Giant Conebill, unlike others in its group, has a fairly long bill.

Flowerpiercers are primarily birds of forest edges, woodlands, and shrubby areas, although some occupy forests. In addition to their nectar eating, many also take small fruits, and some also consume a lot of insects. Various flowerpiercers forage from tree canopies down to small shrubs. They are typically seen singly or in pairs, although some occur in groups of up to twenty or more, and they often join mixed-species foraging flocks. They are lively foragers, and sometimes conspicuously aggressive about defending good feeding sites, chasing away other flowerpiercers as well as hummingbirds and others. Conebills are mainly canopy birds of mountain forests and woodlands, forest edges, and adjacent shrublands; some in the Andes range up to treeline. There are three subgroups: one group of four species favors forest edge or semiopen habitats in low-elevation areas, including swampy forests and mangroves; another, of six species, prefers middle- and high-elevation Andes woodlands; and last, the Giant Conebill occupies only very high-elevation forests. Conebills eat insects, which they take from flowers and other foliage; they also commonly eat small fruits. Occurring singly, in pairs, or in small groups, conebills often join mixed-species foraging flocks that move through the upper parts of forests. The Giant Conebill is a bit different; it creeps along tree trunks and limbs, taking insects from on or under bark.

Flowerpiercers and conebills are presumably monogamous, but relatively little is known about their breeding. The female only or both sexes build a nest in a small tree or shrub. Nests, often bulky and deep, are open and cup-shaped, constructed of grass, leaves, pine needles, fine rootlets, and moss. Only the female incubates eggs; both sexes feed the young.

Two flowerpiercers, one Venezuelan, one Colombian, are endangered, and a single Chilean/Peruvian conebill species is considered vulnerable. Several other species within these groups have extremely small ranges, which places them in jeopardy; extensive habitat destruction in these areas could easily threaten these birds.

CONEBILLS

Distribution:
South America

No. of Living
Species: 11

No. of Species
Vulnerable,
Endangered: 1, 0

No. of Species Extinct
Since 1600: 0

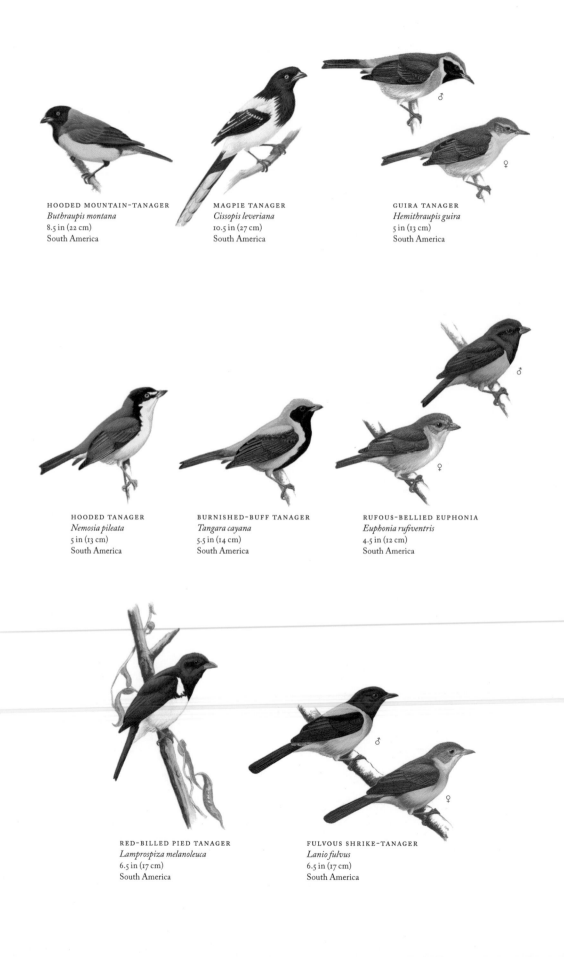

HOODED MOUNTAIN-TANAGER
Buthraupis montana
8.5 in (22 cm)
South America

MAGPIE TANAGER
Cissopis leveriana
10.5 in (27 cm)
South America

GUIRA TANAGER
Hemithraupis guira
5 in (13 cm)
South America

HOODED TANAGER
Nemosia pileata
5 in (13 cm)
South America

BURNISHED-BUFF TANAGER
Tangara cayana
5.5 in (14 cm)
South America

RUFOUS-BELLIED EUPHONIA
Euphonia rufiventris
4.5 in (12 cm)
South America

RED-BILLED PIED TANAGER
Lamprospiza melanoleuca
6.5 in (17 cm)
South America

FULVOUS SHRIKE-TANAGER
Lanio fulvus
6.5 in (17 cm)
South America

Tanagers

TANAGERS comprise a large New World group of beautifully colored, small songbirds, most of which are limited to tropical areas. They are among the American tropics' most visible birds, primarily owing to their habit of associating in mixed-species flocks that gather in the open, often near human habitation, to feed in fruit trees. Bird lovers appreciate them for their diversity of species and wonderful, bright hues. Many are strikingly marked with patches of color that traverse the entire spectrum; indeed, the group is internationally recognized as being among the most fabulously attired of birds. There are some 245 species in the family, Thraupidae, including typical tanagers, honeycreepers, dacnises, and euphonias. (Conebills and flowerpiercers, treated separately in this book, are sometimes also considered tanagers.) Most tanagers are South American, with maximum diversity in the Andes. About 30 species occur in Mexico; only four occur north of the Rio Grande.

Ranging from 3.5 to 11 inches long (9 to 28 cm, most near the smaller end of the range), tanagers are compact birds with medium-long tails. Bill size and shape vary considerably among species, from slender and down-curved to short, stout, and hooked; legs vary in length and thickness. Yellows, reds, blues, and greens predominate in tanager plumage, but a good number are fairly plain, mainly black, brown, or gray. Euphonias are small, stout tanagers, their appearances revolving around a common theme: blue black above, with yellow foreheads, breasts, and bellies. Honeycreepers also sport brilliant plumage and have slender, down-curved bills and brightly colored legs. Dacnises are small tanagers with distinctive short, pointed bills. Tanager sexes typically look alike or nearly so.

Distribution:
New World

No. of Living
Species: 245

No. of Species
Vulnerable,
Endangered: 13, 7

No. of Species Extinct
Since 1600: 0

BERYL-SPANGLED TANAGER
Tangara nigroviridis
5 in (13 cm)
South America

BLUE DACNIS
Dacnis cayana
5 in (13 cm)
Central America, South America

ORANGE-EARED TANAGER
Chlorochrysa calliparaea
5 in (13 cm)
South America

GREEN-AND-GOLD TANAGER
Tangara schrankii
5.5 in (14 cm)
South America

PARADISE TANAGER
Tangara chilensis
5.5 in (14 cm)
South America

SWALLOW TANAGER
Tersina viridis
6 in (15 cm)
South America

BLACK-BACKED BUSH-TANAGER
Urothraupis stolzmanni
6 in (15 cm)
South America

GOLDEN-CROWNED TANAGER
Iridosornis rufivertex
6.5 in (17 cm)
South America

GRASS-GREEN TANAGER
Chlorornis riefferii
8 in (20 cm)
South America

RED-LEGGED HONEYCREEPER
Cyanerpes cyaneus
4.5 in (12 cm)
Mexico, Central America, South America

GREEN HONEYCREEPER
Chlorophanes spiza
5.5 in (14 cm)
Mexico, Central America,
South America

JAMAICAN EUPHONIA
Euphonia jamaica
4.5 in (12 cm)
Jamaica

PUERTO RICAN TANAGER
Nesospingus speculiferus
7.5 in (19 cm)
Puerto Rico

WESTERN STRIPE-HEADED TANAGER
Spindalis zena
6 in (15 cm)
West Indies

GOLDEN TANAGER
Tangara arthus
5 in (13 cm)
South America

BLUE-AND-BLACK TANAGER
Tangara vassorii
5 in (13 cm)
South America

Tanagers inhabit essentially all forested and shrub areas of the American tropics and are particularly numerous in wet forests and forest edges. They prefer the lighter, upper levels of the forest canopy and more open areas, but some favor low, brushy habitats. Most associate in mixed-species tanager flocks, usually together with other types of birds. Finding five or more tanager species in a single group is common, and in the Andes, finding ten or more species together in a single flock is not unusual. A mixed flock will settle in a tree full of ripe fruit and enjoy a meal. Although tanagers' reputations are as fruit-eaters, many also take insects from foliage or even out of the air, and some eat mainly insects. And although most species are arboreal, a few are specialized ground foragers, taking seeds and bugs. Tanagers usually go after small fruits that can be swallowed whole, such as berries, plucking the fruit while perched. After plucking it, a tanager rotates the fruit a bit in its bill, and then mashes it and swallows. (The mashing perhaps permits the bird to enjoy the sweet juice prior to swallowing the fruit.) Some tanagers, such as the ant-tanagers, are frequent members of specialized mixed-species flocks (along with antbirds and others) that spend their days following army ant swarms, feeding on insects that rush from cover at the approach of the devastating ants. Euphonias specialize on mistletoe berries, but eat other fruits and some insects as well. Honeycreepers are specialized for nectar feeding, their bills and tongues modified to punch holes in flower bottoms and suck out nectar; they also take some fruits and insects. Tropical tanagers usually stay in the same area year-round; temperate-zone species, such as those in the United States, are often migratory.

Most tanagers appear to be monogamous, although a number of bigamists have been noted. In many species, male and female stay paired throughout the year. Either the female alone or the pair builds a cup nest (or roofed, in euphonias) of vegetation in a tree or shrub. Only the female incubates eggs; young are fed by both sexes.

Twenty tanager species are currently considered threatened, thirteen are vulnerable and seven are endangered. Two of the latter, both Brazilian, are critically endangered, including one that is known from only a single specimen collected in 1938. Major threats to tanagers are forest clearance and other habitat loss. Several euphonia species are increasingly scarce, apparently because they are captured as prized cage birds in some regions of South America.

New World Sparrows and Old World Buntings

NEW WORLD SPARROWS and OLD WORLD BUNTINGS constitute a diverse group of small seed-eaters, totaling about three hundred species that includes some of North America's most common and visible songbirds, such as Song Sparrows, White-crowned Sparrows, and Dark-eyed Juncos. The group, family Emberizidae, while not closely related to the Old World sparrows (family Passeridae) or "true" finches (Fringillidae), contains many species called sparrows or finches, as well as seedeaters, grassquits, longspurs, and towhees. There are three main subgroups: Eurasian and African buntings; North American sparrows and others; and South American seedeaters, grassquits, and finches. Central America has elements of both the North and South American subgroups. In North America, the sparrow/bunting (emberizid) family is infamous among bird-watchers for its sparrows, many of which are small, brown, and streaked, differing only subtly from one another in plumage pattern, and therefore being extremely difficult to identify to species. Indeed, these birds are undoubtedly the origin of the term LBJ (little brown jobs), which American birders now apply liberally the world over to hard-to-identify small, dark birds. Other celebrated American emberizids are the Song Sparrow, the study of which provided the basis for much of what we know about avian territoriality; the White-crowned Sparrow, the species of choice for many investigations of bird physiology and the relationships between ecology and physiology, especially with regard to the timing of breeding and migration; the Snow Bunting, which breeds farther north than any other land bird; and Galápagos finches, thirteen species of drab black, brown, and

Distribution: Worldwide except Australia and Antarctica

No. of Living Species: 303

No. of Species Vulnerable, Endangered: 19, 17

No. of Species Extinct Since 1600: 0

CUBAN BULLFINCH
Melopyrrha nigra
5.5 in (14 cm)
West Indies

CUBAN GRASSQUIT
Tiaris canora
4.5 in (12 cm)
Cuba

GRASSHOPPER SPARROW
Ammodramus savannarum
5 in (13 cm)
North America, South America

GREEN-BACKED SPARROW
Arremonops chloronotus
6 in (15 cm)
Mexico, Central America

CHESTNUT-BELLIED SEED-FINCH
Oryzoborus angolensis
5 in (13 cm)
South America

COLLARED TOWHEE
Pipilo ocai
8 in (20 cm)
Mexico

PERUVIAN SIERRA-FINCH
Phrygilus punensis
6 in (15 cm)
South America

VARIABLE SEEDEATER
Sporophila corvina
4.5 in (12 cm)
Central America, South America

IMM

SAFFRON FINCH
Sicalis flaveola
5.5 in (14 cm)
South America

WHITE-WINGED DIUCA-FINCH
Diuca speculifera
7.5 in (19 cm)
South America

RED-CRESTED CARDINAL
Paroaria coronata
7.5 in (19 cm)
South America

RUSTY-COLLARED SEEDEATER
Sporophila collaris
4.5 in (12 cm)
South America

WHITE-BELLIED SEEDEATER
Sporophila leucoptera
4.5 in (12 cm)
South America

LARGE GROUND-FINCH
Geospiza magnirostris
6.5 in (17 cm)
Galápagos Islands

DARK-EYED JUNCO
Junco hyemalis
6.5 in (16 cm)
North America

GOLDEN-CROWNED SPARROW
Zonotrichia atricapilla
7 in (18 cm)
North America

SONG SPARROW
Melospiza melodia
6.5 in (17 cm)
North America

RUFOUS-CAPPED BRUSH-FINCH
Atlapetes pileatus
6.5 in (16 cm)
Mexico

SPOTTED TOWHEE
Pipilo maculatus
7.5 in (19 cm)
North America

CHESTNUT-COLLARED LONGSPUR
Calcarius ornatus
6 in (15 cm)
North America

GOLDEN-BREASTED BUNTING
Emberiza flaviventris
6 in (15 cm)
Africa

CAPE BUNTING
Emberiza capensis
6 in (15 cm)
Southern Africa

CINNAMON-BREASTED BUNTING
Emberiza tahapisi
6 in (15 cm)
Africa

LARK-LIKE BUNTING
Emberiza impetuani
5.5 in (14 cm)
Southern Africa

CRIMSON-BREASTED FINCH
Rhodospingus cruentus
4.5 in (11 cm)
South America

mottled birds, the anatomy and geography of which helped Charles Darwin formulate his ideas about evolution via natural selection.

Emberizids are 3.5 to 8.5 inches (9 to 22 cm) long (most about 6 inches [15 cm]), with short, thick, conical bills specialized to crush and open seeds. In some species, the upper and lower parts of the bill can be moved sideways to manipulate seeds. All have relatively large feet, which they use in scratching the ground to find food. Coloring varies greatly but most are dull brown or grayish, with many sporting streaked backs, and some, streaked chests. Buntings are often more colorful than sparrows, many Old World species having yellow in their plumage, or patches of black and white. Emberizid sexes tend to look similar, although in some species, particularly among buntings, males are more brightly colored or patterned than females.

Sparrows and buntings occupy a great variety of habitats. North American birds favor woodlands, grasslands, thickets, meadows, desert areas, and tundra. Many South American species inhabit open areas such as forest edges, parkland, brushy areas, and grassland. Old World buntings typically prefer more open, scrubby, often rocky areas. Many emberizids forage mostly on the ground or at low levels in shrubs or trees. Because they spend large amounts of time in thickets and brush, they can be skulking and inconspicuous. Towhees, juncos, and many sparrows often forage by "double-scratching," that is hopping or jumping backward on the ground, pulling soil back with their feet, exposing hidden food. Sparrows and buntings generally specialize on eating seeds for much of the year, but often feed insects to their young. Some also eat fruit; many are considered almost omnivorous. Most species are strongly territorial, a mated pair aggressively excluding other members of the species from sharply defined areas. In many sparrows, pairs stay together all year; other species often travel in small family groups. Sometimes territories are defended year-round and almost all available habitat in a region is divided into territories. Some species form into large flocks during nonbreeding periods. Songs of emberizids vary from buzzes and trills to pure tones and short, loud melodies.

Most sparrows and buntings are monogamous. The female of a pair, sometimes with the male's assistance, builds a cup-shaped or, often in the New World tropics, domed nest, from grasses, rootlets, and perhaps mosses and lichens. Nests are concealed on the ground or low in shrubs or trees. The female alone or both sexes incubate eggs; both sexes feed the young. Many emberizids are abundant, widespread birds; however, nineteen species are currently vulnerable, seventeen endangered (six of the latter, mainly in South America, critically endangered).

BLACK-FACED GROSBEAK
Caryothraustes poliogaster
7 in (18 cm)
Mexico, Central America

BLUE-BLACK GROSBEAK
Cyanocompsa cyanoides
7 in (18 cm)
Central America, South America

SLATE-COLORED GROSBEAK
Saltator grossus
8 in (20 cm)
Central America, South America

SOUTHERN YELLOW-GROSBEAK
Pheucticus chrysogaster
8.5 in (22 cm)
South America

CRIMSON-COLLARED GROSBEAK
Rhodothraupis celaeno
8.5 in (22 cm)
Mexico

PYRRHULOXIA
Cardinalis sinuatus
8.5 in (22 cm)
North America

YELLOW GROSBEAK
Pheucticus chrysopeplus
9 in (23 cm)
Mexico, Central America

PAINTED BUNTING
Passerina ciris
5.5 in (14 cm)
North America, Central America,
West Indies

Cardinals, Saltators, and Grosbeaks

CARDINALS, SALTATORS, and GROSBEAKS are small to medium-size New World songbirds with short, sturdy, finchlike bills, adapted for powerful seed crunching. They are generally handsome, sometimes crested, and often colorful. The family, Cardinalidae, ranges from central Canada southward to central Argentina and contains forty-three species. Included are a group of buntings, which are generally smaller than other cardinalids (also called, collectively, cardinal-grosbeaks), with smaller bills and brighter plumage, and the Pyrrhuloxia and Dickcissel. Grosbeaks are thick-set, finchlike birds with massive bills. Cardinals are similar, but have conspicuous crests and long, squared tails. Saltators are generally plain-looking cardinalids that favor forest edge and scrub areas. The family is most diverse in the tropics. The most familiar cardinalid to North Americans is undoubtedly the Northern Cardinal, a common, brightly colored (all-red male, reddish brown female) park and garden bird across the eastern United States and Mexico.

Cardinalids, 5 to 9.25 inches (13 to 24 cm) long, vary extensively in coloring. Some are mainly drab olive or gray birds (many saltators) or blue or blue black (some grosbeaks and buntings); others are brighter and two-toned (yellow and black or red and black grosbeaks). At the extreme is the Painted Bunting, considered one of North America's most breathtakingly beautiful little birds, with blue head, green back, and red rump and underparts. Male and female cardinalids look alike or, especially in temperate-zone species, different—the females of many being drab brownish or greenish.

Cardinalids are chiefly arboreal birds, although many will drop to the

Distribution:
New World

No. of Living
Species: 43

No. of Species
Vulnerable,
Endangered: 0, 0

No. of Species Extinct
Since 1600: 0

ground to feed. They occur over a great range of habitat types, from tropical rainforests to forest edges, temperate woodlands, thickets, scrub areas, and gardens. Their diets center often on seeds, the largest-billed species, as one might expect, eating the largest seeds; small species, such as the North American buntings, eat a lot of tiny weed seeds. Grains, soft fruits, berries, and insects are also taken by the group. Although most eat seeds during most of the year, many switch to insects and other small arthropods during the breeding season. Saltators, which tend to occupy open-country habitats such as woodlands, forest edges, clearings, and pastures, favor fruit and berries (and some insect larvae); they will balance a plucked fruit on a horizontal surface and take repeated bites. Cardinalids are typically observed singly or in pairs; in the tropics, many species regularly join mixed-species foraging flocks with such other birds as tanagers and honeycreepers. Tropical species usually remain year-round in the same area; temperate-zone species are often migratory, some, for instance, breeding as far north as southern Canada but wintering in northern South America or the Caribbean. The Northern Cardinal, one of the least migratory of America's birds, remains in the northeastern United States for winter, and is a common visitor to snow-encrusted suburban bird feeders. Many cardinalids are strongly territorial, a male and a female pairing and together defending from others of their species a territory, in which they breed and on which they find all or much of the food for their young. Others, such as some of the tropical grosbeaks, are less territorial, often spending time in loose flocks. Migratory species change their aggressive habits with season. Some American buntings, for instance, are strongly territorial during summer, chasing other buntings away from their real estate, but on their wintering grounds, say in southern Mexico, they congregate in flocks that feed together in agricultural areas. Many members of this family are considered excellent singers, possessing striking, loud, rich songs; some are very persistent singers.

Monogamous breeders, cardinalids have open-cup nests, typically built by females, usually placed in trees and shrubs. Nests are made of such vegetation as twigs, bark, weed stems, and grass, and are usually lined with fine materials like grass, hair, moss, and feathers. The female only or both sexes incubate eggs; in some, males bring food to their incubating mates. Both sexes or the female only (for instance, in the Dickcissel, in which some males have more than one mate), feed the young. None of the cardinalids are currently threatened; four species are considered at risk.

New World Blackbirds

NEW WORLD BLACKBIRDS are mid- to large size songbirds renowned for their ecological diversity and success. They are often abundant, conspicuous birds throughout their range, which encompasses all of North, Central, and South America and the Caribbean. The family, Icteridae, includes about ninety-seven species, variously called blackbirds, caciques, cowbirds, grackles, meadowlarks, orioles, and oropendolas; they vary extensively in size, coloring, ecology, and behavior. Distinguishing them from most other birds is a particular feeding method known as gaping: a bird places its closed bill into crevices or under leaves, rocks or other objects, and then forces the bill open, exposing the previously hidden space to its prying eyes and hunger. Many of these birds, collectively termed icterids, have adapted well to human settlements and are common denizens of gardens, parks, and agricultural areas.

Ranging from 6 to 22 inches (15 to 56 cm) long, icterids generally have sharply pointed, conical bills. Black is their predominant color, but many combine it with bright reds, yellows, or oranges. In some, the sexes are alike (particularly in tropical species), but in others, females look quite different from males, often more cryptically outfitted in browns or grays. Pronounced size differences between the sexes, females being smaller, are common. Grackles, common in urban areas, are primarily black birds with slender bills and, usually, long tails. Blackbirds (the term also used sometimes for the entire family, as in New World Blackbirds) are often marsh dwellers. Caciques, sleek black birds frequently with red or yellow rumps and yellow bills, typically occupy forests, forest edges, or thickets. Orioles are brightly marked birds in yellow or orange mixed with black and white; their preferred habitat is forest. Meadowlarks are

Distribution:
New World

No. of Living
Species: 97

No. of Species
Vulnerable,
Endangered: 4, 7

No. of Species Extinct
Since 1600: 1

PERUVIAN MEADOWLARK
Sturnella bellicosa
8 in (20 cm)
South America

ORIOLE BLACKBIRD
Gymnomystax mexicanus
10.5–12 in (27–30 cm)
South America

TROUPIAL
Icterus icterus
9 in (23 cm)
South America

YELLOW-RUMPED CACIQUE
Cacicus cela
9.5–11.5 in (24–29 cm)
Central America, South America

CRESTED OROPENDOLA
Psarocolius decumanus
14–19 in (36–48 cm)
Central America, South America

RUSSET-BACKED OROPENDOLA
Psarocolius angustifrons
14–19.5 in (35–49 cm)
South America

RED-BREASTED BLACKBIRD
Sturnella militaris
7.5 in (19 cm)
Central America, South America

SCARLET-HEADED BLACKBIRD
Amblyramphus holosericeus
9.5 in (24 cm)
Africa

RED-RUMPED CACIQUE
Cacicus haemorrhous
9–11 in (23–28 cm)
Africa

EPAULET ORIOLE
Icterus cayanensis
8 in (20 cm)
South America

GIANT COWBIRD
Scaphidura oryzivora
12–15 in (31–38 cm)
Central America, South America

SHINY COWBIRD
Molothrus bonariensis
7–8.5 in (18–22 cm)
Central America, South America,
West Indies

BROWN-HEADED COWBIRD
Molothrus ater
7.5 in (19 cm)
North America

YELLOW-SHOULDERED BLACKBIRD
Agelaius xanthomus
8–9 in (20–23 cm)
Puerto Rico

HOODED ORIOLE
Icterus cucullatus
8 in (20 cm)
North America

ORCHARD ORIOLE
Icterus spurius
7 in (18 cm)
North America, South America, West Indies

SCOTT'S ORIOLE
Icterus parisorum
9 in (23 cm)
North America

WESTERN MEADOWLARK
Sturnella neglecta
9.5 in (24 cm)
North America

BOAT-TAILED GRACKLE
Quiscalus major
14.5–16.5 in (37–42 cm)
North America

BOBOLINK
Dolichonyx oryzivorus
7 in (18 cm)
North America, South America

RED-WINGED BLACKBIRD
Agelaius phoeniceus
7–9 in (18–23 cm)
North America, Central America

YELLOW-HEADED BLACKBIRD
Xanthocephalus xanthocephalus
7.5–10 in (19–25 cm)
North America

yellow, black, and brown grassland birds. Oropendolas are spectacular, larger birds of tropical forests and woodlands. Cowbirds, usually quite inconspicuous in shades of brown and black, have a dark secret—they are brood parasites.

Icterids occur in all sorts of habitats, including woodlands, thickets, grassland, marshes, forest edges, and even rainforest canopies, but they are especially prevalent in more open areas. Their regular occupation of marshes is interesting because they are not obviously adapted for living in aquatic environments; they do not have webbed feet, for example, nor can they float or dive. They eat a wide variety of foods including insects and other small animals, fruit, and seeds. Some are fairly omnivorous, as befitting birds that frequently become scavengers in urban settings. A common feature of the group is that seed-eaters during the nonbreeding periods become insect-eaters at breeding time and feed insects to their young. Tropical orioles and caciques commonly join in mixed-species foraging flocks. Outside of the breeding season, icterids, particularly blackbirds and grackles, typically gather in large, sometimes enormous, flocks that can cause damage to roosting areas and agricultural crops.

Icterids pursue a variety of breeding strategies. Some, such as the orioles, breed in classically monogamous pairs, male and female defending a large territory in which the nest is situated. But others, including many caciques and the oropendolas, nest in colonies. Some species, such as Red-winged and Yellow-headed Blackbirds, are strongly polygynous, with a single male during a breeding season mating with several females. Nests, almost always built by females, range from hanging pouches woven from grasses and other plant materials, to open cups lined with mud, to roofed nests built on the ground, hidden in meadow grass. Nestlings are fed by both parents (monogamous species) or primarily by the female (polygynous species). Most cowbirds are brood parasites, building no nests themselves. Rather, females, after mating, lay their eggs in the nests of other species and then let the host species raise their young.

Some of the most abundant birds of the Western Hemisphere are icterids, such as Red-winged Blackbirds and Common Grackles. Brown-headed Cowbirds, open-country birds, have been increasing, actually expanding their range with deforestation. Some tropical orioles have been severely reduced in numbers because they are hunted as prized cage birds. Currently eleven icterids are threatened: four are vulnerable and seven endangered, three of the latter critically so (Montserrat Oriole, Brazil's Forbes' Blackbird, Colombia's Mountain Grackle). The Slender-billed Grackle, a Mexican marsh dweller, became extinct in the early 1900s.

Avian Diversity and Biogeography

AVIAN DIVERSITY

How many kinds of birds are there?

A book seeking to introduce a reader to the birds of the world should be able to say how many different types of birds there are. Unfortunately, we don't know the exact number of living bird species. It is not an easy number to ascertain. The reasons for the difficulty in determining the real number of bird species are interesting, and the explanations afford glimpses into several aspects of ornithology, the study of birds.

Animal species are natural groupings, all individuals of a species reproductively isolated from individuals of other species. One definition of a species is a population or series of populations of organisms that are similar enough to be able to breed freely with each other and produce fertile offspring. So, at least theoretically, we should be able to determine the world's exact number of bird species: simply perform many tests of putting male and female birds together in cages; the ones that mate and produce fertile young are members of the same species; the ones that cannot, are not.

In practice, however, things are not that simple. Captive breeding tests to determine bird species are impractical and probably impossible for many kinds of birds, even if there were people who wanted to attempt them. There are also historical, biological, and scientific issues that prevent, or at least inhibit, easy determination of the number of bird species. Most obviously, a little library research reveals that the total number of bird species changes rather frequently, and not by just a few, but by large increments. Furthermore, no two reference books or other sources seem to provide the same information about bird numbers, and the number of known bird species is growing with passing decades. Indeed, the birds of the world expanded generously during the past century and a half, from a count of 7,500 species in the 1850s, to 8,500 in the mid-1930s, to about 8,600 from the 1950s through the 1970s, to 9,000 during the 1980s, to between 9,600 and 9,900 from the 1990s to today.

What's going on here? Are new species still being discovered and described by science for the first time? Certainly some newly discovered birds have been added to the list of species during the past 150 years, especially earlier in that period. Today, even though most of the terrestrial world has been explored, bird species new to science continue occasionally to appear. Of all the bird families, it is probably among the New World flycatchers (family Tyrannidae) that the most undiscovered species remain. This is because there are many species of flycatchers; they live in nearly every terrestrial habitat; and many of them are drab and inconspicuous. As bird experts reach previously inaccessible locations—hidden valleys, cloud-draped mountain plateaus—in the remotest parts of South America, previously unknown flycatchers are indeed sighted. For instance, one new species was first identified in 1976 in northern Peru, and another was found in 1981 in southern Peru. Two more were first described in the scientific literature in 1997. Birds in other families also have been discovered recently—a new species of barbet was identified in Peru in 1997, a new manakin in Brazil in 1998, a new seedeater in Venezuela in 2001, a new owl in Malaysia in 2001, and a new wren in Colombia in 2003.

But relatively few species have been discovered in, say, the past 50 years, and now, for every few new species discovered, probably a few become extinct. So what accounts for the large recent increases in the number of known living bird species? The answer has to do with avian taxonomists, that is, biologists who study bird classification. One might think that after a couple of hundred years of effort, the scientific classification of birds would be relatively stable. But with each generation of taxonomists, new classifications emerge, sometimes because new methods are used, sometimes because old methods are employed in different ways. With the advent of molecular taxonomy methods (including ones that allow comparative studies of birds' DNA), classifications of birds that were first worked out during the 1800s and early 1900s are undergoing alterations, sometimes radical ones.

The main type of taxonomic change resulting in an increase in the number of described bird species occurs when what used to be considered a single species is "split" into 2 or more separate but closely related species. Usually what happens is that after careful study 2 or more geographic "subspecies," or "races," of a single species, which may look or behave slightly differently, are each given full species status. For example, until the 1990s, the Florida Scrub-jay (*Aphelocoma coerulescens*), Western Scrub-jay (*Aphelocoma californica*), and Island Scrub-jay (*Aphelocoma insularis*), which are now considered separate species that look much alike, were considered subspecies of a single species, the Scrub-jay (*Aphelocoma coerulescens*). In this case, 1 species was split into 3. The splitting was based on genetic, anatomical, behavioral, and fossil evidence indicating that the 3 jays were separate species. Another example: Australia's Green Catbird (*Ailuroedus crassirostris*), a bowerbird, is now often divided into 2 species, the Spotted Catbird (*Ailuroedus melanotis*) in the country's northeast (and in New Guinea) and the Green Catbird (*Ailuroedus crassirostris*) in the southeast. The evidence for the split rests mainly on molecular comparisons of the birds.

Aside from historical and scientific influences on attempts to fix the number of bird species globally (as the number of described species changes over time with more exploration, more research, and improved classification methods), there are also some biological difficulties. Among these is the fact that, although the most accepted definition of a species concerns the inability of a male and a female of different species to mate and produce fertile offspring, in nature this is not always the case. Sometimes 2 species are so alike genetically (because they separated evolutionarily so recently), that they still share most genes, enough so that a male from one species and a female from another can still breed and produce fertile young. Lions breeding with tigers is one example of this phenomenon (the young are called ligers or tiglons; lions and tigers, with nonoverlapping natural distributions, never meet in the wild and so such matings are zoo-only phenomena); wolves breeding with domestic dogs is another. Birds do this also, further complicating the picture for determining species numbers. For instance, a female Island Scrub-jay (found only on Santa Cruz Island, off southern California) might be able to breed successfully with a male Western Scrub-jay from coastal California if the two could meet. We know this kind of interbreeding between very similar species sometimes occurs naturally because there are many cases where such species "collide" and overlap geographically. One example is

North America's Baltimore and Bullock's Orioles, now considered separate species, which nonetheless interbreed and produce hybrid young in parts of their current geographic "zone of overlap" in the American Midwest.

Another biological difficulty in counting species is that some closely related ones look exactly or almost exactly alike, so taxonomists cannot always use direct physical comparison to distinguish species. Good examples are some of the New World flycatchers of genus *Empidonax*, which even expert bird-watchers find difficult to tell apart visually (the birds' songs are more reliable species identifiers), as well as some of the oystercatchers (family Haematopodidae), Eurasian tits (Paridae), larks (Alaudidae), pipits (Motacillidae), and greenlets (Vireonidae).

Finally, some species are known literally from a single specimen; no other evidence of them has ever been found. They are quite mysterious and present problems for counting species: does the specimen in question truly represent a separate species? And if so, are there any more of them left alive? After all, if only a single specimen has been found, perhaps the species is now extinct. For instance, the Negros Fruit-dove is known only from a specimen found on the island of Negros in the Philippines in 1953; the Red Sea Swallow, from a specimen found on the Red Sea coast of Sudan in 1984; and the Cone-billed Tanager, from a specimen found in southern Brazil in 1939. All three of these are considered existing species in most current bird taxonomies.

For all the reasons above, the assignment of species status to various groups (populations) of individuals is an inexact science, and there are many controversial species assignments that bird classification authorities will argue over for decades. This inability to know the exact number of living bird species on Earth reflects a deep and genuine interest on the part of avian taxonomists and others concerned with birds to get the classification "right"—that is, to make it as consistent with nature's groupings as (humanly) possible. In addition to sowing confusion among the uninitiated, reference books issued every few years that give different species totals frustrate bird-watchers, who like to list the bird species they have seen and know the number remaining they might someday spot. For instance, someone who had seen 800 species by 1980 would have calculated he or she had seen 8.9 percent of the globe's birds (based on the then total number of species estimated at 9,000), but in the first decade of the twenty-first century, that person would have seen only 8.2 percent of the birds (based on a total of

9,800 species). In addition, of the 800 species spotted as of 1980, some may have been subsequently split into 2 or more species, or otherwise altered by avian taxonomists, thereby changing the count, and adding confusion to the birder's frustration.

So, how many bird species are there? We don't really know, but I need to pick a number. For this book, I have generally followed a classification scheme that divides the living birds of the world into approximately 9,750 species.

Another way to answer the question "how many types of birds are there?" is to consider bird families. Families of animals, such as birds, are groups containing closely related species (they share a relatively recent common ancestor) that are often very similar in form, ecology, and behavior. For instance, the Blue Jay, Gray Jay, Black-billed Magpie, American Crow, Common Raven, Australian Raven, and Eurasian Jackdaw, which share many anatomical and behavioral traits, are closely related genetically and therefore are all considered members of the globally distributed jay/crow family (Corvidae; all species in the family can be referred to as corvids.) Similarly, all the different kinds of ducks (Mallard, Northern Pintail, Lesser Scaup, etc.) are members of the worldwide duck/geese/swan family, Anatidae (anatids). And all the New World blackbirds (Western Meadowlark, Red-winged Blackbird, Common Grackle, Baltimore Oriole, etc.) are members of family Icteridae (icterids). Families are, in a sense, artificial constructs—ways that people choose to place various species into related groups; there is no absolutely correct way to distribute the world's birds into families, only better and worse ways (that more accurately or less accurately reflect the natural relatedness of various bird groups), and today's family groupings are bound to undergo changes. But the family level of classification is useful for delineating many types of birds, and bird-watchers often organize their sightings by the various families.

So, can I tell you the total number of bird families? Well, not exactly. The problem is that, although many families comprise fairly distinct and obvious aggregations of closely related birds—the three examples above are somewhat representative of this group—others are less distinct, the relationships among a family's members less understood and more controversial. The result is that avian taxonomists argue about the assignment of species to various bird families and, indeed, about the identities, sizes, and even proper number of families. For instance, the Bananaquit, a very common and widespread small bird of the New World tropics, has been variously included in the wood warbler family (Parulidae), allied with the tanager family (Thraupidae), or made the sole member of its own family (Coerebidae). The 2 species of rockfowl, large, terrestrial African songbirds, have, at various times, been considered babblers (family Timaliidae), starlings (Sturnidae), and crows (Corvidae). Now they are usually placed in their own separate family (Picathartidae). The 350 parrot species all look like parrots, and one might assume they would be an uncontroversial group, but they are variously considered a single family (Psittacidae), or 2 families (the 21 cockatoos placed in a separate family, Cacatuidae), or 3 families (the approximately 55 lories and lorikeets also separated out, as family Loriidae). The ioras (4 species), leafbirds (8 species), and fairy-bluebirds (2 species), all confined to Asia, are sometimes considered to comprise a single family (Irenidae), are grouped into 2 families (the leafbirds and fairy-bluebirds comprising Irenidae; the ioras part of another family), or are separated into 3 families (ioras in Aegithinidae; leafbirds in Chloropseidae; fairy-bluebirds in Irenidae). There are many other examples, including among cuckoos, antbirds, cotingas, monarch flycatchers, and Australian chats, to list just a few. A consequence of these shifting family assignments is that there are always several competing global bird classification schemes, or taxonomies, each of which arranges the birds of the world a bit differently; and which change subtly, or sometimes not so subtly, as new research is reported, considered, and acted upon. How many bird families are there? During the past 50 years, taxonomies that divided birds into 145 to 200 or more families have been used, the larger ones mainly in more recent years. For this book, I followed a taxonomy that divides birds of the world into about 200 families.

How does bird diversity compare with that of other vertebrate groups?

How does the total of about 9,800 bird species compare with the numbers for other kinds of terrestrial vertebrate animals? There are about 4,800 mammal species, about 7,800 reptiles, and about 4,800 described species of amphibians. (Almost all birds, mammals, and reptiles living on Earth have been discovered, characterized, and named—that is, scientifically described—but, most experts agree, many more amphibians, because they live partially in water, remain to be discovered. Likewise, the other principal vertebrate group, the fishes, with more than 20,000 recognized species, may still yield many new species as the murky depths of the oceans are explored.)

Why is it that species diversity is greater among birds than among the other groups of mainly terrestrial vertebrates? And, perhaps most curious, why are there roughly twice as many bird as mammal species? After all, mammals, like birds, are "warm-blooded," with efficient respiratory and circulatory systems that allow maintenance of body temperatures typically well above their surroundings and therefore, usually, active year-round lifestyles; and these attributes permit mammals, like birds, to be distributed broadly from the equator to the polar regions. Moreover, mammals have relatively large brains and are considered fairly intelligent. Further, birds, owing to the constraints of flight, are remarkably standardized physically, whereas mammals can be quite diverse in form and still function as mammals (consider how different in form are mice, kangaroos, lions, giraffes, otters, and whales.) But if birds are going to fly, they must look more or less like birds, and have the anatomies and physiologies that birds have. So if birds are in a sense limited in their evolutionary development by flight (and most mammals are not), why are there so many more birds than mammals? There are several reasons, but two main ones have to do with flight, and where birds and mammals live.

Because birds fly and are quite mobile, they are able to make good use of the third dimension of many habitats, height, which many other groups, such as mammals, cannot do as well. Simply put, more species can exist where there are more places (microhabitats) and ecological opportunities (niches) for them to occupy, or "exploit." Birds can easily and rapidly make use of many different places in, say, a forest—some species making their livings on the forest floor, others foraging on tree trunks, others in the lower canopy, and still others in the high canopy. Mammals (or reptiles or amphibians, for that matter) cannot as easily or efficiently utilize all these various levels, so the number of niches open to them is more limited; consequently, the number of their species that can exist at one time in a given region is lower.

In concert with their unsurpassed ability among the terrestrial vertebrates to make use of spatially complex habitats, such as forests and woodlands, and to take advantage of ecological opportunities, birds are most common in the tropics, where highly diverse habitats, such as rainforests, are most available. The relationship between bird diversity and the tropics is discussed below.

Because of their flight abilities, birds can also

- arrive at isolated islands, such as the Galápagos and Hawaiian Islands, where one or a few colonizing species can evolve quickly into many species, as happened with Galápagos finches and Hawaiian honeycreepers;

- cover large geographic ranges to find needed resources, as when albatrosses or penguins leave nesting areas to forage for their young far away from their breeding sites;

- migrate quickly and efficiently over long distances to escape seasonally harsh conditions or food shortage; and

- specialize on eating nectar, like hummingbirds and honeyeaters, because birds, being so light and agile, can readily exploit small flowers near the ground or high in the tree canopy.

These capabilities of birds, all possible because of flight, contribute to the relatively high number of bird species. Some biologists consider the current era (the Cenozoic, the past 65 million years up to the present) as the Age of Mammals (with the previous era, the dinosaur-filled Mesozoic, the Age of Reptiles), meaning mammals diversified recently and have become the dominant terrestrial life form. But given that one measure of ecological success is number of species and there are twice as many bird as mammal species, are mammals truly now the dominant terrestrial animals?

Why is there great variation among bird species in numbers of individuals?

Why are some bird species abundant, others somewhat common, and some rare? Two main factors influence the total size of bird populations: the extent of the geographical range of the species and the density of individuals within the range. The former factor is easy to understand—species with larger ranges, inhabiting more areas, generally can have larger total populations than species with smaller ranges. The latter factor, population density, is more complex. Main influences on population density are available food supply; availability of resources other than food, such as suitable feeding and breeding habitat and nest sites; the impact of predators, parasites, or both; and these days, the impact of negative interactions with people.

So some bird species currently enjoy huge total populations because (1) they have large ranges; (2) they have abundant food and breeding resources; (3) their breeding is often successful (that is, nest predation, parasites, or other factors do not strongly limit nesting success); and (4) they are not under imminent threat from people. In some cases, people have actually aided the success of bird populations, as in the case of introduced birds (such as House Sparrows and European Starlings) and birds that benefited from people's large-scale alterations of habitats (such as New World blackbirds that benefit from conversion of large tracts of forests into more open, agricultural lands, in which blackbirds often thrive). Conversely, some bird species now have very small total populations because (1) their foods or breeding resources are scarce; (2) their breeding success is often strongly depressed by factors such as nest predation or brood parasitism (see p. 166); and/or (3) they are directly or indirectly threatened by people's activities—they are hunted or killed for various reasons; their habitats are altered or destroyed; or they are negatively affected by introduced bird species or other introduced organisms.

Because available food supply is one of the main factors that limits the size of animal populations and because larger birds require more food than smaller birds, smaller bird species are typically more numerous than larger ones. One study found that among North American birds, those weighing about 1.4 ounces (40 grams), sparrow-size species, were the most numerous.

Most critically endangered species have very small total populations because of the actions of people (see "Understanding the Decline in Avian Diversity" below). But some birds are naturally rare, even if they have healthy populations. This is particularly the case with raptors (see p. 71). Although some raptors are common, typically they exist at relatively low densities, as is the case for all "top predators" (predators at the pinnacle of a food chain, preyed upon by no animal, except humans). Usually there is enough food available to support only one or two of a species in a given area. For example, a typical density for a small hawk species, one that may feed on mice and small lizards, is one individual per square mile (2.6 sq km). A large eagle that feeds on monkeys, sloths, and porcupines, such as the massive Harpy Eagle of South and Central America, may be spaced so that a usual density is one individual per 50 square miles (130 sq km), or even more.

How many individual birds are there?

For most species, no true count is possible. But for some highly endangered species, because so few individuals are alive, precise counts are feasible. For instance, when a particular bird species is restricted to a single small island, researchers, through careful surveys, can determine that there are, say, fewer than 1,000 individuals left alive, or fewer than 100. For some, such as the Hawaiian Crow—which occurred only on the Big Island of Hawaii, is on the verge of extinction, and is the subject of much research and conservation effort—fairly precise numbers are known: there were perhaps 75 Hawaiian Crows still in the wild during the mid-1970s, only 11 or 12 in the early 1990s, only 3 in 1999, and none today (some still live as captive birds). Or take the Whooping Crane: wildlife experts know the exact number of living Whooping Cranes (usually fewer than 200 individuals) because this endangered North American species is closely monitored. But for most species, only rough estimates of population sizes are available.

Bird populations (as well as many other types of animals) experience seasonal differences in abundance numbers. Let us say that most birds breed in spring and summer, and most nests successfully fledge, on average, 2 offspring. In this case, if a species has one million members in winter (divided about half and half between males and females), then it will have 2 million members in the summer. Following fledging, and during late summer, fall, and winter, about 50 percent of the total population dies, so that come the next breeding season, about one million individuals are again left alive. In other words, during breeding seasons, the increase in population size via reproduction outpaces mortality rate; during nonbreeding seasons, mortality decreases the population. For example, in an extreme case, it is estimated that each April in North America, when breeding season begins, there are approximately 165 million Red-winged Blackbirds; postbreeding, each July, there are about 350 million of these pretty and prolific birds.

Which are the most abundant birds?

Many seabirds and shorebirds are quite numerous. This is mainly because they have hugely abundant food resources (marine animals—from medium-size fish down to tiny invertebrates) in the ocean and on beaches and mudflats, and they often breed on islands or other relatively inaccessible sites where there are few nest predators. Wilson's Storm-petrel (family Hydrobatidae), with a total population estimated at 100 million or more, is surely one of the most abundant seabirds. Some of its small Southern Ocean breeding islands each have annually about a million breeding pairs. Several petrel and shearwater species (family Procellariidae) are also quite numerous. The Dovekie (or Little Auk; in the puffin family, Alcidae) breeds in the high arctic; there may be, in some years, up to 18 million of its breeding pairs. Among the gulls and terns (family Laridae), the Ring-billed Gull, Common Black-headed Gull, and Black-legged Kittiwake, all have up to a million or more breeding pairs, and the Sooty Tern may have a total world population of up to 50 million individuals. Many penguins (family Spheniscidae) occur in large numbers, with 5 or 6 species having more than a million individuals. The Magellanic Penguin population is between 4.5 and 10 million. Among shorebirds, several species in the sandpiper group (family Scolopacidae) are believed to number, at least in some years, a million or more breeding birds—Common, Green, Wood, Semipalmated, Curlew, and Western Sandpipers, for instance, and the Dunlin, Ruff, and Red Knot. Among aquatic birds, some in the duck/goose family (Anatidae) are hugely abundant, species having total populations of a million or more including the Canada Goose, Common and King Eiders, Eurasian and American Wigeons, Common and Blue-winged Teal, Gadwall, Northern Pintail, Northern Shoveler, Lesser Scaup, and Mallard (the world's most abundant duck, some years with more than 25 million wintering individuals).

Some land birds are also incredibly numerous, and in this case a prime reason is that they specialize on eating seeds, insects, or both, two foods that are usually in almost unlimited supply. Particularly successful in terms of their numbers are those species that have adapted well to people's presence, especially some that were transported by people outside their native ranges and then found success in their new homes. The House Sparrow (family Passeridae)—native to Eurasia and North Africa but introduced to the New World, Southern Africa, Australia, and New Zealand—and the European Starling (family Sturnidae)—also native to Eurasia and North Africa but introduced to North America, the West Indies, Southern Africa, Australia, New Zealand, and Polynesia—are now certainly 2 of the most abundant of the world's birds. The estimated total House Sparrow population is between 120 and 500 million, and for the European Starling, more than 500 million. Likewise, the Rock Dove (Rock, or Feral, Pigeon; family

Columbidae), probably native to parts of Eurasia and northern Africa, through domestication and human transportation, now occurs almost everywhere there are human settlements; the estimated population of the bird worldwide is between 12 and 32 million. Some members of the lark family (Alaudidae)—the Horned Lark, which breeds in Eurasia, Africa, and North America, and the Eurasian Skylark, with a broad Old World distribution and introduced populations in Australia, New Zealand, and Hawaii, among other locations—are incredibly numerous; estimates of the Skylark population range from 70 million to more than 300 million in Europe alone. Also in Europe, the Bank Swallow (also called Sand Martin; family Hirundinidae) often occurs in roosts of up to a million or more individuals. And the Brambling, a small finch (Fringillidae), sometimes occurs in roosts of 10 million or more. Less familiar to North Americans and Europeans, but now widespread and very abundant owing to people transporting them outside their native ranges, are some members of the waxbill family (Estrildidae), such as the Nutmeg Mannikin, native to southern Asia but now also occurring in Australia, Hawaii, and the West Indies. In sub-Saharan Africa, the Red-billed Quelea (a member of the weaver family, Ploceidae), occurs often in flocks of more than a million, causing crop damage; its total population is estimated to be between 500 million and 1.5 billion.

In North America some of the most abundant land birds are probably members of the New World blackbird family (Icteridae). The Red-winged Blackbird is frequently thought of as the most numerous North American land bird, and the Common Grackle and Brown-headed Cowbird likewise have populations in the tens of millions. At one huge winter roost in Louisiana during the mid-1980s, it was estimated that there were 53 million Red-winged Blackbirds, 27 million Common Grackles, and 20 million Brown-headed Cowbirds.

So, how many individual wild birds are there alive on the planet at any one time? Estimates range from 200 billion to 400 billion.

Why are some bird families large, others small?

Some bird families are huge, containing 300 or more species; others are tiny, with 10 or fewer species. In the classification scheme used in this book, 13 families have 200 or more species (table 1 lists the 20 largest); about 20 are single-species families; 36 families contain 2, 3, or 4 species; and 27 others

TABLE 1. The 20 Largest Bird Families

FAMILY	NUMBER OF SPECIES
Tyrannidae (New World flycatchers)	425
Trochilidae (hummingbirds)	335
Psittacidae (parrots)	331
Columbidae (pigeons and doves)	308
Emberizidae (New World sparrows and Old World buntings)	303
Sylviidae (Old World warblers)	279
Muscicapidae (Old World flycatchers)	270
Formicariidae (antbirds)	269
Timaliidae (babblers)	265
Thraupidae (tanagers)	245
Furnariidae (ovenbirds)	240
Accipitridae (hawks, kites, and eagles)	236
Picidae (woodpeckers)	217
Strigidae (typical owls)	188
Turdidae (thrushes)	175
Meliphagidae (honeyeaters and Australian chats)	174
Anatidae (ducks, geese, and swans)	157
Phasianidae (pheasants and partridges)	155
Estrildidae (waxbills)	140
Rallidae (rails, gallinules, and coots)	134

have 5 to 10 species. Why such differences in "species richness" among families? There are a number of factors, which are not mutually exclusive.

First, family age may play a role. Families of animals, just like species or individuals, have a certain lifetime, at the end of which they die out, that is, they become extinct. A family might have many species at some points in its lifetime (when it has expanded its range and, through evolution, diversified and moved into various habitats and achieved various lifestyles) and few species at others, for instance, when it is very old (after many of its constituent species have run their courses and died out). Therefore, older types of birds that today have limited diversity may have been more diverse in the past. For example, the ratites, the ancient group that includes the

Ostrich, Emu, rheas, and so on, were once more diverse and widespread. We know this because there are surviving skeletal materials and written observations of 2 other, recent ratite groups that are now extinct: the moas of New Zealand and the elephant birds of Madagascar and North Africa.

Second, some groups have very limited ranges, which for ecological reasons constrain species diversity. Simply put, if a family is restricted to a small area, especially one that supports only limited habitat types, then a limited number of ecological opportunities, or niches, can be exploited; therefore, a relatively small number of species can exist there. Some examples in this case may be families restricted to isolated islands. The mesites, confined to Madagascar, are ground birds of forests and thickets; the family, Mesitornithidae, has only 3 species. And vangas (family Vangidae) are 14 species of shrikelike birds restricted to Madagascar and the nearby Comoro Islands.

Third, members of some groups are quite successful at exploiting resources relative to members of other groups, and therefore, over time, may become more common. For instance, the most diverse bird family, the New World flycatchers (family Tyrannidae), contains very efficient and successful catchers and consumers of insects. Probably owing to this success and to their occupation of diverse and complex habitats (such as rainforests), the group has been able to spread widely, diversify extensively, and penetrate many available bird niches wherever it occurs. Likewise, tanagers (family Thraupidae), efficient consumers of fruit, a superabundant and, in the tropics, always available food resource, have become very diverse, with about 250 species, mainly in tropical South and Central America. Honeyeaters (distributed in the Australia/New Guinea region and New Zealand; family Meliphagidae), which specialize on eating plant nectar, are very diverse (about 170 species in the family, about 70 in Australia) and hugely abundant in Australia, where many plants produce significant amounts of nectar. (Some evolutionary biologists believe speciation—the creation of new species from those that already exist—is less driven by ecological circumstances than I imply above, and that some animal groups, such as some bird families, are species-rich owing not only to diverse ecological opportunities, but to other factors as well.)

Fourth, a few families have been decimated by people. The best example is the Hawaiian honeycreepers, family Drepanididae. Probably more than 50 species of honeycreepers existed prior to human settlement of the Hawaiian Islands. But from people's first colonization, these birds were hunted for their feathers; their forest homes were cleared for agriculture; their eggs and young were eaten by introduced nest predators such as rats and mongooses; their food was reduced by introduced bird species that competed for food and by introduced farm animals such as pigs, goats, sheep, and cattle, which destroyed native vegetation that provided food; and their immune systems were overwhelmed by introduced diseases. Today, only about 18 of these honeycreeper species remain, and several of the surviving species are endangered, with very small total populations.

How about the species that some taxonomists place in single-species families—what kinds of birds are these?

There are two main reasons a bird is placed in a single-species family. Either it is a species from an ancient group that in the past was more diverse but today is represented by only a single species; or it is a species whose relationships are uncertain. In the latter case, taxonomists have placed the species in a single-species family, keeping it separate, until convincing evidence is found to assign it to another family. Many of the birds that provide taxonomists with classification problems are "oddball" birds—such as the Shoebill (storklike with massive bill, from Central Africa; family Balaenicipitidae), Hoatzin (a large, primitive-looking, Amazonian leaf-eater; Opisthocomidae), Kagu (flightless, from New Caledonia; Rhynochetidae), and Oilbird (large, with batlike habits, from South America; Steatornithidae)—species very different from all other birds.

AVIAN BIOGEOGRAPHY

How are bird species distributed?

We all know that if we traveled to Africa, most of the birds we would see are different from those we would see in South America or Australia. This is because there are geographic patterns in bird distribution. When discussing the distributions of animal (or plant) groups, we are delving into biogeography, a field of study that combines biology and geography. Its aim is to provide scientific explanations for the natural distribution patterns of organisms we see on Earth. It does this by invoking fossil evidence that tells where various kinds of organisms first developed and combining such information with knowledge of continental drift over evolutionary time, and sometimes with knowledge of the climatic histories of continents and ecological relationships among organisms. Two brief avian examples: naturalists in the past were long puzzled by the current distribution of the larger flightless birds—one or a few species on each of the southern continents—tinamous (which can fly modestly) and rheas in South America, Ostrich in Africa, Emu and cassowaries in Australia/New Guinea, kiwis in New Zealand. If they were closely related, how did they become so broadly distributed among continents while at the same time being so isolated from one another? Now, armed with recently acquired knowledge of continental drift, scientists believe these birds represent the few surviving descendants of an ancient group that occurred in Gondwana, the name given to the southern supercontinent that existed many millions of years ago and that broke up (between 180 and 100 million years ago) to form today's various southern continents (including Antarctica). Fossil remains in New Zealand (the moas) and Madagascar (elephant birds) further support the idea that these large flightless birds were once more common and widespread in the Southern Hemisphere. Another example is the todies, 5 species of tiny, plump forest birds restricted to the West Indies. How did it come about that todies occur only in the Caribbean region? A tody fossil found in Wyoming, millions of years old, shows that todies were once much more widespread, indicating that the present todies are a "relict" group, a very narrowly distributed group of a once widespread group.

A few things to consider about bird distribution patterns are the globe's various faunal regions, their characteristic faunas, and their relative species richness; general patterns of animal distribution; bird distributions on islands; and why some species or groups have very limited distributions (endemism).

Zoogeographic Regions: Describing Bird Distributions

Various methods are used to describe the particular regions in which bird groups and species occur. We can divide terrestrial portions of the earth into physically discrete hemispheres or continents and then say which bird groups occur in each hemisphere or continent (for example, waxwings, family Bombycillidae, and accentors, family Prunellidae, are Northern Hemisphere groups; toucans, family Ramphastidae, are a South American group). Old World versus New World, a historical division, also has some utility (for example, the blackbird family, Icteridae, and wood warbler family, Parulidae, are New World groups; the sunbird family, Nectariniidae, is restricted to the Old World). Old World refers to the regions of the globe that Europeans knew of before Columbus, Europe, Asia, Africa. New World refers to the Western Hemisphere—North, Central, and South America. The main climate regions of the earth—tropical, temperate, and arctic—are frequently used to describe bird distributions (for example, parrots are a mainly tropical group). The tropics, always warm, are the regions of the world that fall within the belt from 23.5 degrees north latitude (the Tropic of Cancer) to 23.5 degrees south latitude (the Tropic of Capricorn). Subtropical refers to a region that borders a tropical zone. The world's temperate zones, with more seasonal climates, extend from 23.5 degrees north and south latitude to the Arctic and Antarctic Circles, at 66.5 degrees north and south. Arctic regions, more or less always cold, extend from 66.5 degrees north and south to the poles. Occasionally the terms boreal and austral are used, the former referring to northern, primarily north temperate, regions; the latter, to southern, especially south temperate, regions.

Ever since the 1800s, however, when biologists began traveling regularly among the continents and noticing great similarities in wildlife within broad regions of the world but major differences between and among regions, there has been a scheme to divide the Earth formally into six zoogeographic (or faunal) regions, or realms. These divisions, initially proposed (for birds) by Philip Sclater in 1858 and then generalized for all land animals by Alfred Russel Wallace in 1876, were based on the distributions of bird families known about at that time. (That these divisions are still in common use today to describe animal distributions testifies to the abilities and knowledge of the nineteenth-century naturalists who delineated them.) Unless you take a course in biogeography, it is unlikely you will know about these divisions— but they are how biologically oriented people, such as zoologists, ecologists, and serious bird-watchers, often divide the world.

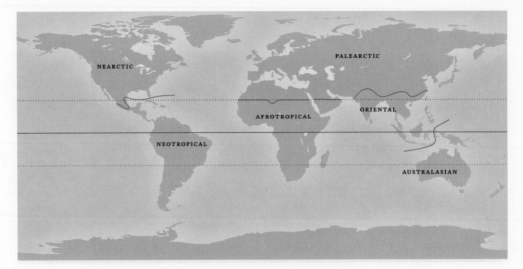

FIGURE I. *Map of the world showing the six zoogeographic regions.*

The first two of the six zoogeographic regions are located north of the tropics: the Nearctic (encompassing North America [north of the tropics] and, sometimes, Greenland) and the Palearctic (Europe, northern Asia, and northern Africa). The next three are largely tropical: the Neotropical (Central and South America and the West Indies), Afrotropical (or Ethiopian; sub-Saharan Africa and Madagascar), and Oriental (all of southern Asia from India to Southeast Asia to Japan) regions. The final one, the Australasian region (or Australian; encompassing Australia, New Guinea, and New Zealand), has both tropical and south temperate components. These 6 regions contain most of the world's terrestrial wildlife. Antarctica is not included because the 7th continent has few terrestrial vertebrates (though penguins, for instance, breed there). For convenience, many bird references include Antarctica with the Australasian region. Other problem places with respect to regional classification are the remote Pacific islands; they have few unique groups of birds or other animals that would set them apart in a separate region, and they have small numbers of groups from several of the continental faunas.

The six zoogeographic regions separate areas of wildlife with common characteristics. Among the birds, various groups are restricted to, concentrated in, or characteristic of the respective regions.

- Birds considered characteristic of the Nearctic region, with about 800 bird species, include New World quail (family Odontophoridae), wrens (Troglodytidae), mockingbirds and thrashers (Mimidae), vireos (Vireonidae), wood warblers (Parulidae), waxwings (Bombycillidae), and sparrows (Emberizidae).

- The Palearctic has about 950 species, with large numbers of ducks, geese, and swans (family Anatidae), pheasants (Phasianidae), larks (Alaudidae), pipits and wagtails (Motacillidae), accentors (Prunellidae), thrushes (Turdidae), Old World warblers (Sylviidae), finches (Fringillidae), and Old World buntings (Emberizidae).

- The Neotropical region (mainly South America), with about 3,100 species, is known for its rheas (family Rheidae), tinamous (Tinamidae), screamers (Anhimidae), guans and chachalacas (Cracidae), trumpeters (Psophiidae), potoos (Nyctibiidae), hummingbirds (Trochilidae), trogons (Trogonidae), motmots (Momotidae), toucans (Ramphastidae), ovenbirds (Furnariidae), antbirds (Formicariidae), cotingas (Cotingidae), manakins (Pipridae), New World flycatchers (Tyrannidae), tanagers (Thraupidae), and blackbirds (Icteridae).

- The Afrotropical region, with about 1,550 species, is rich in guineafowl (family Numididae), turacos (Musophagidae), francolins (Phasianidae), bustards (Otididae), hornbills (Bucerotidae), barbets (Capitonidae), honeyguides (Indicatoridae), larks (Alaudidae), cisticolas (Cisticolidae), shrikes (Laniidae), and waxbills (Estrildidae).

- The Oriental region, with about 1,900 species, is known for its pheasants (family Phasianidae), hornbills (Bucerotidae), bulbuls (Pycnonotidae), sunbirds (Nectariniidae), weavers (Ploceidae), and leafbirds (Irenidae).

- The Australasian region, with between 900 and 1,500 species (depending on which island groups are included), is known for its megapodes (family Megapodiidae), parrots (Psittacidae,

Cacatuidae), kingfishers (Alcedinidae), lyrebirds (Menuridae), honeyeaters (Meliphagidae), birds-of-paradise (Paradisaeidae), and bowerbirds (Ptilonorhynchidae).

The Tropics—Where Most Birds Occur

Many groups of animals, such as insects, lizards, and birds, show a pattern of species number related to latitude. This major pattern, recognized now for almost 2 centuries, is called the global diversity gradient, or latitudinal gradient in species diversity. The higher latitudes (the north and south poles are at 90 degrees latitude; the equator, at 0 degrees latitude) have few species. As one moves to lower and lower latitudes, toward the equator, the number of species increases. Thus, tropical Ecuador, on the equator, has an avifauna (the sum of the species of birds living in an area) of about 1,550 species, twice as large as that of the entire United States, which is more than 30 times larger than Ecuador but is situated at higher latitudes. Often the pattern is continuous. For instance, in the tiny but tropical Central American country of Costa Rica, there are about 600 breeding bird species; moving northward, about 400 occur in southern Mexico, which is still tropical; 200, in temperate, northern Mexico; 180, in the United States' temperate Pacific Northwest; and about 100 in parts of northern Canada and Alaska.

There are a variety of possible explanations for this general trend in species richness, and it is likely that some combination of them accounts for the pattern. Most basically, there is increasing more availability of sunlight energy, higher photosynthetic rates, and therefore more available food closer to the equator. In other words, because animal and plant life is ultimately supported by the energy of the sun, where there is greater sun energy, there will be more species. More sun (together with other factors such as increased moisture and higher temperatures, which are also typical of the tropics) means more photosynthesis (more plant "productivity"), which means more plants, larger plants, and lusher habitats, such as rainforests. Because animals live on plants, both physically and as food sources, more plant material, more kinds of plants, and more complex forests in a region mean more animals can live there. So the greater sunlight energy of the tropics translates to greater plant productivity, which, ultimately, supports more animal species.

Regardless of the causes of the global diversity gradient, the contemporary result among birds is great species richness in the tropics and relatively impoverished bird faunas in temperate and arctic regions. Table 2 lists the 24 countries with the greatest number of bird species. These countries are situated entirely or mainly in the tropics, are very large, or both. Table 3, for comparison, provides bird species information about the United States, Australia, Canada, and the United Kingdom.

Islands—Size and Distance Matter

Another major pattern of species richness concerns islands. The general rules are that the size of an island's avifauna decreases as island size decreases and as distance to the nearest mainland increases. In other words, the smaller and more remote the island, the fewer bird species found there. The reasons for these island biodiversity trends are clear. To explain, we need to consider how organisms first arrive on islands.

If a new island appeared today just off the coast of California, either it split off from the coast and so will have most of the animal species that occur on the adjacent mainland, or it arose from the sea floor, perhaps via volcanic or earthquake activity, and now is pretty much barren of life. In the latter case, many animal species from the mainland, especially birds, which can fly easily to the nearby island, will soon occur on the island. (But not all species of birds on the mainland, or other types of animals, will cross to the new island, thus explaining why even near-shore islands tend to have fewer species than the nearest mainland areas.)

But what about remote islands, such as volcanic ones that arise from the sea floor in midocean? Consider the Hawaiian Islands, volcanic cones in the mid-Pacific that are about as isolated and remote as you can get—2,000 miles (3,200 km) from North America and 3,300 miles (5,300 km) from Japan. Islands such as Hawaii are too far away from mainland areas for mainland species to colonize easily. Only a few organisms will be able to do so, and which species succeed will be governed largely by chance. A large ocean in the way is probably nature's most effective barrier to the spread of terrestrial animal and plant species. Only a few types of terrestrial organisms have the ability to cross oceans (and even then, by chance, only a few species of each type will). How do these species cross oceans?

1. They fly—birds and bats arrive under their own power. (Some colonizers, such as snakes or mammals, could swim to previously unsettled islands, but only over short distances; this would not apply to the Hawaiian Islands, with no near sources of colonizing immigrants.)

2. They arrive passively, not under their own power, by drifting ashore. Tiny plant seeds and small spiders and bugs may drift in on the wind; seeds and fruit that can survive saltwater immersion may float and drift through the ocean—coconuts, for instance, appear to be adapted for long-term ocean drifting, which is why coconut palms line beaches worldwide.

3. They come by "boat," definitely tourist class, rafting in on floating bits of vegetation, for instance, or on trees washed off of mainland areas during storms. Eggs or even adults of certain animals may reach remote islands this way, including spiders, insects, lizards (but not amphibians, whose thin wet skins needed for breathing do not mix well with saltwater), even small rodents. (Remember that for the colonization to succeed, more than a single mouse or lizard must arrive on an island; generally you must have at least 2, a male and a female. Actually, according to genetic principles, if you want the new population to have long-term success in its new home, you probably need more than 2 individuals in order to increase the genetic diversity of the population so it is not too inbred.) Finally, because seabirds and ducks cover vast distances in their ocean-crossing feeding and migratory flights, some organisms could travel over thousands of miles of open ocean in mud on birds' feet (plant seeds; insect, fish, and amphibian eggs) or even in bird intestines, eventually being excreted on isolated islands (plant seeds). Many native plants of Hawaii may have first arrived in the islands as seeds in bird intestines.

So the methods by which birds and other animals colonize islands explain the relationship between distance from mainland and number of species present: the greater the distance, the fewer species that manage to colonize. But what about the relationship between island size and number

TABLE 2. Countries with 850 or More Total Bird Species

COUNTRY	AREA SQUARE MILES (× 1,000)	AREA SQUARE KM (× 1,000)	TROPICAL ENVIRONMENT?	NUMBER OF BIRD SPECIES BREEDING	NUMBER OF BIRD SPECIES TOTAL
Colombia	440	1,139	Yes	1,700	1,721
Peru	496	1,285	Yes	1,541	1,710
Brazil	3,286	8,512	Mostly	1,500	1,635
Ecuador	109	284	Yes	1,388	1,559
Indonesia	735	1,905	Yes	1,530	1,539
Venezuela	352	912	Yes	1,296	1,360
Bolivia	424	1,099	Mostly	?	1,275
China	3,705	9,597	Mostly not	1,103	1,244
India	1,269	3,288	Partly	926	1,219
Democratic Republic of Congo	906	2,345	Yes	929	1,086
Tanzania	365	945	Yes	827	1,076
Kenya	224	580	Yes	847	1,068
Mexico	756	1,958	Partly	772	1,054
Cameroon	184	475	Yes	690	1,000
Myanmar	262	679	Partly	867	999
Uganda	91	236	Yes	830	992
Argentina	1,068	2,767	Mostly not	897	983
Sudan	967	2,506	Yes	680	937
Panama	30	77	Yes	732	929
Thailand	198	513	Yes	616	915
Angola	481	1,247	Yes	765	909
Nigeria	357	924	Yes	681	862
Ethiopia	436	1,128	Yes	626	861
Costa Rica	20	51	Yes	600	850

Note: Species numbers presented here are estimates only, because the total number of species in a country or region varies slightly depending on the method of counting and the source consulted. For example, sources may or may not include seabirds in their counts. Some avifauna totals include only birds that breed within a given region; others include all birds in the region, including nonbreeding migrants. Some include species seen only occasionally in a given country, while others do not. Also, species numbers change when various new systems of classification come into vogue. Therefore, most important to note in the table is, for instance, not that 1,721 species have been seen in Colombia, but that Colombia, a moderate-size tropical South American country, supports a huge number of bird species, more than 1,700.

TABLE 3. Number of Bird Species in the United States, Australia, Canada, and the United Kingdom

	AREA			NUMBER OF BIRD SPECIES	
COUNTRY	SQUARE MILES (× 1,000)	SQUARE KM (× 1,000)	TROPICAL ENVIRONMENT?	BREEDING	TOTAL
United States	3,619	9,373	No	650	768
Australia	2,968	7,687	Partly	649	751
Canada	3,582	9,976	No	426	578
United Kingdom	94	243	No	230	590

Note: Species numbers presented here are estimates only, because the total number of species in a country or region varies slightly depending on the method of counting and the source consulted. For example, sources may or may not include seabirds in their counts. Some avifauna totals include only birds that breed within a given region; others include all birds in the region, including nonbreeding migrants. Some include species seen only occasionally in a given country, while others do not. Also, species numbers change when various new systems of classification come into vogue.

of species? Smaller islands tend to have fewer species, say of birds, because smaller islands have fewer habitat types and so there are fewer niches that bird species can occupy. Think about it this way: A large island may have almost all the kinds of habitats that a mainland region has: low, middle, and high elevation zones, beach, scrub, forests, grasslands, wetlands, and so on, and each habitat can support a characteristic avifauna. But small islands may have just a few habitats—some coastal islands are comprised of just rocks along the water and forested interiors, and atolls, barely rising from the ocean's surface, often consist simply of coral beach, some grasses and shrubby vegetation, and coconut palms. Another factor contributing to larger islands having more species is that they support greater population sizes than smaller islands. Because large populations, on average, are less likely to become extinct than small populations, species on large islands are more likely to persist. Therefore, even if similar numbers of species become established on both large and small islands, large islands, over long periods of time, will retain more of them.

Endemics: Some Birds Occur in Very Limited Areas

The word *endemic*, when used to describe animal or plant distributions, means "peculiar to a particular place." An endemic species or group of species is one that occurs in a certain area and nowhere else. But the size or type of area referred to is variable. A given species of sparrow, say, may be endemic to the Western Hemisphere, to a single continent such as South America, to a coastal region of Argentina, or to a speck of an island off Argentina's coast. (As far as anyone knows, all species on our planet are endemic to Earth; and if the only life in the universe is ours, then all life is endemic to Earth.) Therefore, a modifier is often used with the word endemic to clearly indicate what is meant. There are hemispheric endemics, continental endemics, regional endemics, country endemics, state endemics, and island endemics. The term *near-endemic* is used to indicate a species or group that occurs in a particular place and perhaps one or a few other areas nearby. For instance, the Palm Cockatoo occurs on the northeastern tip of Australia but also across the Torres Strait in New Guinea. It can be considered an Australian near-endemic (and a New Guinea near-endemic); it is endemic to the Australasian region (which encompasses Australia, New Zealand, New Guinea, and some of the Pacific islands).

Not only species are endemic. Many genera (plural of genus), families, and even entire orders of birds are endemic to particular regions of the world. For instance, the American blackbird family, Icteridae, with 25 genera and about 97 species, is endemic to the New World; the tody family, Todidae, with only 1 genus and 5 species, is endemic to the Caribbean region; toucans (family Ramphastidae, 6 genera, about 40 species), motmots (Momotidae, 6 genera, 10 species), cotingas (Cotingidae, 26 genera, about 70 species), and manakins (Pipridae, 15 genera, about 50 species) are endemic to South and Central America and Mexico; and turacos (Musophagidae, 6 genera, 23 species), rockfowl (family Picathartidae, 1 genus, 2 species), bushshrikes (Malaconotidae, 7 genera, 43 species), sugarbirds (Promeropidae, 1 genus, 2 species), and indigobirds and whydahs (Viduidae, 1 genus, 19 species) are endemic to Africa. Many bird families are confined totally or mainly to tropical regions; a few smaller families are concentrated in or nearly restricted to temperate regions (such as waxwings, family Bombycillidae; long-tailed tits, Aegithalidae; and accentors, Prunellidae). Bird orders with restricted distributions include the tinamous (order Tinamiformes; limited to South and Central America and Mexico), the mousebirds (Coliiformes; limited to Africa), and the Hoatzin (Opisthocomiformes; limited to South America).

Why are some species endemic to only small areas?

Why are some species endemic to small areas while others are spread over huge regions such as multiple continents or even most of the world (the Barn Swallow, European Starling, House Sparrow, Peregrine Falcon, and Osprey are more or less globally distributed)? Another way of asking this is: What determines a species' present distribution? History is the answer. When a species' distribution is confined to a restricted or small area, (1) there are one or more barriers to further spread (an ocean, a mountain range, a thousand miles of tropical rainforest in the way); (2) the species evolved only recently and has not yet had time to spread; or (3) the species evolved long ago, spread long ago, and now has become extinct over much of its prior range. A history of isolation also matters. The longer a group of animals and plants are isolated from their close relatives, the more time they have to evolve by themselves and to change into new, different, even unique groups. The best examples are on islands. Some islands once were attached to mainland areas, but continental drift or changing sea levels (or some combination) led to their isolation in the middle of the ocean; other islands arose wholly new via volcanic activity beneath the seas. For instance, the island of Madagascar was once attached to Africa and India. The organisms stranded on its shores when it became an island had probably 100 million years in isolation to develop into the highly endemic fauna and flora we see today. About 80 percent of the island's plants and animals are thought to be endemic, including half the bird species and essentially all the mammals and reptiles. Other examples of islands or island nations with high concentrations of endemic animals are Indonesia, where about 15 percent of the world's bird species occur, a quarter of them endemic, and New Guinea, where 45 percent of birds are endemic.

Where are the greatest numbers of endemic bird species?

Countries with high numbers of endemics are listed in Table 4.

TABLE 4. Countries with More Than 20 Endemic Bird Species

NUMBER OF ENDEMIC SPECIES	COUNTRY OR ISLAND
408	Indonesia (includes Sulawesi, 96; Moluccas, 67; Lesser Sundas, 46; Java and Bali, 28; Irian Jaya, 39; Borneo, which is shared with Malaysia, 31)
313	Australia (mainland)
185	Brazil
185	Philippines
114	Peru
105	Madagascar
92	Mexico
94	Papua New Guinea (includes Bismarck Archipelago, 38)
74	New Zealand
70	China
67	Colombia
67	United States (mainland 9; Hawaii 36)
58	India
43	Solomon Islands
40	Venezuela
37	Ecuador
28	Ethiopia
26	Jamaica
25	French Polynesia
25	Sao Tomé and Principe
24	Fiji
24	Democratic Republic of Congo
24	Tanzania
24	Sri Lanka
22	New Caledonia
21	Cuba
21	Japan

Note: The number of endemic species for each area varies depending on source consulted. Factors that affect different counts include the date of the count (more species being discovered or delineated in later years); whether all outlying islands that are political possessions of the country in question are included in the count; and the avian taxonomy used for the count.

*Which bird families have the most narrowly endemic species
and which have the least?*

One way to look at this is to examine species that have relatively small,
or restricted, ranges, defined as a species having a total range of less than
50,000 square kilometers (about 19,300 square miles; about the same size of
Costa Rica). About 25 percent of all bird species and about 75 percent of all
threatened species have restricted ranges by this definition. Table 5 lists bird
families with the most and least restricted-range species.

UNDERSTANDING THE DECLINE IN AVIAN DIVERSITY

How many birds are threatened?

Although currently there are vast numbers of individual birds and a large
number of types of birds, many species are now threatened with extinction.
We know (from historical documents and, for many species, surviving body
materials) that at least 128 bird species became extinct during the past 500
years, 103 of them in the past 200 years. Currently about 12 percent of bird
species (about 1,180 of them) are considered by authoritative conservation
organizations (for example, International Union for Conservation of Nature,
and BirdLife International) to be globally threatened, that is, in danger of
total extinction. After years of careful field surveys, consultation of scientific
literature, and gathering data from international sources, an exhaustive
compilation of information about threatened birds was produced in 2000
(BirdLife International's *Threatened Birds of the World*); it presents current
ranges and population sizes, threats, and conservation information for each
threatened species. According to this comprehensive source, about 680
species are vulnerable (facing a high risk of extinction in the medium-term
future); about 320 are endangered (facing a very high risk of extinction in the
near future); and about 180 are critically endangered (facing an extremely
high risk of extinction in the immediate future). After considering available
evidence and analysis of population trends, this source predicts that, unless
significant conservation measures are undertaken, perhaps 460 of these 1,180
threatened species will become extinct by 2100.

TABLE 5. Families with Most and Least
Restricted-Range Species

FAMILY	TOTAL SPECIES	RESTRICTED- RANGE SPECIES	PERCENTAGE OF SPECIES
MOST			
Drepanididae (Hawaiian honeycreepers)	18	18	100
Mesitornithidae (mesites)	3	3	100
Zosteropidae (white-eyes)	94	79	84
Todidae (todies)	5	4	80
Paradisaeidae (birds-of-paradise)	44	29	66
Megapodiidae (megapodes)	19	11	58
Tytonidae (barn owls)	16	9	56
LEAST			
Hirundinidae (swallows)	89	8	9
Paridae (tits and chickadees)	55	5	9
Anatidae (ducks, geese, and swans)	157	11	7
Ardeidae (herons, egrets, and bitterns)	63	4	6
Meropidae (bee-eaters)	25	1	4
Threskiornithidae (ibises and spoonbills)	33	1	3
Otididae (bustards)	25	0	0
Ciconiidae (storks)	19	0	0
Gruidae (cranes)	15	0	0
Remizidae (penduline tits)	13	0	0
Recurvirostridae (stilts and avocets)	11	0	0

What are the main threats to birds?

The sources of threats to birds are primarily linked, as might be expected, to their interactions with humans and to the consequences of human activities, such as agricultural, residential, and economic development. In fact, 99 percent of the 1,180 bird species now threatened are so classified owing to human activities (the remaining 1 percent are thought threatened because of natural disasters—volcanic or cyclone damage to small islands, for instance—or other natural phenomena). The three major human-induced threats that account for most of the declines of bird populations are habitat loss, direct exploitation, and invasive species.

Habitat loss, simply put, consists of humans damaging or destroying the natural habitats in which birds live, forage, and breed. Birds are specialized to live in their native habitats and most, in the absence of adequate amounts of such habitat, cannot obtain sufficient nutrition or shelter or cannot breed, and so die out. Main causes of habitat loss are ever-expanding agricultural activities (small-holder farming, shifting slash-and-burn farming, crop plantations), logging for timber and firewood, and livestock ranching and grazing. New human settlements and the draining and filling in of marshlands also take their toll. All major habitat types—forests, woodlands, shrublands, grasslands, and wetlands—are affected by these forces. Habitat loss is implicated as sole, chief, or contributing threat for about 1,000 (85 percent) of all currently threatened birds; 74 percent of those species are forest-dwellers impacted by recent loss of tropical forests.

Birds are hunted, trapped, or killed for a number of reasons. Chief among these is hunting for food, which impacts about 365 of the 1,180 threatened species. The second most common form of direct exploitation of birds (affecting about 230 threatened species) is capture for the pet trade, either locally as cage birds or for the illicit international trade in exotic pets. Birds are killed in large numbers because farmers or ranchers believe the birds damage their crops, orchards, or livestock. They are also killed for sport and for their parts (for instance, for their wings, bills, or feathers, to make decorative or ceremonial objects). Generally, but not always, larger birds are more affected by hunting and trapping than smaller birds (chickenlike birds for food; parrots, especially, for the pet trade).

Invasive, or "alien," species are those transported by people to sites they would probably never reach on their own; they are said to be *introduced* to the new areas. People may, for various reasons, intentionally introduce organisms. Many others, such as rats, mice, ants, and cockroaches, are accidentally introduced when, for example, they are transported in food-storage areas of ships. Invasive species wholly or partly caused many of the bird extinctions that occurred in the past 200 years, principally when predators such as rats, cats, and mongooses were introduced to islands that had been generally predator-free. Island species are so sensitive to alien introductions, and can be so quickly harmed and endangered, because they have lived in isolation for thousands of years. In essentially predator-free environments, they never evolved appropriate defenses or behaviors to deal with dangerous situations, or if their colonizing ancestors had such defenses, they were lost through generations of nonuse. Introduced species can also negatively affect native birds indirectly, causing their populations to crash and pushing them closer to extinction. They might, for example, compete with native species for the same foods; damage or destroy plants that native birds depend on; and/or bring with them diseases for which native species have no defenses or immunity. For example, many of the remaining Hawaiian honeycreepers (p. 439) are threatened now because some introduced bird species compete with native honeycreepers for the same foods, including fruit, nectar, and insects; because introduced pigs and goats eat or otherwise damage or destroy plants that some native honeycreepers depend on; and especially because introduced mosquitoes transmit diseases such as avian malaria, to which most of the honeycreepers quickly succumb. Currently, approximately 300 (about 25 percent) of globally threatened species are being impacted by introduced predators, and 200 or so are being affected negatively by introduced species in other ways (competition, disease, etc.).

Which kinds of birds are threatened, and where are they located?

Many types of birds are threatened with extinction but there are some general patterns in what kind they are and where they live. Some types of birds are more susceptible than others to the negative impacts of humans and to natural catastrophes. Species with small ranges are particularly vulnerable. When and if their numbers fall, these species or groups face a greater chance of extinction than others because they lack other places "to go," other populations in far-off places that might survive. Narrowly endemic birds, such as the restricted-range species referred to in Table 5, are subject to such extinctions. Particularly good examples are species that are endemic to islands. If a species of bird occurs only on a single island or on a small group of tiny islands near to each other, all its eggs are, so to speak, in one basket. If there is a calamity there—a powerful hurricane, a volcanic eruption—the entire species could become extinct, because all individuals there die and there are no others elsewhere. This type of species extinction—but caused by people instead of natural catastrophes—has apparently happened often to island-bound birds over the past 500 years as people colonized remote islands. People caused habitat destruction and brought animal predators that the native birds had no fear of or experience with. In one of the worst examples, 99 percent of forests were destroyed on the Philippine island of Cebu (mainly to clear land to create sugarcane plantations), which caused the extinction of many of Cebu's endemic forest birds. Worldwide, about 130 bird species have probably become extinct in the last 500 years, and about 100 of them were island endemics. (And the problem persists: more than 900 of the 9,800 or so living bird species are island endemics, and so continually vulnerable.)

Flightless birds, for obvious reasons, are also highly vulnerable to people and to mammalian predators (rats, cats, dogs, mongooses) that people introduce to areas in which they live. The best examples are among the rails, a group of often-skulking small and medium-size swamp and dense-vegetation birds that inhabit most parts of the world, including many isolated islands. Five or more rail species from tropical Pacific islands and several from New Zealand became extinct as people colonized their islands. Some of these rails, such as the Guam Rail (now extinct in the wild but surviving in captivity), were flightless. In fact, fully a quarter of the world's rail species are almost or completely flightless. They are thought to have lost the power of flight because it is energetically expensive to maintain (especially the large chest muscles) and unnecessary where strong natural selection pressures do not favor it (such as oceanic islands that lack mammalian predators). Many of the surviving flightless rails are endangered.

Birds that occur in certain kinds of habitats are more often threatened with extinction than others. For instance, tropical forest birds are increasingly threatened as tropical forests are cut and burned for logging, fuel wood, or agricultural and other land development. And grassland birds, such as some larks and pipits, are threatened because their relatively flat, open, grass-growing homes are ideal for various kinds of agriculture (crop farming, ranching, grazing)—so grasslands in good, natural condition are increasingly scarce.

Some birds, of course, not only survive amid the massive environmental

TABLE 6. Countries with at Least 30 Threatened Bird Species

COUNTRY	TOTAL THREATENED	VULNERABLE	ENDANGERED	CRITICALLY ENDANGERED
Indonesia	114	70	30	14
Brazil	113	54	36	23
Colombia	77	40	24	13
Peru	73	46	23	4
China	73	59	11	3
India	68	52	9	7
Philippines	65	41	12	12
Ecuador	62	43	13	6
New Zealand	62	40	16	6
United States	53	29	10	14
Mexico	39	19	12	8
Argentina	39	31	4	4
Russia	38	27	8	3
Malaysia	35	29	3	3
Thailand	35	26	5	4
Myanmar	33	26	4	3
Vietnam	33	21	10	2
Australia	33	20	11	2
Tanzania	33	23	9	1
Papua New Guinea	32	29	2	1
Japan	32	22	8	2

changes wrought by people, but even thrive, increasing their ranges and numbers. Those that have benefited from people's alterations of habitats and other developments include some members of the heron and egret family (Ardeidae), some gulls and terns (Laridae), some nightjars (Caprimulgidae), some pigeons and doves (Columbidae), some thrushes (Turdidae), some swallows (Hirundinidae), some blackbirds and cowbirds (Icteridae), some in the crow family (Corvidae), and many finches and sparrows (Fringillidae, Estrildidae, Emberizidae).

Threatened bird species are not evenly distributed among the various regions of the world. They are concentrated in the Neotropics and Southeast Asia, areas with very high species diversity and many endemic birds. Countries with the most threatened species are listed in Table 6. Some small places have large numbers of threatened birds. For example, the Hawaiian Islands, which occupy only about 6,425 square miles (16,641 sq km) of usually dry land, support 27 threatened species (11 vulnerable, 6 endangered, 10 critically endangered [a few of which may now be extinct]).

Roughly half the globe's bird species are thought to be experiencing declining overall populations, and many undoubtedly will become extinct in our lifetimes and over the next few generations. But with increasing education about environmental threats and with conservation measures, most will survive and thrive for long periods of time; and these beautiful, splendid animals will be available for our watching far, far into the future.

Bibliography

ALI, S., AND S. D. RIPLEY. 1978–1999. *Handbook of the Birds of India and Pakistan.* Vols 1–10. Oxford University Press, Dehli.

ALSTRÖM, P., AND K. MILD. 2003. *Pipits and Wagtails.* Princeton University Press, Princeton.

BEAMAN, M., AND S. MADGE. 1998. *The Handbook of Bird Identification for Europe and the Western Palearctic.* Princeton University Press, Princeton.

BEEHLER, B. M., T. K. PRATT, AND D. A. ZIMMERMAN. 1986. *Birds of New Guinea.* Princeton University Press, Princeton.

BIRDLIFE INTERNATIONAL. 2000. *Threatened Birds of the World.* Lynx Edicions and BirdLife International, Barcelona and Cambridge, Eng.

BREWER, D., AND B. K. MACKAY. 2001. *Wrens, Dippers, and Thrashers.* Yale University Press, New Haven.

BRAZIL, M. A. 1991. *The Birds of Japan.* Smithsonian Institution Press, Washington, DC.

BROWN, J. H., AND M. V. LOMOLINO. 1999. *Biogeography.* 2nd ed. Sinauer Associates, Sunderland, MA.

BROWN, L. H., E. K. URBAN, AND K. NEWMAN, EDS. 1982–2003. *The Birds of Africa.* Vols. 1–7. Academic Press, London and San Diego.

BYERS, C., J. CURSON, AND U. OLSSON. 1995. *Sparrows and Buntings: A Guide to the Sparrows and Buntings of North America and the World.* Houghton Mifflin, New York.

CAMPBELL, B., AND E. LACK, EDS. 1985. *A Dictionary of Birds.* T. and A. D. Poyser, Calton, Eng.

CHANTLER, P., AND G. DRIESSENS. 2000. *Swifts: A Guide to the Swifts and Treeswifts of the World.* 2nd ed. Yale University Press, New Haven.

CLEERE, N., AND D. NURNEY. 1998. *Nightjars: A Guide to the Nightjars, Nighthawks, and Their Relatives.* Yale University Press, New Haven.

CLEMENT, P. 2000. *Thrushes.* Princeton University Press, Princeton.

CLEMENT, P., A. HARRIS, AND J. DAVIS. 1993. *Finches and Sparrows: An Identification Guide.* Princeton University Press, Princeton.

CLEMENTS, J. F. 2000. *Birds of the World: A Checklist.* Ibis Publishing, Vista, CA.

COATES, B. J., AND K. D. BISHOP. 1997. *A Guide to the Birds of Wallacea.* Dove Publications, Alderley, Australia.

DAVIES, S. J. J. F. 2002. *Bird Families of the World: Ratites and Tinamous.* Oxford University Press, Oxford.

DEMEY, R. 2002. *A Guide to the Birds of Western Africa.* Princeton University Press, Princeton.

DOUGHTY, C., N. DAY, AND A. PLANT. 1999. *Birds of The Solomons, Vanuatu and New Caledonia.* A & C Black, London.

FEARE, C., AND A. CRAIG. 1999. *Starlings and Mynas.* Princeton University Press, Princeton.

FERGUSON-LEES, J., AND D. A. CHRISTIE. 2001. *Raptors of the World.* Houghton Mifflin, New York.

FLINT, V. E., R. L. BOEHME, Y. V. KOSTIN, AND A. A. KUZNETSOV. 1984. *A Field Guide to Birds of the USSR.* Princeton University Press, Princeton.

FORSHAW, J. M., AND W. T. COOPER. 1977. *The Birds of Paradise and Bowerbirds.* Collins, Sydney and London.

FRITH, C. B., AND B. M. BEEHLER. 1998. *Bird Families of the World: The Birds of Paradise.* Oxford University Press, Oxford.

FRY, C. H., K. FRY, AND A. HARRIS. 1992. *Kingfishers, Bee-eaters, and Rollers.* Princeton University Press, Princeton.

FULLER, E. 2000. *Extinct Birds.* Oxford University Press, Oxford.

GASTON, A. J. 1998. *Bird Families of the World: The Auks.* Oxford University Press, Oxford.

GIBBS, D., E. BARNES, AND J. COX. 2001. *Pigeons and Doves: A Guide to the Pigeons and Doves of the World.* Yale University Press, New Haven.

GILL, F. B. 1994. *Ornithology.* 2nd ed. W. H. Freeman, New York.

GOODWIN, D. 1982. *Estrildid Finches of the World.* British Museum of Natural History, London.

GRIMMET, R., C. INSKIPP, AND T. INSKIPP. 1999. *A Guide to the Birds of India, Pakistan, Nepal, Bangladesh, Bhutan, Sri Lanka, and the Maldives.* Princeton University Press, Princeton.

HARRAP, S., AND D. QUINN. 1995. *Chickadees, Tits, Nuthatches, and Treecreepers.* Princeton University Press, Princeton.

HARRISON, C. S. 1990. *Seabirds of Hawaii: Natural History and Conservation.* Cornell University Press, Ithaca, NY.

HARRISON, P. 1983. *Seabirds: An Identification Guide.* Houghton Mifflin, New York.

HAYMAN, P., J. MARCHANT, AND T. PRATER. 1991. *Shorebirds: An Identification Guide.* Houghton Mifflin, New York.

HEATHER, B., AND H. ROBERTSON. 1997. *Field Guide to the Birds of New Zealand.* Oxford University Press, Oxford.

HIGGINS, P. J., J. M. PETER, AND W. K. STEELE. 1990–2001. *Handbook of Australian, New Zealand, and Antarctic Birds.* Vols. 1–5. Oxford University Press, Oxford.

HILTY, S. L., AND W. L. BROWN. 1986. *A Guide to the Birds of Colombia.* Princeton University Press, Princeton.

HOWARD, R., AND A. MOORE. 1994. *A Complete Checklist of the Birds of the World.* Academic Press, London.

HOWELL, S. N. G., AND S. WEBB. 1995. *A Guide to the Birds of Mexico and Northern Central America.* Oxford University Press, New York.

HOYO, J. DEL, A. ELLIOTT, AND D. A. CHRISTIE, EDS. 1992–2003. *Handbook of the Birds of the World.* Vols. 1–8. Lynx Edicions, Barcelona.

ISLER, M. L., AND P. R. ISLER. 1999. *The Tanagers: Natural History, Distribution, and Identification.* Smithsonian Institution Press, Washington, DC.

JARAMILLO, A., AND P. BURKE. 1999. *New World Blackbirds: The Icterids.* Princeton University Press, Princeton.

JOHNSGARD, P. A. 1978. *Ducks, Geese, and Swans of the World.* University of Nebraska Press, Lincoln.

———. 1981. *The Plovers, Sandpipers, and Snipes of the World.* University of Nebraska Press, Lincoln.

———. 1993. *Cormorants, Darters, and Pelicans of the World.* Smithsonian Institution Press, Washington, DC.

———. 2000. *Trogons and Quetzals of the World.* Smithsonian Institution Press, Washington, DC.

JONES, D. N., R. W. R. J. DEKKER, AND C. S. ROSELAAR. 1995. *Bird Families of the World: The Megapodes.* Oxford University Press, Oxford.

JUNIPER, T., AND M. PARR. 1998. *Parrots: A Guide to Parrots of the World.* Yale University Press, New Haven.

KEMP, A. C. 1995. *Bird Families of the World: Hornbills.* Oxford University Press, Oxford.

KENNEDY, R. S., P. C. GONZALES, E. C. DICKINSON, H. C. MIRANDA, JR., AND T. H. FISHER. 2000. *A Guide to the Birds of the Philippines.* Oxford University Press, Oxford.

KING, B., M. WOODCOCK, AND
E. C. DICKINSON. 1975. *A Field
Guide to the Birds of South-East Asia.*
Collins, London.

KÖNIG, C., F. WEICK, AND J.-H.
BECKING. 1999. *Owls: A Guide to the
Owls of the World.* Yale University
Press, New Haven.

LAMBERT, F., AND M. WOODCOCK.
1996. *Pittas, Broadbills, and Asities.*
Pica Press, London.

LANGRAND, O. 1990. *Guide to the
Birds of Madagascar.* Yale University
Press, New Haven.

LEFRANC, N., AND T. WORFOLK.
1997. *Shrikes: A Guide to the Shrikes
of the World.* Yale University Press,
New Haven.

LEKAGUL, B., AND P. D. ROUND.
1991. *A Guide to the Birds of Thailand.*
Saha Karn Bhaet, Ltd., Bangkok.

LEVER, C. 1994. *Naturalized Animals:
The Ecology of Successfully Introduced
Species.* T. and A. D. Poyser, London.

MACKINNON, J., AND K. PHILLIPPS.
1993. *A Field Guide to the Birds of
Borneo, Sumatra, Java, and Bali.*
Oxford University Press, Oxford.

———. 2000. *A Field Guide to the
Birds of China.* Oxford University
Press, Oxford.

MADGE, S. 1992. *Waterfowl: An
Identification Guide to the Ducks,
Geese and Swans of the World.*
Houghton Mifflin, New York.

MADGE, S., AND H. BURN.
1994. *Crows and Jays.* Princeton
University Press, Princeton.

MATTHYSEN, E. 1998. *The Nuthatches.*
T & A. D. Poyser, London.

MEFFE, G. K., AND C. R. CARROLL.
1997. *Principles of Conservation
Biology.* 2nd ed. Sinauer Associates,
Sunderland, MA.

MONROE, B. L., AND C. G. SIBLEY.
1997. *World Checklist of Birds:
English Names and Systematics.* Yale
University Press, New Haven.

MULLARNEY, K., L. SVENSSON, D.
ZETTERSTRÖM, AND P. J. GRANT.
1999. *Birds of Europe.* Princeton
University Press, Princeton.

NATIONAL GEOGRAPHIC. 2002. *Field
Guide to the Birds of North America.*
4th ed. National Geographic
Society, Washington, DC.

NEWMAN, K. 1983. *The Birds of
Southern Africa.* Macmillan,
Johannesburg.

NEWTON, I. 1998. *Population
Limitation in Birds.* Academic Press,
San Diego.

OLROG, C. C. 1984. *Las Aves
Argentinas.* Administración de
Parques Nacionales, Buenos Aires.

PEÑA, M. DE LA, AND M. RUMBOLL.
1998. *Collins illustrated checklist:
Birds of southern South America and
Antarctica.* HarperCollins, London.

PERRINS, C. M., AND A. L. A.
MIDDLETON. 1985. *The Encyclopedia
of Birds.* Facts on File Publications,
New York.

PIZZEY, G., AND F. KNIGHT. 1997. *A
Field Guide to the Birds of Australia.*
HarperCollins Publishers, Sydney.

POOLE, A., AND F. B. GILL, EDS.
1992–2003. *The Birds of North
America.* Birds of America Inc.,
Philadelphia.

PORTER, R. F., S. CHRISTENSEN,
AND P. SCHIERMACKER-HANSEN.
1996. T. and A. D. Poyser, London.

PRATT, H. D., P. L. BRUNER, AND
D. G. BERRETT. 1987. *A Field
Guide to the Birds of Hawaii and the
Tropical Pacific.* Princeton University
Press, Princeton.

RAFFAELE, H., J. WILEY, O.
GARRIDO, A. KEITH, AND J.
RAFFAELE. 1998. *A Guide to the
Birds of the West Indies.* Princeton
University Press, Princeton.

RIDGELY, R. S., AND P. J.
GREENFIELD. 2001. *The Birds
of Ecuador: Field Guide.* Cornell
University Press, Ithaca, NY.

RIDGELY, R. S., AND J. A. GWYNNE,
JR. 1989. *A Guide to the Birds of
Panama.* Princeton University Press,
Princeton.

RIDGELY, R. S., AND G. TUDOR. 1989.
The Birds of South America. Vol 1.
University of Texas Press, Austin.

———. 1994. *The Birds of South
America.* Vol 2. University of Texas
Press, Austin.

ROBSON, C. 2000. *A Guide to the
Birds of Southeast Asia.* Princeto n
University Press, Princeton.

ROWLEY, I., AND E. RUSSELL. 1997.
*Bird Families of the World: Fairy-
Wrens and Grasswrens.* Oxford
University Press, Oxford.

SCHODDE, R., AND I. J. MASON.
1999. *The Directory of Australian
Birds. I. Passerines.* CSIRO
Publishing, Collingwood, Victoria.

SCOTT, J. M., S. MOUNTAINSPRING,
F. L. RAMSEY, AND C. B.
KEPLER. 1986. "Forest Bird
Communities of the Hawaiian
Islands: Their Dynamics, Ecology,
and Conservation." *Studies in Avian
Biology* 9: 1–431.

SHORT, L. L., AND J. F. M. HORNE.
2001. *Bird Families of the World:
Toucans, Barbets, and Honeyguides.*
Oxford University Press, Oxford.

SIBLEY, C. G., AND B. L. MONROE,
JR. 1990. *Distribution and Taxonomy
of Birds of the World.* Yale University
Press, New Haven.

SIBLEY, D. A. 2000. *The Sibley Guide
to Birds.* Alfred A. Knopf, New
York.

SICK, H. 1993. *Birds in Brazil.*
Princeton University Press,
Princeton.

SIMPSON, K., AND N. DAY. 1996.
*The Princeton Field Guide to the
Birds of Australia.* 5th ed. Princeton
University Press, Princeton.

SINCLAIR, I., P. HOCKEY, AND W.
TARBOTON. 1997. *Birds of Southern
Africa.* 2nd ed. Struik, Cape Town,
South Africa.

SNOW, D. W. 1982. *The Cotingas.*
British Museum of Natural History,
London.

STANGER, M., M. CLAYTON, R.
SCHODDE, J. WOMBEY AND
I. MASON. 1998. *CSIRO List of
Australian Vertebrates: A Reference
with Conservation Status.* CSIRO
Publishing, Collinngwood, Victoria.

STATTERSFIELD, A. J., M. J.
CROSBY, A. J. LONG, AND
D. C. WEGE. 1998. *Endemic Bird
Areas of the World: Priorities for
Biodiversity Conservation.* BirdLife
International, Cambridge, Eng.

STEVENSON, T., AND J. FANSHAWE.
2002. *Field Guide to the Birds of East
Africa.* T. and A. D. Poyser, London.

STILES, F. G., AND A. F. SKUTCH.
1989. *A Field Guide to the Birds of
Costa Rica.* Cornell Univeristy Press,
Ithaca, NY.

TAYLOR, B., AND B. VAN PERLO.
1998. *Rails: A Guide to the Rails,
Crakes, Gallinules and Coots of the
World.* Yale University Press, New
Haven.

UNITED NATIONS DEVELOPMENT
PROGRAMME, UNITED NATIONS
ENVIRONMENT PROGRAMME,
WORLD BANK, WORLD
RESOURCES INSTITUTE. 2000.
*World Resources, 2000–2001: People
and Ecosystems, the Fraying Web of
Life.* Elsevier Science, Oxford Eng.

WELLS, D. R. 1999. *The Birds of
the Thai-Malay Peninsula.* Vol 1.
Academic Press, London.

WILLIAMS, T. D. 1995. *Bird Families
of the World: The Penguins.* Oxford
University Press, Oxford.

WINKLER, H., D. A. CHRISTIE,
AND D. NURNEY. 1995. *Woodpeckers.*
Pica Press, London.

ZIMMERMAN, D. A., D. A. TURNER,
AND D. J. PEARSON. 1996. *Birds
of Kenya and Northern Tanzania.*
Princeton University Press,
Princeton.

General Index

A

Acanthisittidae, 238
Acanthizidae, 275
accentors, 423–5, 487
Accipitridae, 71
accipitrids, 71–5
Aegithalidae, 371, 493
Aegithinidae, 474
Aegothelidae, 179
African buntings, 455
African warblers, 389–90
Afrotropical region, 486–7
Ailuroedus, 471
akalats, 347
Alaudidae, 401, 472, 480, 487
albatrosses, 31–2, 476
Alcedinidae, 195, 488
Alcidae, 141, 479
alcids, 141–2
alethes, 341
alien species, 498
amakihis, 441
Anatidae, 7, 63, 473, 479, 487
Anhimidae, 62, 487
anhingas, 49–50
Anhingidae, 49
anis, 167–9
Antarctica, 8, 21–2, 139, 486
antbirds, 245–7, 487
antpeckers, 413
antpittas, 245
antshrikes, 245
antthrushes, 245
antvireos, 245
antwrens, 245
anvil-headed storks, 56
apalises, 389–90
Aphelocoma, 471
Apodidae, 181
Apodiformes, 181, 185
Apterygiformes, 14
aracaris, 223
Aramidae, 103
arboreal, 9
Ardeidae, 51, 501
Artamidae, 309
asities, 238
astrapias, 305
Atrichornithidae, 267
at-risk species, 11
attilas, 257
auklets, 141
auks, 141–2
Australasia, 8
Australasian region, 486–7
Australasian babblers, 289–90
Australasian mudnesters, 290
Australasian robins, 283–4
Australasian treecreepers, 263–4
Australian chats, 277–81
avadavats, 413
avian biogeography, 483–95
avian distribution, 483–95
avian diversity, 469–74, 496–502
avifauna, 488
avocets, 129–30

B

babblers, 289, 291, 397–400
Balaenicipitidae, 56, 483
bananaquits, 443, 446
barbets, 219–21, 470, 487

bare-eyes, 245
barn owls, 171–4
barwings, 397
batises, 335–7
bazas, 71
becards, 257
bee-eaters, 205–6
beeswax, 225
bellbirds, 251, 285
bell-magpies, 311
belly-wetting, 145
berrypeckers, 427–8
biodiversity, 7, 474–6
bird abundance, 478–80
bird biogeography, 483–95
bird distribution, 483–95
bird diversity, 469–74, 496–502
bird families, 480–3
birds-of-paradise, 305–7, 487
bird species, number of, 469–73
bird-watching, 1–3
bishops, 407–11
bitterns, 51–3
blackbirds, 463–7, 487, 501
bluebills, 413
bluebirds, 341
bobwhites, 87
Bombycillidae, 339, 485, 487, 493
boobies, 45–7
boubous, 329
bowerbirds, 269–70, 487
bowers, 269–70
Brachypteraciidae, 209
breeding: polyandrous, 10, 20, 86, 111, 118, 124, 423; polygamous, 10, 110; polygynous, 10, 423, 467; promiscuous, 10, 89, 307
bristlebills, 381
bristlefronts, 249
bristleheads, 309–10
broadbills, 237
bronze-cuckoos, 163
brood-parasitism, 163, 166, 225–7, 467
brownbulls, 381
brush-turkeys, 81
Bucconidae, 217
Bucerotidae, 211, 487
buffalo-weavers, 407
bulbuls, 381–4, 487
bullfinches, 435
buntings, 455–9, 461
Burhinidae, 123
bushbirds, 245
bushshrikes, 329–30
bustards, 97, 109–10, 487
butcherbirds, 311–3
buttonquail, 83–6, 97
buzzards, 71

C

Cacatuidae, 157, 474, 488
caching, 304
caciques, 463–7
Callaeidae, 290
camaropteras, 389
Campephagidae, 321
canaries, 435–8
canasteros, 244
Capitonidae, 219, 487
Caprimulgidae, 175, 501
Caprimulgiformes, 175
caracaras, 77

Cardinalidae, 461
cardinalids, 461–2
cardinals, 461–2
cardinal-grosbeaks, 461
Cariamidae, 104
cassowaries, 13, 15–6, 484
Casuariiformes, 13
catbirds, 357–8
Cathartidae, 69
Certhiidae, 263
chachalacas, 91–2, 487
chaffinches, 435
Charadriidae, 119
Charadriiformes, 111, 115, 119, 123, 125, 129, 131, 135, 137
chats, 277, 347, 443
chatterers, 397
chat-tyrants, 257
chickadees, 375–6
chiffchaffs, 391
Chionidae, 139
Ciconiidae, 55
Ciconiiformes, 51, 55–7
Cinclidae, 345
cinclodes, 244
Cinclosomatidae, 291
cisticolas, 389–90, 487
Cisticolidae, 389, 487
class, 5
classification, 3–5
clay licks, 155
Chloropseidae, 474
Climacteridae, 263
cochoas, 347
cockatoos, 157–8
cocks-of-the-rock, 251
Coerebidae, 447, 474
colies, 189
Coliidae, 189
Coliiformes, 189, 494
Columbidae, 147, 479, 501
Columbiformes, 143, 147
comets, 185
condors, 69–70
conebills, 447–9
Conopophagidae, 250
conservation status, 10
cooperative breeding, 400
coots, 97–8
Coraciidae, 207
Coraciiformes, 195, 199, 201, 205, 207, 211
Corcoracidae, 290
cordonbleus, 413
cormorants, 49–50
Corvidae, 7, 301, 473–4, 501
corvids, 301–4
cotingas, 251–3, 487
Cotingidae, 251, 487, 493
couas, 166
coucals, 166
countershading, 25
coursers, 131–3
courtship displays, 10, 26, 29, 32, 56, 101, 256, 403
cowbirds, 463–7, 501
crab-plovers, 127
Cracidae, 91, 487
Cracticidae, 311
crakes, 97
cranes, 97, 99–101
creepers, 263–4
crescent-chests, 249
crimson-wings, 413

crombecs, 391
crop, 136
crossbills, 435–8
Crotophagidae, 167
crowned-pigeons, 147
crows, 301–4, 501
crow-shrikes, 311
crytic coloration, 110, 175
cuckoo-rollers, 209
cuckoos, 163–6
cuckoo-shrikes, 321–2
Cuculidae, 163, 167
Cuculiformes, 159, 163
curassows, 91–2
curlews, 115
currawongs, 311–3

D

dacnises, 451–4
DDT, 78
Dendrocolaptidae, 239
Dicaeidae, 428
Dicruridae, 297
dikkops, 123
Diomedeidae, 31
dippers, 341–5
distraction display, 121, 133
distribution, 8, 483–95
diversity, 3, 7, 469–74
diving birds, 21–2, 25–6, 27–9, 44, 47, 50, 67, 141–2
diving-petrels, 37–8
domestic ducks, 63
double-scratching, 459
doves, 147–50, 501
dowitchers, 115
Drepanididae, 439, 482
dreps, 439
Dromadidae, 127
drongos, 297–9
ducks, 63–7, 487
Dulidae, 340
dynamic soaring, 31, 33

E

eagles, 71–5
eared-nightjars, 177
earthcreepers, 244
ecological diversity, 7
egrets, 51–3
elaenias, 257
elephant birds, 482, 484
Emberizidae, 455, 487, 501
emberizids, 455–9
emeralds, 185
Empidonax, 472
emu-wrens, 271–3
endangered species, 10, 11
endemic species, 8, 493–5
Estrildidae, 413, 480, 487, 501
estrildid finches, 413–6
Ethiopian region, 486–7
Euphonies, 451–4
Eurasian buntings, 455
Eurasian tits, 472
Eurylaimidae, 237
Eurypygidae, 107
extinct species, 20, 29, 38, 127, 142, 150, 155, 232, 281, 287, 290, 304, 472, 441, 446, 467, 496, 500